REF 920 OAK
Watertown High School
1300205
A to Z of STS scientists

D1573347

Library Media Center
Watertown High School

A to Z
of
STS Scientists

NOTABLE SCIENTISTS

A TO Z
OF
STS SCIENTISTS

Elizabeth H. Oakes

Library Media Center
Watertown High School

Facts On File, Inc.

A TO Z OF STS SCIENTISTS

Notable Scientists

Copyright © 2002 by Facts On File

All rights reserved. No part of this book may be reproduced or utilized in any form or by any means, electronic or mechanical, including photocopying, recording, or by any information storage or retrieval systems, without permission in writing from the publisher. For information contact:

Facts On File, Inc.
132 West 31st Street
New York NY 10001

Library of Congress Cataloging-in-Publication Data

Oakes, Elizabeth, 1964–
    A to Z of STS scientists / Elizabeth Oakes.
        p.   cm.
    Includes bibliographical references and index.
    ISBN 0-8160-4606-9 (acid-free paper)
    1. Scientists—Biography. 2. Engineers—Biography. I. Title.

Q141 .O23 2002
509.2′2—dc21
[B]   2002023535

Facts On File books are available at special discounts when purchased in bulk quantities for businesses, associations, institutions, or sales promotions. Please call our Special Sales Department in New York at (212) 967-8800 or (800) 322-8755.

You can find Facts On File on the World Wide Web at http://www.factsonfile.com

Text design by Erika K. Arroyo
Cover design by Cathy Rincon
Chronology by Sholto Ainslie

Printed in the United States of America

VB Hermitage 10 9 8 7 6 5 4 3 2 1

This book is printed on acid-free paper.

Contents

List of Entries   vii
Acknowledgments   ix
Introduction   xi

Entries A to Z   1

Entries by Field   329
Entries by Country of Birth   333
Entries by Country of Major Scientific Activity   337
Entries by Year of Birth   341
Chronology   345
Bibliography   351
Index   355

List of Entries

Abbe, Cleveland
Ackerman, Thomas P.
Agricola, Georgius
Aiken, Howard Hathaway
Alfvén, Hannes
Allen, Paul
Alvarez, Luis Walter
Ampère, André-Marie
Apgar, Virginia
Appert, Nicolas
Archimedes
Arkwright, Sir Richard
Armstrong, Edwin
Avery, Oswald Theodore
Babbage, Charles
Baekeland, Leo Hendrik
Baird, John Logie
Bakewell, Robert
Banting, Sir Frederick G.
Barton, Clara
Becquerel, Antoine-Henri
Benz, Karl Friedrich
Berliner, Emile
Bessemer, Sir Henry
Best, Charles Herbert
Binet, Alfred
Boulton, Matthew
Brindley, James
Brunel, Isambard Kingdom
Brunel, Sir Marc Isambard
Bunsen, Robert Wilhelm Eberhard von
Caldicott, Helen
Campbell-Swinton, Alan Archibald
Carlson, Chester
Carothers, Wallace Hume
Carson, Rachel Louise
Chain, Sir Ernst Boris
Cooke, Sir William Fothergill
Cousteau, Jacques-Yves
Crick, Francis
Crookes, Sir William
Daily, Gretchen
Daimler, Gottlieb Wilhelm
Dalton, John
Daniell, John Frederic
Darby, Abraham
Darwin, Charles Robert
Davy, Sir Humphry
De Laval, Carl Gustav Patrik
De Vries, Hugo
Diesel, Rudolf
Domagk, Gerhard
Drake, Edwin Laurentine
Dunlop, John Boyd
Eastman, George
Edison, Thomas
Einstein, Albert
Einthoven, Willem
Elion, Gertrude Belle ("Trudy")
Farnsworth, Philo
Fermi, Enrico
Ferranti, Sebastian Ziani de
Fleming, Sir Alexander
Florey, Howard Walter
Forrester, Jay
Franklin, Benjamin
Franklin, Rosalind Elsie
Freud, Sigmund
Frisch, Otto Robert
Gabor, Dennis
Galilei, Galileo
Gilbert, Walter
Goddard, Robert Hutchings
Goodyear, Charles
Greenfield, Susan
Gurdon, Sir John Bertrand
Haber, Fritz
Hancock, Thomas
Hargreaves, James
Harrison, John
Hero of Alexandria
Hertz, Heinrich Rudolf
Hodgkin, Dorothy Crowfoot
Hollerith, Herman
Holmes, Arthur
Hoobler, Icie Gertrude Macy
Hopkins, Donald

Hopper, Grace Murray
Hounsfield, Godfrey Newbold
Huygens, Christiaan
Hyatt, Gilbert
Hyatt, John Wesley
Jacquard, Joseph-Marie
Jenner, Edward
Jobs, Steven
Joliot-Curie, Irène
Joule, James Prescott
Kapitsa, Pyotr Leonidovich
Kay, John
Kelly, William
Kelvin, Baron (William Thomson)
King, Mary-Claire
Kipping, Frederic Stanley
Kolff, Willem J.
Krupp, Alfred
Lancefield, Rebecca Craighill
Lanchester, Frederick William
Lavoisier, Antoine-Laurent
Lawes, Sir John Bennet
Lear, William
Leblanc, Nicolas
Lebon, Philippe
Leclanché, Georges
Leibniz, Gottfried Wilhelm
Leopold, Aldo
Lister, Joseph
Love, Susan
Lucid, Shannon W.
Lumière, Auguste and Louis
Magellan, Ferdinand
Maiman, Theodore
Marconi, Guglielmo
Marcy, Geoffrey
Matzinger, Polly Celine Eveline
Mauchly, John William
Maxwell, James Clerk
McClintock, Barbara
Mead, Margaret
Medawar, Peter Brian
Meitner, Lise
Mendel, Johann Gregor
Mercator, Gerardus
Mestral, George de
Michaux, Pierre
Morgan, Thomas Hunt
Napier, John
Newcomen, Thomas
Newton, Sir Isaac
Niepce, Joseph
Nipkow, Paul Gottlieb
Nobel, Alfred Bernhard
Oppenheimer, J. Robert
Ørsted, Hans Christian
Ortelius, Abraham
Otis, Elisha Graves
Otto, Nikolaus August
Papin, Denis
Pappus
Parsons, Sir Charles Algernon
Pascal, Blaise
Pasteur, Louis
Pavlov, Ivan Petrovich
Pennington, Mary Engle
Perkin, William Henry
Piaget, Jean
Planck, Max
Playfair, Lyon
Pliny the Elder
Polhem, Christopher
Polo, Marco
Poncelet, Jean-Victor
Popov, Alexander Stepanovich
Quimby, Edith H.
Richards, Ellen Swallow
Röntgen, Wilhelm Conrad
Rowland, F. Sherwood
Sagan, Carl Edward
Salk, Jonas Edward
Sanger, Frederick
Savery, Thomas
Shockley, William
Sikorsky, Igor
Silbergeld, Ellen Kovner
Singer, Isaac Merrit
Skinner, B. F.
Soddy, Frederick
Sperry, Elmer Ambrose
Sperry, Roger
Spode, Josiah
Starley, James
Stephenson, George
Sutherland, Ivan Edward
Swan, Sir Joseph Wilson
Swedenborg, Emanuel
Swinburne, James
Telford, Thomas
Tesla, Nikola
Thomson, Elihu
Trevithick, Richard
Tull, Jethro
Venter, J. Craig
Volta, Count Alessandro
Von Neumann, John
Watson-Watt, Robert Alexander
Watt, James
Wells, Horace
Westinghouse, George
Wheatstone, Sir Charles
Whitehead, Robert
Wilkinson, John
Wozniak, Stephen
Wright, Wilbur and Orville
Young, James
Zeiss, Carl Friedrich
Zeppelin, Ferdinand
Zworykin, Vladimir

Acknowledgments

Many thanks to Bill Baue, head writer and researcher, for his devotion to accuracy and his willingness to dig for the details that bring these scientists and inventors to life. Bill's contribution to this project was immense. For assistance with photographs, I thank those scientists who graciously responded to my requests and the many librarians and archivists who helped, especially Heather Lindsay with the Emilio Segrè Photo Archives at the American Institute of Physics. I would also like to express my gratitude to the University of Montana Mansfield Library, where much of the research for this book was completed, and to the authors of the many science reference books I consulted.

As always, my sincerest thanks go to Frank K. Darmstadt, my editor, as well as to the rest of the staff at Facts On File for their incredible help and support.

INTRODUCTION

Certain discoveries and inventions in history, both ancient and contemporary, have impacted society in irreversible ways. Here are the biographies of 208 scientists and inventors—men and women from all periods of history, as far back as the third century B.C.—whose work has had this kind of reverberating impact on society. You will meet Edith Quimby, who developed the earliest therapeutic uses for radiation; the television pioneer Philo Farnsworth, who invented the mechanism for the electronic transmission of images; and the Russian Pyotr Kapitsa, famous for his research in cryogenics as well as for his determined defense of intellectual freedom in scientific research. The even more famous, such as Louis Pasteur and the Wright brothers, are included as well. Diverse as their work and backgrounds may be, these men and women share in common a significant role in the interplay between science and society.

THE SCIENTISTS

A to Z of STS Scientists brings together an array of well-known and lesser-known scientists and inventors, providing the basic biographical details of their lives. The focus, however, is on their work, with their scientific achievements situated in their proper social context and presented in everyday language that makes even the most complex concepts accessible.

Because the field of Science, Technology, and Society Studies is interdisciplinary, A to Z of STS Scientists presents scientists and inventors from backgrounds as diverse as early aviation technology and modern computer science. No one discipline rules. There are stories of biophysicists, chemists, physicians, inventors, electrical engineers, psychologists, pathologists, and more.

Some are well-known scientific greats; others are contemporary scientists, whose work is just verging on greatness. Among these are minority scientists and inventors, who have often been excluded from books such as this.

To compile the entrant list, I relied largely on the judgment of other scientists, consulting established reference works, periodicals, and Web resources. Despite this process, I cannot claim to present the "most important" historical and contemporary figures. Time constraints and space limitations prevent the inclusion of many deserving scientists and inventors.

THE ENTRIES

Entries are arranged alphabetically by surname, with each entry given under the name by which the entrant is most commonly known. The typical entry provides the following information:

Entry Head: Name, birth/death dates, nationality, and field of specialization.

Essay: Essays range in length from 750 to 1,200 words, with most averaging around 900 words. Each contains basic biographical information—date and place of birth, family information, educational background, positions held, prizes awarded, etc.—but the greatest attention is given to the entrant's work. Names in small caps within the essays provide easy reference to other people represented in the book.

In addition to the alphabetical list of scientists and inventors, readers can search for entrants by scientific field, country of birth, country of major scientific activity, and year of birth. These indexes are located at the back of the book.

Abbe, Cleveland
(1838–1916)
American
Meteorologist

Cleveland Abbe ushered in the modern era of meteorology by instituting a national system of daily weather reports and forecasts that served as the prototype for the U.S. Weather Bureau, which he also helped to organize. Abbe helped transform the reporting of weather from a highly localized phenomenon based on conjecture into a coordinated system based on observed facts and informed projections of potential weather developments. Abbe's "probabilities," as he initially called them, acted as the precursor to the present-day weather forecast

Abbe was born on December 3, 1838, in New York City, brother of Robert Abbe, the pioneer in plastic surgery who introduced radiation therapy to the United States. Growing up in the city, he became enthralled with weather by reading articles by Merriam, Espy, and Joseph Henry (among others) in the daily newspapers. In the summer of 1857, he read William Ferrel's classic article on the theories of storms and winds in the *Mathematical Monthly,* which guided him into the study of meteorology. That year, he graduated from the Free Academy (now the College of the City of New York) and proceeded to conduct graduate studies in astronomy under F. Brunow at Ann Arbor, Michigan, until 1860, and then under B.A. Gould at Cambridge, Massachusetts, until 1864.

Abbe spent the next two years studying and working as an assistant under astronomer Otto Struve at the Observatory of Pulkova in Russia. Upon his return to the United States, he worked briefly at the Naval Observatory before taking up the directorship of the Cincinnati Observatory.

In his inaugural address on May 1, 1868, he outlined his intention of establishing a system of weather reports. John Gano, president of the Cincinnati Chamber of Commerce, pledged his support for such a project, and the Western Union Telegraph Company donated transmissions over its telegraph lines of weather reports from the 40 volunteer meteorological correspondents enlisted by Abbe. The first *Cincinnati Weather Bulletin* was dispatched on September 1, 1869. In October 1869, Abbe devised a code of cipher for abbreviating the weather reports.

Abbe's *Cincinnati Weather Bulletin* served as the prototype for the nationalization of a weather-reporting system, which Smithsonian observer Increase Allen Lapham of Milwaukee urged Congress to establish under the auspices of the Signal Corps of the Army. The U.S. Congress

Cleveland Abbe instituted the first national system of daily weather reports and forecasts. *(AIP Emilio Segrè Visual Archives)*

announced a joint resolution supporting the measure on February 2, 1870, and on February 9, President Grant signed the initiative into law, charging the secretary of war with establishing it under the Army Signal Service. Abbe married on May 10, 1870, in the midst of preparations for the institution of the weather report, which went into effect in November 1870.

On January 3, 1871, Abbe was appointed civilian assistant to the chief signal officer, General Albert J. Myer. Together, they organized the Weather Bureau of the Army Signal Service, which oversaw the national weather reports. The reports consisted of thrice daily synopses of current weather conditions, along with "probabilities," or forecasts of possible atmospheric developments. Abbe devised a system to reduce traffic on the electromagnetic telegraph wires by having all of the reporters at the major stations opening up their lines at specific appointed times, each to give their report and then listen to others' reports, thereby disseminating all the necessary information in a mere 20- to 30-minute interchange. Despite the efficiency of such a system, Western Union refused to dispatch all weather reports on March 4, 1871, forcing the Weather Bureau to use competing telegraph companies for their transmissions.

Abbe continued to impose order on the system he innovated, determining the altitude above sea level of all Signal Service barometers in 1872. The next year, he launched the *Monthly Weather Review*, a slim bulletin of weather statistics that expanded some 20 years into one of the most respected meteorological journals in the world under Abbe's editorship. Also in 1873, the International Meteorological Congress established the "Daily Bulletin of Simultaneous International Meteorological Observations," based on Abbe's national system.

Abbe published prolifically. His most important papers included "Treatise on Meteorological Apparatus and Methods," published in 1887, and "Preparatory Studies for Deductive Methods in Storm and Weather Prediction," published in 1889. Other important titles included *Solar Spots and Terrestrial Temperature; A Plea for Terrestrial Physics; Atmospheric Radiation;* and *Treatise on Meteorological Apparatus.*

Abbe was duly recognized for his contributions to science. For example, he was elected to the National Academy of Sciences in 1879. Perhaps the most distinguished honor was his receipt of the Marcellus Hartley medal for Eminence in the Application of Science to the Public Welfare on April 17, 1916. He was unable to attend the ceremony, however, due to ill health. Half a year later, Abbe died at his home in Chevy Chase, Maryland, on October 28, 1916. In his honor, flags in front of the Department of Agriculture and the Weather Bureau in Washington, D.C., were flown at half-mast on the day of his funeral.

Also in his memory, the American Meteorological Society named the Cleveland Abbe Award for Distinguished Service to Atmospheric Sciences by an Individual after him.

⊠ Ackerman, Thomas P.
(1947–)
American
Meteorologist

The theory of nuclear winter, or the catastrophic atmospheric consequences wrought by nuclear war, elicited a sea change in the public perception of the viability of employing nuclear weapons tactically. Thomas Ackerman participated on the team that proposed a scientific model for a nuclear winter scenario in the early 1980s. The theory's reception varied along political lines: antinuclear activists embraced it as evidence of the insanity of maintaining nuclear arsenals, while the conservative contingent attacked its scientific limitations.

Thomas P. Ackerman attended the University of Washington throughout his academic career, earning his master of science degree in physics in 1971 and his Ph.D. in atmospheric science in 1976. After receiving his doctorate, he went to work as a research scientist at the National Aeronautics and Space Administration (NASA) Ames Research Center.

In 1982, the Swedish environmental journal *Ambio* published an article in which Paul Crutzen and John W. Birks coined the term *nuclear winter* to describe the aftereffects of a nuclear war. Interestingly, they theorized that the resulting environmental effects would eclipse the destructiveness of the actual explosions, as carbon soot from the resulting fires would blanket the atmosphere, preventing sunlight from reaching the earth's surface. When CARL EDWARD SAGAN read this account, he grasped the political implications of such a theory, and he realized that the scientific community could offer the antinuclear movement the ultimate deterrent: a description of mutual assured destruction, or global suicide.

Sagan set out to create a scientific model of nuclear winter, using computer software to extrapolate the effects of a nuclear holocaust. He enlisted Ackerman, along with Richard P. Turco, Owen B. Toon, and James B. Pollack, to form the team later known by the acronym TTAPS. The group developed a one-dimensional model projecting the likely outcomes of significant nuclear events. In their report, "Nuclear Winter: Global Consequences of Multiple Nuclear Explosions," published in the December 23, 1983, issue of *Science*, they proposed that nuclear weapons exploding over 100 cities, releasing an explosive power totaling as little as 100 megatons, would send so much dust and smoke into the atmosphere that the temperature would drop anywhere from 5 to 15 degrees, an outcome that could have catastrophic environmental consequences.

The nuclear winter theory galvanized the political community: the antinuclear movement used it as an apocalyptic rallying cry to discontinue the stockpiling of nuclear arms, and indeed to reach disarmament treaties. However, the conservative faction seized upon the theory's limitations, pointing out that it did not take into account the division of the earth's surface into water and land (which would create heat transfer), the difference between daytime and nighttime sunlight (TTAPS postulated 24-hour sunlight at one-third strength), and the limitations of existing computers to take into account the multiple variables factoring into a realistic scenario. Conservatives further accused the TTPAS team of sacrificing scientific integrity in order to advance a political agenda, a position confirmed by the opinions of leading scientists (including Nobel laureate Richard Feynman) who criticized the study's methodologies.

The TTAPS team, along with Crutzen and Birks, received the 1985 Leo Szilard Lectureship

Award from the American Physical Society, reaffirming their scientific integrity. In 1988, Ackerman joined the faculty of Pennsylvania State University as a professor of meteorology and associate director of Earth System Science Center, and he then held a concurrent position on the NASA's MISR (Multi-angle Imaging SpectroRadiometer) science team and as a site scientist for the Tropical Western Pacific site in the Department of Energy's Atmospheric Radiation Measurement Program.

Ackerman also continued to collaborate with the TTAPS team, conducting further research on the nuclear winter question. In 1990, the group published a follow-up article in *Science*, in which they defended their original theory by offering more sophisticated modeling (available due to more sophisticated computer programs) and taking into account more realistic variables.

Since then, with the demise of the Soviet Union and the melting of the cold war, little research has been applied to the nuclear winter theory. However, the theory lodged itself in the collective conscious, exerting a significant influence on public policy as well as personal angst. The reception of the theory demonstrated the necessity of maintaining impeccable scientific integrity, especially when scientific findings carry political implications. Ultimately, the theory's influence eclipsed the question of its scientific validity, as it forced a more considered approach to the question of the destructive capacity of nuclear weapons, and the wisdom of maintaining vast nuclear arsenals in a state of readiness.

⊗ Agricola, Georgius
(1494–1555)
German
Mineralogist, Geologist, Metallurgist

Though his exhaustive knowledge of diverse subjects earned him the title "The Saxon Pliny," Georgius Agricola was best known as the author of *De re metallica libri XII* (*On the Subject of Metals*), a seminal text in the understanding of metallurgy and the mining and smelting processes of the time. Living in the mining capitals of St. Joachimsthal in Bohemia (Czechoslovakia) and Chemnitz, Germany, Agricola had extensive exposure to every aspect of mining, including the management of the mines and the machinery used, such as pumps, windmills, and waterwheels, which he incorporated into his books.

Born on March 24, 1494, in Glauchau, Saxony, to Gregor Bauer, a dyer and wool draper, Georg Bauer latinized his name to Georgius Agricola as was the custom at the time. His youngest and favorite brother followed his footsteps to Chemnitz in 1540 to become a metallurgist, and his oldest brother entered the priesthood in Zwickau and later in Glauchau. In 1526 Agricola married the widow of Thomas Meiner, the director of the Schneeberg mining district, and she died in 1541. In 1542 Agricola

Georgius Agricola's great knowledge of mining and metallurgy was commemorated by future president Herbert Hoover in 1912 when he prepared an English edition of Agricola's masterwork, *On the Subject of Metals*. (Smith Collection, Rare Book & Manuscript Library, University of Pennsylvania)

remarried, this time to Anna Schütz, the daughter of the guild master and smelter owner Ulrich Schütz, who entrusted his wife and children to Agricola's care upon his death in 1534.

In 1514, at the late age of 20, Agricola entered Leipzig University, earning his B.A. in 1515. The university retained him as a lecturer in elementary Greek for the next year, after which Agricola went to Zwickau where he organized a new Schola Graeca in 1519 and wrote his first book in 1520, *De prima ac simplici institutione grammatica,* which described new humanistic methods of teaching. Agricola then fled from the radicalism of the Reformation back to Leipzig, where he studied medicine between 1523 and 1526. During this time, he served on the editorial board in Bologna and Venice for the Aldina editions of texts by Galen and Hippocrates, interests he maintained later in life. After earning his M.D., he left Italy via the mining districts of Carinthia, Styria, and the Tyrol bound for Germany, where he stayed briefly until the mining city of St. Joachimsthal elected him town physician and apothecary in 1527.

In 1534 Agricola departed for Chemnitz, yet another city renowned for its mining, which elected him burgomaster, or mayor, in 1545. The combination of chronically sick miners and heavy metals allowed Agricola to investigate the pharmaceutical uses of minerals. Agricola published on a wide range of topics, including politics and economics. In 1554 he published *De peste libri III,* based on his experiences administering to sufferers of the black plague that swept through Saxony between 1552 and 1553.

De re metallica libri XII (*On the subject of metals*), his crowning achievement, did not appear until 1556, four months after his death. The text surveyed all aspects of mining at the time, from working conditions to metallurgy to smelting processes. Agricola had finished writing it during his return visit to St. Joachimsthal, where he had started drafting it 20 years earlier. While there, he met designer Blasius Weffring, who spent the next three years creating 292 woodcuts to illustrate the text. A year later Phillipus Bech translated the work into Old German but retained the woodcuts, creating an edition so fine that it survived 101 years in seven editions.

Agricola died on November 21, 1555, in Chemnitz. Mining engineer and future U.S. president Herbert Hoover revived Agricola's legacy by preparing an English edition of his masterwork, *On the Subject of Metals,* in 1912, a testament to the primacy of Agricola's work.

Aiken, Howard Hathaway
(1900–1973)
American
Computer Engineer

Howard Aiken helped usher in the computer age by inventing the Harvard Mark I and Mark II, the precursors to modern digital computers. The *New York Times* hailed the significance of his invention: "At the dictation of a mathematician, it will solve in a matter of hours equations never before solved because of their intricacy and the enormous time and personnel which would be required to work them out on ordinary office calculators." Aiken himself did not fully comprehend the potential of his invention, estimating in 1947 that only "six electronic digital computers would be required to satisfy the computing needs of the entire United States."

Howard Hathaway Aiken was born on March 9, 1900, in Hoboken, New Jersey. Soon after his birth, though, his family moved to Indianapolis, Indiana, where he attended Arsenal Technical High School. While studying electrical engineering at the University of Wisconsin, he worked for the Madison Gas and Electric Company. After earning his bachelor's degree in 1923, the company promoted him to chief engineer.

In 1927, Aiken moved to Chicago to work for Westinghouse Electric Manufacturing Company. Four years later, he took up a research position in the physics department at the University of Chicago. He conducted doctoral study there and at Harvard University, focusing his dissertation on a theory of space-charge conduction in vacuum tubes. This topic required calculations that would have taken him a lifetime to complete, so in 1937 he proposed the design and construction of a calculating machine. "The desire to economize time and mental effort in arithmetical computations, and to eliminate human liability to error is probably as old as the science of arithmetic itself," he wrote, jovially adding that the computer was "only a lazy man's idea."

Harvard physics department chair Frederick Saunders pointed out that lab technician Carmelo Lanza had already worked on such a machine, stored in the Science Center attic: a set of brass wheels from CHARLES BABBAGE's analytical engine. The prospect of completing Babbage's unfinished task (the existing nineteenth-century technology could not actualize his design) inspired Aiken, who kept the wheels in his office thereafter: "There's my education in computers, right there," he would say of them.

Aiken sought to build a machine that answered the multiple demands of scientists and mathematicians: ". . .whereas accounting machines handle only positive numbers, scientific machines must be able to handle negative ones as well; that scientific machines must be able to handle such functions as logarithms, sines, cosines and a whole lot of other functions; the computer would be most useful for scientists if, once it was set in motion, it would work through the problem frequently for numerous numerical values without intervention until the calculation was finished; and that the machine should compute lines instead of columns, which is more in keeping with the sequence of mathematical events."

Harvard granted Aiken his Ph.D. in 1939, appointing him an instructor in physics and communication engineering. That year, the Navy Board of Ordnance contracted Harvard to conduct research in preparation for World War II. At the same time, Aiken was searching for financial support from the private sector—he first appealed to the Monroe Calculating Machine Company, which declined but referred him to International Business Machines (IBM) president Thomas J. Watson, who promised support of $200,000.

IBM engineer Robert V. D. Campbell supervised construction of the Automatic Sequence Controlled Calculator (ASCC), as the computer was first called, at the Endicott, New York, IBM plant. Measuring 51 feet long, two feet wide, and eight feet high, it weighed more than 30 tons, with its 530 miles of wires and 760,000 moving parts—including 2,200 counter wheels and 3,300 relay components. Operators fed information in by tape or punch card, with output returning on punch cards or by electronic typewriter. It sounded like a "roomful of ladies knitting" when running.

The computer could manipulate positive and negative numbers to 23 decimal places, adding them in three-tenths of a second and multiplying in four seconds; it could also subtract, divide, and store tabulations in its 72 storage registers. GRACE MURRAY HOPPER, who collaborated with Aiken to develop these library functions and later invented the COBOL computer language, discovered the first computer "bug"—a moth squished by a relay switch.

The computer went into operation in May 1944. In accordance with the original agreement, IBM donated the computer to Harvard, which dedicated it on August 14, 1944, earning it its lasting name—the Harvard Mark I, which remained functional for the next 14 years and now resides (in sections) in the Harvard Science Center lobby, at the Smithsonian's

National Museum of History and Technology, and in the IBM historical collection. After finishing the Mark I, Aiken continued to advance his design; the Navy posted him at the Naval Proving Ground at Dahlgren, Virginia, where he finished the Mark II in 1947. Whereas the Mark I combined electronic with mechanical workings, the Mark II was fully electronic. Run by 13,000 electronic relays, the Mark II could add in two-tenths of a second and multiply in seven-tenths of a second, storing up to 100 ten-digit figures and signs.

By 1952, Aiken had designed and built Marks III and IV. In recognition of his work, the U.S. Navy promoted him to the rank of commander in its research department, and Harvard promoted him to a full professorship in applied mathematics. He also founded Harvard's computer science program, one of the first of its kind. He retired from Harvard in 1961, moving to Fort Lauderdale, Florida, to take up a professorship in information technology at the University of Miami.

In 1964, Aiken received the Harry M. Goode Memorial Award from the Computer Society. He died on March 14, 1973, in St. Louis, Missouri, before the personal computer revolution brought the digital computers that he invented into households worldwide.

Alfvén, Hannes
(1908–1995)
Swedish
Astrophysicist

Although Hannes Alfvén's name graces many physical phenomena that he identified—Alfvén waves and the Alfvén speed, Alfvén layers, Alfvén critical points, Alfvén radii, and Alfvén distances—his actual role in the advancement of astrophysics has been obscured by the belated acceptance accorded many of his discoveries. He received the 1970 Nobel Prize for "his contribu-

Hannes Alfvén was the first space scientist to be awarded the Nobel Prize. *(AIP Emilio Segrè Visual Archives)*

tions and fundamental discoveries in magnetohydrodynamics, and their fruitful applications to different areas of plasma physics." Early in his career, he theorized that plasma, or highly ionized gases, filled interstellar space (instead of a vacuum, as was then believed), an assertion that created much controversy until sophisticated experimentation proved him right. Alfvén attributed this dynamic to the politics of science, which ironically resist the overturning of the status quo (despite the fact that science bases itself on the notion of objectivity and the notion of progress).

Hannes Olof Gösta Alfvén was born on May 30, 1908, in Norrköping, Sweden. His mother, Anna-Clara Romanus, was one of the first female physicians in Sweden, and his father, Johannes Alfvén, was also a doctor. Two early experiences inspired his later interest in astrophysics and wave theories: the gift of an astronomy book by Camille Flammarion; and his building of a radio receiver as a member of his high school radio club. He studied mathematics and experimental and theoretical physics at the University of Uppsala, writing his doctoral dissertation on "Ultra-Short Electromagnetic Waves"—which combined his two childhood interests. The university granted 26-year-old Alfvén his Ph.D. in 1934 and retained him as a docent in physics (as did the Nobel Institute for Physics in Stockholm).

The year before, Alfvén had published the genesis of his theory on the origin of cosmic radiation in *Nature*. In it, he noted the insufficiency of contemporary cosmic ray theories, which did not "seem to be in accordance with the latest experimental results," instead insisting on the necessity of explaining "the origin of the cosmic rays, introducing no new hypotheses, and only applying the kinetic gas theory to the conditions of world space." In 1937, his proposal of a galactic magnetic field met with widespread resistance (if not scorn), as it directly contradicted the prevailing wisdom that a vacuum filled interstellar space. As with much of his work, it took several decades before experiments verified his theories, and by then, his name had often been forgotten in conjunction with the discovery. In this case, the study of cosmic magnetism is now an accepted discipline within astrophysics.

In 1939, Alfvén submitted a paper explaining magnetic storms and the auroras to the *Journal of Geophysical Research*, which rejected it because it flew in the face of accepted theories propounded by Sydney Chapman, Alfvén's scientific nemesis, in favor of a return to the previously accepted theories of Kristian Birkeland. Chapman's theories gained predominance due to their mathematical elegance, but satellite measurements in 1974 proved Alfvén and Birkeland correct.

In 1940, the Royal Institute of Technology in Stockholm appointed 32-year-old Alfvén professor of electromagnetic theory and electric measurements, which he considered a greater honor than his later Nobel Prize. Two years later, he published a succinct paper, "Existence of Electromagnetic-Hydrodynamic Waves" in *Nature*. Although this theory of electromagnetic wave propagation through plasma was not immediately adapted upon its publication, six years later, Alfvén waves gained almost instant acceptance when the renowned physicist ENRICO FERMI nodded in agreement ("Of course," he reputedly commented) at Alfvén's presentation of the theory in a lecture to the University of Chicago; "the next day, every physicist said, 'of course,'" according to Alfvén.

In 1946, Alfvén changed titles to professor of electronics, indicative of a change in his research focus. In 1950, he published *Cosmical Electrodynamics*, a book that introduced many of his theories to the scientific community for the first time, as so many international scientific journals shied away from publishing his controversial ideas, relegating them to more obscure journals, often those published only in Swedish. In 1963, the Royal Institute created a chair of plasma physics for him.

In 1967, Alfvén joined the University of California at San Diego as a professor of electrical engineering, a position that he split with his Royal Institute post, crossing the Atlantic each equinox. That year, the Royal Astronomical Society granted him its Gold Medal, and three years later, he shared the 1970 Nobel Prize in physics with French physicist Louis Néel. The next year, he received both the 1971 Gold Medal of the Franklin Institute and the Lomonosov Medal of the U.S.S.R. Academy of Sciences.

In 1990, the Royal Institute established the Alfvén Laboratory in his honor, and the next year, he retired. Alfvén died at his home in Djursholm, Sweden, on April 2, 1995.

Allen, Paul
(1953–)
American
Computer Engineer

Paul Allen collaborated with Bill Gates to found Microsoft, the computer software company that provided the programming for almost all personal computer applications. Allen and Gates wrote the programming that launched the company, though as their company expanded, they exerted their visionary influence by wedding hardware with the appropriate software. Although Allen did not actually write the programming, he is largely responsible for delivering the operating systems that drove IBM's personal computers and later for the development of the Windows interface, which facilitated amateur users' interactions with personal computers. In this sense, Allen helped fuel the personal computer boom that ushered in the information age.

Allen was born on January 21, 1953, in Seattle, Washington. His father was head librarian at the University of Washington. His mother, Faye, hosted "science club" meetings in the family's home to encourage her 10-year-old son's burgeoning interest in science.

In 1965, Allen entered Lakeside, a private college-preparatory school in Seattle, which proved to be a seminal event when he met his future business partner, Bill Gates, three years later. The pair indulged their interest in computers by programming in BASIC, and Allen mentored junior high school students by teaching computer courses as a high school student. Allen has said about his high school years that he "wanted to look at every computer [he] could, understand the software, the neat things this computer could do that other computers couldn't."

In 1971, Allen graduated from Lakeside and matriculated at Washington State University. The next year, he and Gates collaborated to construct a computer for measuring traffic from a $360 Intel 8008 chip, and the pair launched their first company, Traf-O-Data, to market the technology. In 1974, Honeywell of Boston hired Allen as a programmer. In January of the next year, the cover of *Popular Mechanics* magazine featured a picture of the Altair, a computer kit based on the then-new Intel 8080 chip. This first personal computer captured Allen's imagination, as both he and Gates had foreseen the day when personal computers would become available to the general consumer, and computers would grace the desktops not only at corporate offices but also in homes. They also realized that many companies stood poised to manufacture the hardware, but precious few companies, if any, were prepared to produce programming, or software, for these personal computers.

Allen and Gates approached Model Instrumentation and Telemetry Systems (MITS), the Albuquerque, New Mexico–based maker of the Altair, promising delivery of a version of BASIC for the Altair that they had yet to write. After winning the contract, Allen and Gates spent the next eight weeks frantically composing the code in a computer lab at Harvard University (where Gates was studying), and delivered on their promise. Impressed by the programming, MITS hired Allen as its associate director of software in 1975. Gates accompanied Allen to Albuquerque, where the pair founded a company called Microsoft to develop and market software that would fill the technology gap they had identified.

By 1977, both Apple and Radio Shack had commissioned Microsoft to produce versions of BASIC for their Apple II and TRS-80 computers (respectively). In November of that year, Allen resigned from his MITS position to take up a full-time position at Microsoft, in which he owned a 36 percent share. Allen headed research and development. The next year, Microsoft, buoyed by sales of more than one million dollars, moved to Bellevue in the founders' home state of Washington.

As the 1980s approached, the computer giant International Business Machines (IBM) realized the potential of the personal computer (PC) market, but it needed an operating system to drive its computers. IBM turned to Microsoft, and again Gates and Allen promised delivery of an operating system they had not built yet. Allen contacted Tim Patterson of Seattle Computer Products to ask for the rights to its "Quick and Dirty Operating System" (QDOS) for use in an unnamed client's computers. Patterson sold Allen the rights for less than $100,000, and Microsoft became the supplier of the operating system for IBM PCs, which dominated the market within three years.

That year, 1983, Microsoft introduced the graphical interface, or the mode by which users interacted with their computers, called "Windows," a program that revolutionized personal computing by hiding the programming code behind visual icons. Also in 1983, Allen learned that he had developed Hodgkin's disease, prompting him to resign from Microsoft. However, he held onto his Microsoft stock, making him a billionaire after Microsoft went public in 1986.

Allen shifted his focus to investment, and he wielded tremendous power through his financial strength, though some have criticized his investments for lacking vision. He owned three different companies to handle his portfolio, which included controlling interest in such businesses as TicketMaster, the Portland Trail Blazers basketball team, and the Seattle Seahawks football team, as well as in technology-based companies such as Asymetrix, Starwave, and Dream Works SKG.

Allen indulged his passion for rock and roll music by funding the Experience Music Project, which grew out of his idea for a Jimi Hendrix museum, and located the postmodern museum on the site of the Seattle World's Fair that he attended in 1962. In September 2000, Allen resigned from the Microsoft board to focus his attention on his numerous other enterprises, though he remained a senior strategy adviser.

Alvarez, Luis Walter
(1911–1988)
American
Physicist

In his first years as a research physicist at the University of California at Berkeley in the 1930s, Luis Alvarez earned the title "prize wild idea man," which acknowledged both his wide-ranging investigations and his ability to identify important questions in need of solutions. A member of the Manhattan Project that developed the atomic bomb during World War II, Alvarez followed the *Enola Gay*, the plane that dropped the atomic bomb on Hiroshima, Japan, in an another B-29 bomber from which he witnessed

A member of the Manhattan Project, Luis Alvarez witnessed the destruction of Hiroshima from a B-29 bomber that followed the *Enola Gay*, the plane that dropped the atomic bomb. *(AIP Emilio Segrè Visual Archives)*

the destruction of Hiroshima. Alvarez also modified the bubble chamber technique invented by University of Michigan physicist Donald Glaser for use in conjunction with particle accelerators to identify previously unknown elementary particles. This work earned him the Nobel Prize in physics in 1968, though he never rested on these laurels, remaining active and innovative in physics for the next 20 years.

Alvarez was born on June 13, 1911, in San Francisco, California, to Dr. Walter Clement Alvarez, a research physiologist at the University of California at San Francisco, and Harriet Skidmore Smythe, an Irish woman whose family instituted a missionary school in Foochow, China. Alvarez had two children, Walter and Jean, with his first wife, Geraldine Smithwick, a fellow University of Chicago student, but their relationship ended over the strain of wartime separation. Alvarez had two more children, Donald and Helen, with his second wife, Janet Landis, whom he married in 1958.

Alvarez entered the University of Chicago in 1928, earned his B.S. in 1932, and graduated in 1936 with his Ph.D. in physics. He appreciated his graduate adviser, Nobel laureate Arthur Compton, not for guiding him but rather for staying out of his way while he immersed himself in his research. At UC Berkeley he conducted research alongside two other Nobel laureates, Ernest Orlando Lawrence and Felix Bloch.

Alvarez earned his reputation as the wild idea man quickly at UC Berkeley with three significant discoveries in the 1930s. Within his first year there he discovered how atomic nuclei decayed and orbital electrons absorbed them, a process known as K-electron capture. He also invented a mercury vapor lamp with a student, and with Bloch, he developed a process for determining the magnetic moment of neutrons by slowing their motion in a beam. In the early 1940s he conducted research for the military at the Massachusetts Institute of Technology on radio detecting and ranging systems, developing three new types of radars in three years.

Alvarez never suffered from guilt over his role in the development of the atomic bomb and continued to support nuclear technology, as he believed the benefits far outweighed the problems. After the war Alvarez devoted his research to particle physics, building larger and faster particle accelerators. He transformed Glaser's bubble chamber design to use liquid hydrogen instead of diethyl ether as the liquid through which the particles passed, leaving their bubble tracks. By this method he discovered dozens of new elementary particles.

In the final years of his career, Alvarez projected high-energy muon rays (the subatomic particles of cosmic radiation) at King Chephren's pyramid in Gaza, Egypt, to determine that it had no secret chambers. The Warren Commission called upon his physics expertise to verify the fact of a lone assassin in the murder of President John F. Kennedy. Finally, he teamed up with his son, UC Berkeley geologist Walter Alvarez, to hypothesize that the dust created by the earth's collision with an asteroid caused the "winter" that exterminated dinosaurs, a widely accepted theory based on iridium deposits in Italian sedimentary rock.

Alvarez remained on the faculty of UC Berkeley until his retirement in 1978. His 1968 Nobel Prize and the 22 patents he held on his inventions attested to the significance of his work. He died of cancer on September 1, 1988, in Berkeley, California.

Ampère, André-Marie
(1775–1836)
French
Physicist, Mathematician

A mathematical child prodigy self-educated according to the principles of Rousseau, André Ampère established his scientific significance

The "amp," the unit for measuring electrical current, is named for André-Marie Ampère. *(AIP Emilio Segrè Visual Archives)*

after a fit of inspiration in September and October of 1820, when he developed the science of electrodynamics. Both Ampère's Law, which established the mathematical relationship between electricity and magnetism, and the ampere, or amp, a unit for measuring electrical current, were named after him. Though Ampère maintained wide-ranging interests in mathematics, chemistry, metaphysics, philosophy, and religion, his main source of fame derived from his work with electricity and magnetism.

Ampère was born in Lyon, France, on January 22, 1775, to Jean-Jacques Ampère, an independent merchant who provided his son with a complete library, and Jeanne Desutières-Sarcey Ampère, a devout Catholic who instilled her faith in her son. Ampère spent the rest of his life reconciling his reason with his faith. His personal life amounted to a series of disasters, starting with the execution by guillotine of his father on November 23, 1793, in the midst of the French Revolution. Tragedy took a respite during his relationship with Julie Carron, whom he courted against all odds and married on August 7, 1799. Tragedy revisited him when she died on July 13, 1803, of an illness contracted during the birth of their son, Jean-Jacques, on August 12, 1800. Ampère then entered into an ill-advised marriage to Jeanne Potot on August 1, 1806. The only positive outcome of the wedding was the birth of his daughter, Albine.

Though Ampère did not hold a degree, he taught mathematics at the Lyceum in Lyon before his appointment as a professor of physics and chemistry at the école centrale in Bourg-en-Bresse in February 1802. In 1808 Napoleon appointed him inspector-general of the new university system, a post Ampère held until death. In 1820 the University of Paris hired him as an assistant professor of astronomy, and in August 1824 the Collège de France appointed him as the chair of the experimental physics department.

In early September 1820 François Arago reported to the Académie des Sciences on the discovery by Danish physicist HANS CHRISTIAN ØRSTED that a magnetic needle deflected when current in nearby wires varied, thus establishing a connection between electricity and magnetism. In less than a month, Ampère presented three papers to the Académie that established the science of electrodynamics by positing that magnetism was simply electricity in motion. Specifically, Ampère worked with two parallel wires with electrical current flowing through them: he discovered that the currents attracted each other when heading in the same direction and repelled one another when traveling in opposite directions. The implications of his experiments suggested a whole new theory of matter. Ampère published a comprehensive overview of his findings in 1827 in his *Memoir on the Mathe-*

matical *Theory of Electrodynamic Phenomena, Uniquely Deduced from Experience*. The scientific community did not embrace his theories until Wilhelm Weber incorporated them into his theory of electromagnetism later in the century.

Ampère died on June 10, 1836, while alone on an inspection tour in Marseilles, France. Despite his tragic personal life, Ampère led a successful scientific life, contributing an entire field to the study of science.

⊗ Apgar, Virginia
(1909–1974)
American
Physician

Physician Virginia Apgar's best-known contribution to the medical sciences was the development of a test designed to evaluate the health of a newborn. This breakthrough system helped save the lives of many infants and became a standard procedure in hospitals worldwide. Apgar made numerous other advances in obstetric anesthesia and raised the awareness Americans had regarding birth defects.

Born on June 7, 1909, in Westfield, New Jersey, Apgar was raised in a family environment that encouraged the love of music as well as science. Apgar's mother was Helen Apgar and her father, Charles Apgar, was a businessman who had many hobbies, including astronomy and wireless telegraphy. Charles Apgar influenced his daughter to take an interest in science.

In the 1920s, Apgar entered Mount Holyoke College, where she was a very active student. Apgar not only waitressed and worked in the school library and other facilities to fund her tuition, but she also participated in numerous extracurricular activities, such as the college newspaper, orchestra, theater, and tennis. After graduating in 1929, Apgar entered medical school at Columbia University's College of Physicians and Surgeons. She received her medical degree in 1933 and hoped to pursue a career in surgery. A professor encouraged Apgar to instead investigate anesthesiology, then a new field. Apgar became only the 50th physician to achieve certification in the new specialty.

After completing her education, Apgar took a teaching position at Columbia University's medical school in 1936. Two years later, she became the director of the anesthesia division at Presbyterian Hospital. By 1949, Apgar had successfully built an anesthesiology department at the school, thereby lending the new field credibility, and was named full professor. In her research, Apgar concentrated on the effects of anesthesia during birth. She observed that newborn infants were generally taken directly to the hospital nursery with no examination—primary attention was given instead to the mothers. Because of the cursory treatment given to the newborns, health problems generally were not noticed until they became critical or fatal. Apgar thus developed a simple test, which came to be known as the Apgar Score System, that nurses could administer immediately after birth. Using a scale ranging from zero to two, the test

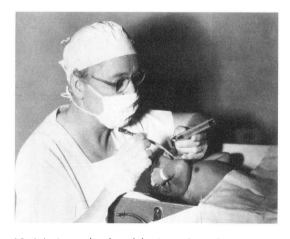

Virginia Apgar developed the Apgar Score System, a series of quick tests used in hospitals around the world to determine a newborn's health. *(National Library of Medicine, National Institutes of Health)*

evaluated the newborn's skin color, muscle tone, breathing, heart rate, and reflexes. The resultant total score, known as the Apgar Score, indicated the physical condition of the infant. A low score was indicative of problems requiring immediate attention, while a high score signified a healthy baby. Apgar first introduced the system in 1952 and published it in 1953.

After many years researching obstetric anesthesia and working as a physician, Apgar returned to school, earning a master's degree in public health from Johns Hopkins University in 1959. Also that year Apgar became a senior executive with the charitable organization National Foundation—March of Dimes. Apgar's job was to raise funds to support research on birth defects and to increase public support and awareness. In 1967, Apgar became the director of the organization's research department.

In addition to her pioneering work related to the Apgar Score System, Apgar also conducted research on neonatal acid-base status and introduced the anterior approach to the stellate ganglion. Among the many honors Apgar received were the Elizabeth Blackwell Citation from the New York Infirmary in 1960, the Distinguished Service Award of the American Society of Anesthesiologists in 1961, and the Gold Medal for Distinguished Achievement in Medicine, awarded in 1973 by Columbia University's College of Physicians and Surgeons. Apgar was also named Woman of the Year in science by *Ladies' Home Journal* in 1973 and was honored with a U.S. stamp in her name in 1994. Apgar died from liver disease at the age of 65 in 1974.

Appert, Nicolas
(1752–1841)
French
Food Technologist

Nicolas Appert invented the food-preserving process of boiling (and thereby sterilizing) food in glass jars—the basis of modern canning. His invention freed people from the need for fresh food, allowing them to preserve it by an easy process (as opposed to salting or smoking), enabling food to be kept over long periods of time. Appert originally developed his process at the behest of the French military, and canning played an important role in subsequent military history, though it also played a significant role in the broader society. Canning remains an integral part of society, as the pace of modern life allows less time for buying and preparing fresh food.

Nicolas-François Appert was born on October 23, 1752, at Châlons-sur-Marne, just east of Paris, France. His father was an innkeeper who educated his son in the arts of brewing and pickling, tasks that were an integral part of running a hotel at that time. Appert apprenticed as a chef and confectioner at the Palais Royal Hotel in Châlons and then served the Duke and Duchess of Deux-Ponts in these capacities. He subsequently moved to Paris, where he established himself as a renowned confectioner by 1780. He was also a champagne bottler, a practice that may have influenced his future in canning.

During the Napoleonic era, with the French military spreading its influence geographically, the challenge of feeding the army and navy became such an acute problem that the French Directory offered a prize in 1795 for a practical solution for preserving food. Appert rose to the challenge, experimenting over the next 14 years in search of a satisfactory food-preservation method.

With little scientific understanding, Appert proceeded intuitively, relying on trial-and-error as his primary methodology. Eventually, he happened upon his preserving technique of heating foods to above the boiling point of water in an autoclave of his own design, then submerging glass jars and bottles filled with food, which he then corked and sealed with wax. He experimented with about 70 different types of food before he was satisfied with his results in preservation.

By 1804, Appert had devised a method refined enough to open the world's first canning factory, thanks to the financial support of de la Reynière. Aptly called the House of Appert, it was located in Massy, just south of Paris. By 1809, he had succeeded in preserving food in jars, which he submitted to the Directory for trials. Testing by the French Navy and the Consulting Bureau on Arts and Manufacturing confirmed the success of Appert's canning process, as the food stayed preserved. On January 30, 1810, Appert received the 12,000-franc prize, which he invested in his business. The next year, he published an account of his findings, entitled *The Art of Preserving all Kinds of Animal and Vegetable Substances for Several Years*, in fulfillment of one of the contest requirements that he disseminate his methods and research.

Appert's contributions to society did not end with the canning process. Also in 1810, he invented peppermint schnapps as an ice-cream topping. Napoleon's wife, Marie-Louise, the duchess of Parma, brought the recipe back to Austria, where it became a hit not only as a condiment but also as a drink. He also invented the bouillon cube, a dried concentrate that could be rehydrated into a soup base, and devised an acid-free method of extracting gelatin from bones.

In 1812, Appert received a gold medal from the Society for the Encouragement of National Industry, and a decade later, he was named a "Benefactor of Humanity." That same year, in 1822, Appert improved upon his canning method by switching from the use of glass jars sealed with cork and wax to using cylindrical tin-plated steel cans, the method inherited by the modern canning industry. The House of Appert continued to can food until 1933, and Appert's basic technique remains in use currently.

The collapse of the Napoleonic empire left Appert bankrupt, as enemies of the old regime trashed his factories. He died in Massy on June 3, 1841. After he died, LOUIS PASTEUR established the scientific basis for much of Appert's work, revealing the biological and chemical means of food decay as well as the science of food preservation.

Archimedes
(287 B.C.–212 B.C.)
Sicilian/Greek
Physicist, Mathematician

Karl Friedrich Gauss considered Archimedes the greatest mathematician ever, with only SIR ISAAC NEWTON as his equal. Archimedes' estimation for the numerical value of pi survived as the best approximation available into the Middle Ages. However, Archimedes was most renowned for his practical applications of mathematical and physical theories. Two of his innovations, Archimedes' principle and the Archimedes screw, involved the displacement of water, and he was considered the founder of hydrostatics.

Though little is known about Archimedes' personal life, his own writings revealed the identity of his father, the astronomer Phidias. Archimedes was born in Syracuse on the island of Sicily in about 287 B.C. Evidence suggests that he traveled to Alexandria, where he studied mathematics under the successors of Euclid. He returned to Sicily, which was under the rule of King Hieron II, supposedly a relative or at least a friend (Archimedes dedicated *The Sand-Reckoner* to Hieron's son, Gelon).

In this late text, Archimedes contrived a notation system for very large numbers, and in another text, *Measurement of the Circle*, he estimated the value of pi as 22/7, a relatively accurate figure. However, his applications of mathematical concepts proved even more profound than his abstract realizations. In *De architectura*, Vitruvius told the dubious story of Archimedes' solution to a problem posed by Hieron—how to test whether a gift crown was indeed pure gold, as claimed the giver, or alloyed with less precious metals, as Hieron suspected. As Archimedes pondered the problem in the bathtub, he noticed that the farther he immersed his body into the tub, the more water spilled over the edge. He hypothesized that the density of the displaced water equaled the density of his submerged body. In his excitement, he rushed through the streets naked shouting in Greek, *"Heurçka!"* While testing the authenticity of the crown, he noticed that a block of pure gold equal in weight to the crown displaced less water than the crown, thus casting doubt on its veracity.

This test, which hinged on relative density and buoyancy, became known as Archimedes' principle. Archimedes described this principle, along with his understanding of buoyancy (or the upward force exerted on solids by liquids), in *On Floating Bodies*, a text that established him as the founder of hydrostatics. Of even greater practical value was his invention of what became known as the Archimedes screw, a device that could draw water along an ascending helix.

Archimedes was apparently most proud of his formulation of the volume of a sphere as two-thirds that of the cylinder in which it is inscribed, as discussed in one of his most famous works, *On the Sphere and the Cylinder*. When Cicero was the Quaestor of Sicily in 75 B.C., he tracked down Archimedes' grave and verified that it was indeed inscribed with a sphere and cylinder as well as the formula for their intersection.

Hieron called upon Archimedes to invent weapons to stay the Roman invasion of Sicily in 215 B.C., led by Marcellus. Experts doubt that Archimedes invented the weapon of mirrors that ignited distant ships with focused sunlight, though he did invent various catapults. Marcellus ultimately captured Sicily in 212 B.C., and though the general himself admired Achimedes' work, his soldiers put the mathematician to death, supposedly while he was making calculations in the sand.

Arkwright, Sir Richard
(1732–1792)
English
Textile Industrialist

Sir Richard Arkwright is credited with inventing the spinning frame, a machine that transformed cotton-spinning not only by improving the quality of production but also by multiplying exponentially the output of production. He instituted many of the policies and procedures that contributed to the success (and the oppression) of the Industrial Revolution: factory housing, child labor, worker exploitation, extended workdays, and mass production. Arkwright thus created a living legacy, as many of these business practices are still in use.

Arkwright was born on December 23, 1732, in Preston, in Lancashire, England, the youngest of 13 children. His cousin Ellen educated him, as his parents could not afford to send him to school. As a teenager, he apprenticed to a barber, then established his own barbershop, which was successful until he opened a pub that lost money. He had one son with his first wife. After she died, he remarried in 1761. In 1762, he started a wig-making business and traveled around the countryside gathering women's hair for wigs.

On his travels, he met John Kay, a Warrington clockmaker who had designed a spinning machine in partnership with Thomas Highs of Leigh. However, the pair had run out of funds, forcing them to abandon the project. Arkwright financed Kay to continue building the machine; legend has it that they worked in a room rented from a teacher in a house secluded behind gooseberry bushes—neighbors reported that the racket they made building the machine sounded like the devil tuning his bagpipes.

In 1768, Arkwright moved from Preston, where he feared that local spinners would attack him for robbing them of their handicraft, to Nottingham, where he took up a new partnership with John Smalley and filed a patent for his spinning frame. The genius of Arkwright's design, and what distinguished it from JAMES HARGREAVES's spinning jenny and other spinning machines of the day, was the order of operation: instead of stretching and spinning the cotton fibers simultaneously, the spinning frame drew out the cotton roving with three sets of paired rollers turning at increasing speeds, which stretched the roving to quadruple its original length before twisting it. Furthermore, the spinning frame spun 128 threads at a time and required only one unskilled laborer to operate it. The spinning frame had the necessary features to implement mass production.

Arkwright first built a mill run by horsepower, but he could not afford the expansion necessary to grow the business, so he sought new partnerships. In 1769, the Derby hosiers Samuel Need and Jedediah Strutt agreed to finance Arkwright's new design, the water frame, essentially a spinning frame driven by waterpower. In August 1771, Arkwright leased land on the Bonsall Brook just above the River Derwent in Cromford, Derbyshire, where he built a five-story factory. The key to this location was the Cromford Sough, which drained hot springs water into the Bonsall Brook just above the factory site, thus preventing the river water from freezing in the winter and allowing Arkwright to run his factory year-round.

The population of Cromford could not provide enough workers to run the factory, so Arkwright advertised throughout Derbyshire for large families, enticing them by providing housing. Indeed, he'd built a village of cottages specially constructed with a weaving shed on the top floor for the weavers to turn into cloth the yarn that their wives and children had milled at the factory. Children, who started work as early as age six, comprised his primary workforce (some two-thirds of Arkwright's 1,900 workers), as their dexterity made them up to 10 times more productive than adults. A brilliant manager, he persuaded workers to put in 13-hour days, and he ran several shifts so that his mills operated 23 hours a day.

In 1773, Arkwright's mills produced the first cloth made completely from cotton (previously only the weft was cotton, and the warp was linen), but a 1736 act of Parliament had imposed a double duty on calicoes; in 1774, Parliament enacted a special exemption for Arkwright, enabling him to continue his all-cotton fabric production. He filed a patent in 1775 covering all the preliminary stages of his production process—including processes invented by others, who in 1783 disputed the patent, which was annulled in 1785.

Arkwright was knighted in 1786, and the next year he was made the high sheriff of Derbyshire. He continued to expand his business, establishing factories throughout Derbyshire, Lancashire, Staffordshire, and even extending into Scotland. He tended to follow the same blueprint at all his factories, constructing stone buildings 30 feet by 100 feet or longer, and five-to-seven stories tall. His factory system became the model for the Industrial Revolution and was followed throughout the industrializing world. He continued to innovate, using steam power to run some of his factories after 1790. When Arkwright died on August 3, 1792, in Cromford, he had amassed a fortune of some half-a-million pounds.

Armstrong, Edwin
(1890–1954)
American
Engineer, Communications Industry

"Major" Edwin Armstrong could be considered one of the fathers of radio, as he invented three of the key elements to radio broadcasting: the regenerative oscillating circuit, which amplified the signal to audible levels; the superheterodyne circuit, which also amplified weak signals; and frequency modulation (FM), which prevented static while creating a clearer signal. Although the scientific community recognized Armstrong's achievements, the radio industry tried to escape its duty to pay him his due, forcing him to bankrupt himself in the process of defending his patents. Armstrong died not knowing that his inventions, specifically FM, would eventually become the standard in radio broadcasting.

Edwin Howard Armstrong was born on December 18, 1890, in New York City, the first child of a pair of native New Yorkers. His mother, Emily Smith, was a former public school teacher, and his father, John Armstrong, was vice president of the U.S. branch of Oxford University Press. While in London on business, his father picked up a copy of the *Boy's Book of Inventions*, and the entry on GUGLIELMO MARCONI inspired the 14-year-old to research the science of wireless communication and become an inventor.

Two years earlier, the family had moved to Yonkers, where Armstrong attended Public School Six until 1905, when he entered Yonkers High School. For his 1910 graduation, his father gave him a red motorcycle, on which he commuted to Columbia University to study electrical engineering. In the summer of 1912, he made the first of his major discoveries—what he called the regenerative circuit, which amplified a radio signal by bouncing it within the receiving tube to generate an oscillation that actually created electromagnetic waves, thus transforming the receiver into a kind of transmitter. Simply stated, he used feedback to amplify the signal, increasing its power as much as 20,000 times a second.

In 1913, Armstrong patented his invention and graduated from Columbia, where he remained as an instructor and assistant to the inventor and professor Michael Pupin, whose chair Armstrong inherited upon Pupin's retirement. His patent was issued on October 6, 1914, and he licensed use of the regenerative circuit to Marconi's company that year.

When the United States entered World War I, Armstrong was commissioned as a captain in the U.S. Army Signal Corps as head of the airplane radio section of the Research and Inspection Division, stationed in Paris. He outfitted the U.S. military forces with radios and was responsible for researching means of intercepting enemy radio transmissions. Toward this end, he conducted experiments from the Eiffel Tower to invent what he called the superheterodyne circuit, which converted high incoming radio frequency into a lower radio frequency before amplifying it, thus allowing the receiver to pick up barely audible signals. From a technical perspective, he achieved higher gain with fewer tubes, without creating any oscillation. The army promoted him to the rank of major (a title he retained for the rest of his life, even after his discharge from the military), and he received a ribbon from the French Legion of Honor.

In 1920, Armstrong sold the rights to his two circuit patents to Westinghouse Electric and Manufacturing Company for $335,000, and he later sold superregenerative circuit rights to Radio Corporation of America for a large block of stock. While negotiating with RCA president David Sarnoff, Armstrong fell in love with his secretary, Marion MacInnes, and the couple married on December 1, 1923.

Unfortunately, Armstrong became involved in a protracted legal battle over his regenerative circuit patent with Lee DeForest, who had invented the audion tube upon which Armstrong's improvement was based; DeForest also filed a patent for the regenerative circuit a year after Armstrong received the patent. The suit bounced through the courts from 1922 through 1934, when the U.S. Supreme Court judged in DeForest's favor on a technical misunderstanding. Interestingly, RCA failed to assist Armstrong, as its president, David Sarnoff, realized that if DeForest won the suit, RCA would retain its patent for another 10 years. On the other hand, the Institute of Radio Engineers supported Armstrong by refusing to rescind its first Medal of Honor, which it had awarded Armstrong in 1918, insisting that Armstrong was the rightful inventor.

Ironically, Armstrong's regeneration created the problem of static in some ways, because it amplified not only the signal but also any interference, heard as static. Armstrong solved this problem, despite expert opinion that static was as inevitable as poverty, by flip-flopping the existing standard, which modulated the amplitude (AM) on a fixed frequency, by holding the amplitude constant while modulating the frequency (FM). Armstrong filed four patents for frequency modulation (FM) in 1933. On June 9, 1934, the first trial of FM transmitted an organ recital from the top of the Empire State Building, with AM sending "hundreds of thousands times more static" than FM, according to Armstrong's friend who picked up both signals.

Despite the clear superiority of FM over AM broadcasting, the industry was entrenched in the existing technology and could not afford to shift its standard in the middle of the Depression. In 1938, Armstrong spent $300,000 of his own money to build a 425-foot tower and radio station on Hudson River Palisades at Alpine, New Jersey, to transmit FM signals. By 1939, there were 40 FM stations, and between May and July 1940, when the Federal Communications Commission opened up more of the broadcast spectrum, it received more than 500 FM applications.

In 1941, the Franklin Institute awarded Armstrong its Franklin Medal, thus confirming the Institute of Radio Engineers' conferral of recognition on Armstrong as the rightful father of radio. During World War II, Armstrong altruistically allowed the U.S. government to use his patents (he'd accumulated 42 royalty-free). The period after the war found him embattled with the industry, which had forced him to file 21 infringement suits to honor his patents. Bankrupted by legal fees, Armstrong got into a violent fight with his wife, who fled to her sister's in Connecticut. Distraught, Armstrong committed suicide by walking out the window of his 13th-floor Manhattan apartment on January 31, 1954.

The next year, Columbia University established the Armstrong Memorial Research Foundation in his honor. Between 1954, when she prevailed over RCA, and 1967, when she defeated Motorola, Armstrong's widow won all 21 infringement suits, recouping some $10 million and regaining the honor of her husband.

Avery, Oswald Theodore
(1877–1955)
Canadian
Bacteriologist, Physician

Oswald Avery was the first scientist to identify deoxyribonucleic acid, or DNA, as the material responsible for genetic transfer. With this discovery, Avery helped usher in the age of biogenetics, a branch of science that has wrought profound changes on society, from the use of DNA in courts as evidence to confirm the guilt or innocence of criminals to the cloning of sheep. A decade after Avery's 1944 discovery of DNA as the substance of genetics, James Watson and FRANCIS CRICK established the structure of DNA, further confirming the genetic function of the substance.

Oswald Theodore Avery was born on October 21, 1877, in Halifax, Nova Scotia, in Canada. His mother was Elizabeth Crowdy, and his father, Joseph Francis Avery, was a Baptist clergyman. When he was 10 years old, Avery and his family immigrated to New York City, where his father became pastor of the Mariners'

Temple in the Bowery on the lower east side of Manhattan—a mission church supported by the wealthy philanthropist John D. Rockefeller. In 1892, both Avery's father and his older brother, Ernest, died, leaving Avery to father his younger brother, Roy, while his mother went to work at the Baptist City Mission Society.

In 1893, Avery traveled to upstate New York, where he enrolled in Colgate Academy, and three years later matriculated into Colgate University to study humanities. In 1900, he graduated with a bachelor of arts degree and returned to New York City to study medicine at Columbia University's College of Physicians and Surgeons, where he earned his medical degree in 1904. He proceeded to practice general surgery but soon became disillusioned with his field's inability at the time to treat serious diseases, leading him into medical research as a means of advancing the field. In 1907, he landed a position as associate director of the bacteriological division of Brooklyn's Hoagland Laboratory, where his student nurses dubbed him "The Professor."

Avery studied how bacteria species caused infectious diseases. He excelled in practical applications: for example, he developed a quick and simple method for differentiating human from bovine streptococcus hemolyticus. He also focused much of his attention on the pneumococcus bacteria, determining the optimum and limiting hydrogen-ion concentration for pneumococcus growth. At this same time, Rufus Cole, director of the newly established hospital of the Rockefeller Institute for Medical Research, was focusing his laboratory's research efforts on finding a pneumococcus serum to fight lobar pneumonia. Then known as the "captain of the men of death," it claimed the lives of some 50,000 people per year in the United States.

One of Avery's papers on secondary infection in tuberculosis patients, which demonstrated a highly organized approach to clinical investigation, caught Cole's attention, prompting him to drive from Manhattan to Brooklyn to visit the bacteriologist. Impressed, Cole wrote Avery two letters offering him a position at the Rockefeller Institute Hospital; Avery was too busy to reply to either letter, requiring Cole to drive to Brooklyn again to make the offer in person. "The Professor" finally accepted in 1913, and after working at Rockefeller, his colleagues shortened his nickname to "Fess."

Avery, a recluse who never married, shared an apartment and a laboratory for many years with Alphonse Raymond Dochez. Together, the pair focused their research on the clear gelatinous coating encapsulating virulent pneumococci, discovering in it what they dubbed a "specific soluble substance." Avery and Dochez subsequently found this "SSS" in lobar pneumonia patients' urine and blood, thus verifying that virulence resides in pneumococcal coating.

Furthermore, Avery and Dochez discovered that culture fluids derived from this coating reacted with pneumococcal antisera in a type-specific manner—in other words, type I pneumococci culture fluids reacted only with type I antisera, type II cultures with type II antisera, and so forth. This marked the discovery of immunological specificity, or the reaction of specific bacteria with specific antigens. Avery next set about to analyze the chemical makeup of the capsule, enlisting the assistance of Michael Heidelberger. In 1922, the young chemist identified the coating's active ingredients as polysaccharides, or sugar molecules, earning pneumococcus Avery's nickname as "the sugar-coated microbe."

In 1928, the British Ministry of Health medical officer Frederick Griffith hypothesized that pneumococci might switch types upon losing their polysaccharides, based on his experimental injection of mice with live, capsule-deficient (hence harmless) pneumococci bacteria mixed with heat-killed virulent pneumococci of a different immunological type, which resulted in the death of the mice. At first, Avery was dubious of

Griffith's "transformation," but subsequent duplication of his results inspired Avery to pursue the identity of the transforming substance.

Avery enlisted two young scientists—Colin M. MacLeod and Maclyn McCarty—to assist in the complex process of isolating which component of the pneumococcal bacteria was responsible for the immunological transference. After more than a decade of frustrating work, the trio finally published their findings in a milestone paper entitled "Studies on the Chemical Nature of the Substance Inducing Transformation of Pneumococcal Types. Induction of Transformation by a Deoxyribonucleic Fraction Isolated from Pneumococcus Type III." The paper appeared in a 1944 edition of the *Journal of Experimental Medicine*. Its daunting title translates into one of the most significant biological findings of the 20th century—that deoxyribonucleic acid, or DNA, serves as the material by which genetic transferal takes place. A decade later, in 1953, James Watson and Francis Crick published the structure of DNA, spurring subsequent research that confirmed DNA's role as the genetic substance.

Avery retired in 1943, the year before the publication of the landmark DNA paper. He moved to Nashville, Tennessee, where his brother, Roy, taught bacteriology at Vanderbilt University. His career was honored with the receipt of the Paul Ehrlich Gold Medal and the Copley Medal of the Royal Society of London, a prestigious organization to which he belonged as a foreign member. Avery died of cancer on February 20, 1955. Rockefeller University erected and named a granite gateway on its northwest corner in Avery's honor.

Babbage, Charles
(1792–1871)
English
Mathematician, Computer Scientist

Charles Babbage's frustration with the persistence of human errors in mathematical tables gave birth to the notion of mechanical computation, an idea that led to the advent of modern computer technology. Once Babbage recognized the potential for the mechanization of mathematics, he devoted himself obsessively to its realization. He designed and attempted to build three mechanical computers in his lifetime, though it was not until May 1991 that Doron Swade successfully completed building a Babbage computer at the Science Museum in London for the equivalent of $500,000. Part of the driving force behind Babbage's project was his belief in the connection between science and culture, especially industry, which stood to progress on the shoulders of scientific discovery.

Babbage was born to affluent parents on December 26, 1792, in Teignmouth, England. He attended Cambridge University from 1810 to 1814, though he found that many of his professors could not match his intellect. In an attempt to bolster British mathematical standards to the level of those in continental Europe, he joined with John Herschel and George Peacock to campaign against British intellectual isolationism. Toward this end, he helped found the Analytical Society in 1815. He was instrumental in the founding of many other organizations, among them the Royal Astronomical Society in 1820 and the Statistical Society of London in 1834, a testament to his wide-ranging interests. His theory of "operational research" helped with the establishment of the British postal system in 1840. His diverse scientific interests included cryptanalysis, probability, geophysics, astronomy, altimetry, ophthalmoscopy, statistical linguistics, meteorology, actuarial science, and lighthouse technology.

In 1827 Cambridge University honored Babbage with an appointment as the Lucasian professor of mathematics, a position he held through 1839. The position would have served as a perfect platform from which to preach against and perhaps reform the substandard teaching of mathematics in Britain. Ironically, he did not use the position as a bully pulpit because of his obsession with mechanized computation and an analytical machine. In 1923 he commenced work on what became known as his Difference Engine No. One, based on addition rather than multiplication. The construction of this eight-foot-tall machine took a decade and cost the equivalent of $85,000, though he never completed it due to a loss of funding. He followed up in 1847 with his Difference Engine No. Two, a streamlined version of No. One, though he never constructed this design as the government refused to finance it.

Babbage's concept acted as a harbinger of modern computers, though his engine was an analog decimal machine as opposed to the modern computer, which is a binary digital machine. Lord Byron's daughter Ada Lovelace created programs for the prototypical computer, establishing the field of computer programming in conjunction with Babbage's designs for his computer. These two cohorts devised systems to predict winning horses but lost much money as these systems were never foolproof.

Although Babbage criticized the Royal Society for its conservatism, he was elected as one of its members in 1816. Babbage died on October 18, 1871, in London.

Baekeland, Leo Hendrik
(1863–1944)
Belgian/American
Chemist

Leo Baekeland discovered the first fully synthetic substance that could be molded into almost any shape imaginable—he named his innovation "Bakelite," partly after his own name, but its generic nomenclature is plastic. Other scientists had already discovered the same polymer from the reaction of carbolic acid, or phenol, with formaldehyde, but Baekeland devised a means of transforming it from a gooey resin into an infinitely transmutable substance by heating it under pressure, and he thereby ushered in the plastic age.

Baekeland was born on November 14, 1863, in Ghent, Belgium. His mother worked as a maid, and his father was a shoe repairman. He taught himself photography, then attended

"Bakelite," an early plastic, was discovered by Leo Hendrik Baekeland, a prominent chemist and businessman in the early 20th century. *(Smith Collection, Rare Book & Manuscript Library, University of Pennsylvania)*

night classes at the Municipal Technical School of Ghent to learn the chemistry of film developing. Unable to afford the chemicals necessary for the process, he recycled his own by melting down a silver watch given to him as a gift and figuring out how to extract and purify silver nitrate from it for use in exposing his pictures.

Baekeland entered the University of Ghent at the age of 17 to study chemistry under F. Swarts, whose daughter, Celine (known as "Bonbon"), he later married. The couple had two children. Upon graduation with a bachelor of science degree in 1882, he was retained by the university as a professor of chemistry while he pursued doctoral studies. He earned his doctorate *maxima cum laude* two years later and served as a professor of chemistry and physics at the Government Higher Normal School of Science in Bruges from 1885 through 1887. He then returned to the University of Ghent until 1889, when he immigrated to the United States after using a traveling fellowship to study in France and Britain.

In the United States, Baekeland first worked as a chemist for the E. & H. T. Anthony Company, a photographic paper manufacturer, before setting out on his own as a consultant and inventor of what the field lacked. He invented the first contact developing paper—an unwashed silver chloride emulsion—that succeeded on the commercial market. He then "invented" (the validity of this assertion is dubious) Velox, otherwise known as "gaslight" paper because it could be developed with artificial light (which was gas at the time) and thus did not rely on direct sunlight for development. While his scientific claim may have been hyperbolic, his business sense was very accurate, as he established the Nepera Chemical Company in partnership with Leonardo Jacobi to manufacture Velox, and GEORGE EASTMAN bought the company out in 1899 for $1 million.

Baekeland moved his family into Snug Rock, an estate on the banks of the Hudson River north of Yonkers, New York, converting the barn into a lab to continue his experimentation in search of his next big hit—a synthetic substitute for shellac, the natural resin that served as an electrical insulator, among other things. He followed in the footsteps of scientific predecessors, such as Adolf Von Baeyer, who combined phenol (a coal-tar derivative) and formaldehyde (an embalming fluid derived from wood alcohol). However, Baekeland hit upon a means of controlling pressure and temperature perfectly with his "bakelizer," a kind of industrial-strength pressure cooker that transformed the viscous, shellac-like by-product of the phenol-formaldehyde reaction into a hard-but-moldable substance, the perfect malleable synthetic.

After applying for a patent on July 13, 1907, and securing patent number 942,699 soon thereafter, Baekeland introduced polyoxybenzylmethylenglycolanhydride to the scientific community at the 1909 meeting of the New York chapter of the American Chemical Society and to the world thereafter as "Bakelite." He founded and served as president of the General Bakelite Corporation in 1910 to manufacture the plastic, but he again exercised his business sense by licensing the use of his process to other manufacturers, who made such imitations as Redmanol and Condensite, while he simultaneously pushed Bakelite as the genuine article. By 1911, Bakelite had set up plants in the United States as well as in Germany. By 1922, Baekeland had orchestrated the consolidation of several plastics makers into the Bakelite Corporation, which he continued to serve as president until 1939, when the company folded into Union Carbide and Carbon Company.

Baekeland also served as president of the three major chemical associations in the United States—the American Chemical Society in 1924, the American Electrochemical Society, and the American Society of Chemical Engineering. The first organization awarded him its Nichols Medal in 1909, and the third society granted him its Perkin Medal in 1916; the Franklin Institute bestowed its Franklin Medal on him in 1940. Baekeland died on February 23, 1944, in Beacon, New York.

Baird, John Logie
(1888–1946)
Scottish
Engineer, Communications Industry

John Logie Baird ushered in the age of television by transmitting the first television broadcast in

1923, across his small apartment. Over the next three years, he demonstrated his innovation to a fascinated public. However, electronic television systems, using cathode-ray tubes, supplanted his mechanical systems in popular use, due more to corporate pressure than to true technological superiority. Baird thought so far ahead of his time that many of his innovations are only now coming into use—for example, high-definition television, or HDTV, with over a thousand lines of resolution per screen, a system that Baird developed in the 1940s, became commercially available in 1990.

Baird was born on August 13, 1888, in Helensburgh, Scotland. He was the youngest of four children born to Jessie Morrison Inglis and the Reverend John Baird, a Presbyterian minister. He studied electrical engineering at Glasgow's Royal Technical College and then at the University of Glasgow, but these undergraduate studies were interrupted by World War I, for which he was deemed physically unfit to fight. At about this time, he read about PAUL GOTTLIEB NIPKOW's 1884 patenting of the *electrisches teleskop,* or a set of synchronizing spinning disks that sent and received light signals through a photosensitive selenium cell—this theory captured Baird's imagination.

During the war, Baird worked as an engineer with the Clyde Valley Electrical Power Company. He quit this job in 1919 to establish his own company, manufacturing his patented "undersocks." He sold this enterprise after a year and moved to the Caribbean island of Trinidad to set up a jam and chutney manufacturing company, which he likewise sold after a year. Back in Britain in 1920, he ran into Captain O. G. Hutchinson, who agreed to finance Baird's attempt to build a television, the term meaning "to see from a distance," coined by Constantin Perskyi at the 1900 International Electricity Congress in Paris. After establishing a soap-making company in London in 1921 and then selling it in 1922, he moved to Hastings, a sea-coast town 60 miles south of London, to conduct research.

Baird constructed his prototypical television from an old tea chest, discarded hat and biscuit boxes, darning needles, and sealing wax. He filed a patent for what he called a "televisor" in July 1923 (which was granted in May 1924) and soon thereafter transmitted a silhouette of a Maltese cross several yards across his apartment. He conducted his first public demonstration at Selfridge's department store in London on March 25, 1925. In October of that year, he transmitted the first image of a human being—William Taynton, a teenage office boy terrified by the hot lights cast upon him. Then, on January 26, 1926, he demonstrated the first fully functioning mechanical television to members of the British Royal Institution, transmitting moving images in gradations of tone.

Buoyed by these successes, Baird established the Baird Television Development Company and obtained a transmitting license from the British Post Office, whose telephone lines he utilized in 1927 to make the first "cable" transmission—the 438 miles from London to Glasgow. The next year, he performed the first transatlantic television transmission to an amateur radio operator in Hartsdale, New York. However, he ran into resistance from the British Broadcasting Corporation (BBC), whose general manager, J. C. W. Reith, viewed him as a competitor.

Baird reacted to the BBC's refusal to grant him a transmitting license by threatening to send "pirate" broadcasts, a threat he carried out by transmitting "unauthorized" television signals from Berlin in 1929. This pressured Reith into signing an agreement with Baird in September of that year for the transmission of a series of experimental television broadcasts, though few people had receiving televisions at the time. The BBC broadcast the Derby (the popular horse race) live on June 3, 1931, a first. Also in 1931, Baird married the concert pianist

Mary Albu, and the couple eventually had one son and one daughter.

The 1930s brought intense competition between rivals vying for the adaptation of their television systems as the standard. Specifically, the Electric and Musical Industries, Ltd. (EMI) in Britain and the Radio Corporation of America (RCA) poured their corporate monies into research and development. RCA unveiled the "iconoscope," invented by VLADIMIR ZWORYKIN, in 1933, and EMI followed suit two years later by introducing the "emitron" camera tube—an iconoscope clone, as RCA and EMI shared cross-licensing agreements in an effort to create a corporate monopoly over the airwaves.

In an effort to keep up with the corporate joneses, Baird licensed the manufacturing of his televisor to the Gaumont British conglomerate, which produced some 20,000 sets. Gaumont sought to hybridize Baird's mechanical system with the superior electronic systems by licensing PHILO FARNSWORTH's "image dissector," which the youngster had conceived of while still in high school in the early 1920s. Despite these advances, the BBC abandoned Baird's television system in February 1937, in favor of the EMI system developed by the Marconi Company.

Ironically, EMI triumphed through its corporate strength more than through technical superiority, as Baird's expertise allowed him to compete with superior technologies—viewers who compared EMI's fully electronic transmissions (boasting 405 lines per screen) to Baird's electromechanical broadcasts (transmitting 240 lines) could detect no difference. Baird's testimony in 1943 before the British government's Hankey Committee, charged with deciding the best course for television standardization, convinced them to endorse systems that he developed—including stereoscopic color television and high-definition television of more than 1,000 lines (a goal that was not achieved until Japanese manufacturers produced high-definition television technology in 1990).

Baird's work presaged the development of prerecorded television (which accounts for some 95 percent of broadcasting nowadays), the design on the Sony Trinitron tube (based on Baird's single electronic gun CRT), and the Rank-Cintel film scanning system (an extension of Baird's Cinema Television system) that much contemporary television utilizes. However, many of Baird's technological advances did not come into use during his lifetime. Baird died on June 14, 1946, in Bexhill, in the Sussex region of England.

⊗ **Bakewell, Robert**
(1725–1795)
English
Agricultural Scientist

Robert Bakewell established the practice of agricultural science, experimenting with his land and his animals to discover the best farming methods. He innovated new techniques of flooding and manuring his lands, and he manipulated his animal-breeding to retain the most desirable traits while avoiding negative traits. His breeding methods helped to feed the population explosion accompanying the Industrial Revolution and established a preoccupation in farmers for the genetic disposition of their animals.

Bakewell was born on May 23, 1725, at Dishley Grange, just north of Loughborough, in Leicestershire, England. He was the third child born to Rebecca Bakewell; Robert Bakewell, his father, was a tenant farmer renting 440 acres of farmland at Dishley. Bakewell grew up farming, foregoing formal education in favor of an agricultural tour of Europe to systematically observe the farming techniques of different cultures. He apprenticed to his father upon his return from Europe and inherited the run of Dishley upon his father's death in 1760.

Bakewell ran his farm as a kind of working experiment in agricultural methods. For example, he divided his land into arable land and grassland, devoting one-quarter to the former and the rest to the latter. Of the arable land, he apportioned 15 acres for growing wheat, 25 for spring corn, and 30 for turnips. He transported turnips to storage or market in a flat-bottomed boat until he realized in 1789 that turnips float, after which he simply threw the root into the canal and collected them, clean and ready to store, at the far end of the canal.

Bakewell took a holistic approach to farming, treating his animals humanely (he kept calves with their mothers, for example) and taking advantage of natural synergies. He housed his cattle over the winter in stalls specially designed on platforms that allowed manure to seep into collecting ditches, thus conserving straw for use as fodder instead of for lining stalls. Bakewell even wintered neighboring cattle herds for free in exchange for their manure, as he fertilized his arable land liberally and experimented with different manuring techniques on specially marked plots.

Bakewell's most significant contributions to the agricultural sciences were his animal-breeding techniques, which he guarded carefully for fear that his laboriously gathered information would be stolen (he never published his findings.) He perfected the technique of "breeding out-and-in," whereby he would out-breed his animals to pick up desired traits from afar, then in-breed within his own herd to fix the positive traits in his stock. He also maintained unusually large herds, raising 400 sheep, 160 cattle, and 60 horses without overgrazing his land.

Contrary to the practice of the day, Bakewell controlled breeding on his farm, separating males from females and allowing them to intermingle only to mate in hand-selected pairs. By these methods, Bakewell created the New Leicester (or Dishley) breed of sheep from the old Lincolnshire breed. Dishley sheep featured fattier mutton, better for nourishing the poor. Also, Bakewell instituted the practice of ramletting, or leasing out studs for the mating season to introduce his favorable traits into other herds. In 1786, Bakewell earned 1,000 guineas by letting 20 rams, and in 1789, he made 1,200 guineas on three rams alone.

Bakewell bred longhorn cattle to produce extra meat while consuming less food, thus creating a much more efficient breed. Bakewell's longhorn breed became extremely popular (despite the fact that they were no good at milk production), until an apprentice of his bred shorthorn cattle. Bakewell treated his workers well, retaining them for a minimum of four years (in part to prevent the loss of his trade secrets). He further minimized the migration of his secrets by feeding his sheep on flooded fields before selling them, as he had discovered that this practice infected them with liver fluke, killing them within a year, thus preventing other farmers from gaining his breeding secrets.

Though stingy with his agricultural knowledge, Bakewell was generous with the money he made from his farming innovations, treating houseguests like royalty (in fact, many were from royalty in Europe). This practice eventually bankrupted Bakewell in 1776. Bakewell never married but lived with his single sister, Hannah. Childless, he passed the farm on to his nephew, Honeybourn. Bakewell died on October 1, 1795, at Dishley Grange.

Banting, Sir Frederick G.
(1891–1941)
Canadian
Physician

Sir Frederick Banting conceived of the experiment that led to the discovery of insulin, the pancreatic substance that regulates blood sugar levels in diabetics. He carried out the experiment in collaboration with CHARLES HERBERT BEST and James Bertram Collip, under the supervision of John James Rickard Macleod, with whom he shared the 1923 Nobel Prize in physiology or medicine. The discovery of insulin revolutionized the lives of diabetics, extending their life expectancy and improving their quality of life significantly.

Frederick Grant Banting was born on November 14, 1891, near Alliston, Ontario. He was the youngest of five children born to Margaret Grant and William Thompson Banting, a farmer. In 1911, Banting entered the University of Toronto's Victoria College to study theology for the Methodist ministry. Within a year, however, he shifted his major to medicine, and in preparation for World War I, the university speeded up his course of study to grant him his medical degree in December 1916, whereupon he departed for England to serve as a lieutenant in the Canadian Army Medical Corps. He was assigned to the orthopedic hospital at Ramsgate, where he performed surgery under Clarence L. Starr.

In the Battle of Cambria near Haynscourt in France in September 1918, Banting caught shrapnel in his right forearm, though he continued to attend to wounded soldiers despite his own wound. The next year, he received the Military Cross from the British government for heroism under fire.

Upon his return to Canada after the war, Banting interned as the resident surgeon under Starr at Toronto's Hospital for Sick Children. In 1920, he established a private surgical practice in London, Ontario, where his fiancée, Edith Roach, was teaching (though the couple never married). To supplement the meager income from his practice (only one patient visited in his first month), Banting taught orthopedics as a demonstrator at the University of Western Ontario for two dollars an hour.

On the evening of October 30, 1920, preparing for a lecture on carbohydrate metabolism in the pancreas, Banting read an article by Moses Baron entitled "The Relation of the Islets of Langerhans to Diabetes, With Special Reference to Cases of Pancreatic Lithiasis" in the November issue of the journal *Surgery, Gynecology, and Obstetrics*. The article discussed the case of a blocked pancreas that had shriveled up, all except for the islets of Langerhans, which Banting suspected held the key to unlocking the mystery of diabetes. Unable to sleep thereafter, Banting got up at 2:00 A.M. to write in his journal: "Diabetus [sic] Ligate pancreatic ducts of dogs. Keep dogs alive till acini degenerate leaving Islets. Try to isolate the internal secretion of these to relieve glycosurea [sic]." Banting thus conceived of the experiment that would eventually lead to the isolation of insulin, the chemical mechanism for controlling diabetes.

Later that morning, Banting discussed his experiment idea with the neurophysiologist Frederick R. Miller, head of the laboratory he was working in, who recommended that Banting contact the University of Toronto's J. J. R. Macleod, an expert in carbohydrate metabolism. At their early-November meeting, Macleod reacted skeptically at first, assuming that Banting would follow in the footsteps of numerous other scientists who had researched the role of the pancreas in diabetes fruitlessly, until Banting revealed his intention of preparing a pancreatic extract, an approach Macleod met with enthusiasm. Macleod promised Banting 10 dogs, an assistant versed in blood and urine analysis techniques, and eight weeks in his laboratory, starting in mid-May 1921.

Macleod supervised Banting's experiment for a month before taking off for the summer to his home in Scotland. In his absence, Banting and his assistant, Charles Best, a 22-year-old medical student who had only just received his bachelor's degree in physiology and biochemistry, proceeded with the experiment. They met failure at first, as the catgut sutures tying off the pancreatic ducts disintegrated, prompting them to substitute silk suturing. On July 30, 1921, they removed the withered pancreas, chopped it and ground it in a cold mortar, mixed it with saltwater, filtered it through cheesecloth, then injected it into the diabetic dog, whose blood-sugar level decreased from 0.2 to 0.12—Banting and Best cautiously considered their extract, which they called "isletin," a success.

Banting and Best refined their extract by using the pancreases of fetal calves, which contained higher concentrations of islet cells. They were joined by biochemist J. B. Collip, who purified their extract chemically. When Macleod returned from his vacation, Banting (who had abandoned his surgical practice and teaching to pursue this experiment) requested a salary, which Macleod grudgingly granted him by appointing him as a senior demonstrator at the University of Toronto. Banting presented a preliminary report of his and Best's findings to the Physiological Journal Club in Toronto on November 14, 1921.

In January 1922, Banting and Best acted as human guinea pigs, injecting themselves with insulin, the name they adopted for the extract. Finding no adverse side effects, they injected the extract into 14-year-old diabetic Leonard Thompson, who was hovering on the brink of death. The boy's blood-sugar level decreased, and it dropped even further with the administration of a purer, more potent extract 12 days later (he lived another 13 years, eventually dying not of diabetes but of pneumonia contracted after a motorcycle accident). The scientists reported their success in the *Journal of the Canadian Medical Association*, and the Eli Lilly pharmaceutical company began production and distribution of insulin, the new diabetes-regulating drug. Insulin radically transformed the lives of diabetics, allowing them to regulate their blood-sugar levels and lead comparatively normal lives.

The granting of the 1923 Nobel Prize in physiology or medicine to Macleod and Banting infuriated the latter, who considered the former a mere overseer, while the overlooked Best had collaborated as a primary experimenter. Banting therefore split his prize money with Best, while Macleod split his with Collip. The Canadian Parliament also granted Banting an annuity of $7,500 from 1923 on. Also that year, the University of Toronto appointed Banting to an endowed chair named after him—the Banting and Best Chair of Medical Research. The next year, Banting married Marion Robertson, who bore him one son, William, in 1928. After a long separation, the couple divorced in 1932, and in 1937, Banting married Henrietta Ball, a technician in the Banting and Best Department of Medical Research, which he headed for the entirety of his career.

In 1934, King George V, himself a diabetic, knighted Banting as a baronet of the British Empire. At the outbreak of World War II, Banting volunteered as a medical consultant acting as a liaison between England and Canada. He died on February 21, 1941, when his transatlantic flight crashed in Newfoundland. Banting's memory survived, however, as diabetics the world over pay tribute to him as the discoverer of insulin, the miracle drug that regulates their lives. The Canadian Diabetes Foundation bought the house in London, Ontario, where he conceived of his famous experiment, establishing the Banting Museum in it. In 1989, Queen Elizabeth lit the Flame of Hope in the square beside the museum, which will remain burning until Banting's work is extended with the discovery of a cure for diabetes.

Barton, Clara
(1821–1912)
American
Nurse, Humanitarian

Best known for establishing the American Red Cross, the U.S. arm of the International Red Cross, Clara Barton was a humanitarian who tirelessly worked to help victims of natural disasters. She was affiliated with the American Red

Cross for more than two decades. Barton also worked as a battlefield nurse during the Civil War, and her fearless dedication resulted in the moniker "Angel of the Battlefield."

Born Clarissa Harlowe Barton on December 25, 1821, in the small town of North Oxford, Massachusetts, Barton was the youngest of five children born to a middle-class farming family. Barton's father, Stephen Barton, was not only a farmer but also a businessman and state legislator. Her father largely influenced Barton's compassion for her fellow humans. Barton's mother, Sarah Stone, was an intense and eccentric woman with a fierce temper. With many daily family chores, Barton developed a strong work ethic at an early age. She was also active in the community, tutoring underprivileged children and helping to nurse the sick.

Because teaching was one of the few respectable professions for women in the early 1800s, Barton became a teacher in 1839. After several years of teaching, she journeyed to New York to attend the Clinton Liberal Institute. When she finished her schooling, Barton returned to teaching, working at the first public school in New Jersey. In 1854, however, Barton, fighting depression and a nervous breakdown, left teaching for good. She moved to Washington, D.C., where she secured a job as a copy clerk in the U.S. Patent Office, making her the first female clerk in the federal government.

Though her job as a copyist was not necessarily fulfilling, Barton's life was changed forever with the outbreak of the Civil War in 1861. Hoping to help the troops, she requested permission to become a battlefield nurse. Women were not allowed near the battlefield, but Barton was extremely persistent, and in 1862 she was given a pass to the front lines. For three years, Barton traveled to battlefields with wagons of medical and food supplies. She fed wounded soldiers while also tending to their injuries. In 1865, the Civil War ended, and Barton spent the year attempting to identify dead soldiers and locate missing soldiers. Through her tireless efforts, she managed to identify 22,000 men.

In the mid to late 1860s, Barton traveled across the United States to deliver lectures about her experiences during the Civil War. Though Barton's lectures were well received, the physical and emotional pressures pushed her dangerously close to a breakdown in 1868. On the advice of her doctor, she traveled to Europe to recuperate. In 1869, Barton was in Geneva, Switzerland, where she met a group affiliated with the International Red Cross. One of the group, Dr. Louis Appia, asked for Barton's assistance in persuading the U.S. government to endorse the Treaty of Geneva. Among the provisions included in the treaty was the establishment of the Red Cross, an international wartime relief organization. Barton agreed, but before she could return to the United States,

Clara Barton, who is best known for establishing the American Red Cross. (National Library of Medicine, National Institutes of Health)

the Franco-Prussian War commenced, and she joined the Red Cross workers on the front lines.

Barton returned to the United States in 1873, but she soon suffered a nervous breakdown and was unable to work for two years. By 1877, Barton's health was stabilized, and she began her efforts to form a U.S. branch of the Red Cross. Dr. Appia appointed Barton the U.S. representative, and she worked to gain support from the government and generate publicity and national support. Barton also expanded the scope of the Red Cross to include peacetime efforts, such as helping victims of natural disasters, diseases, and major accidents. In 1881, Barton was elected president of the American Red Cross and succeeded in forming the first chapters. A year later, the U.S. government ratified the Treaty of Geneva, and from 1883, Barton dedicated her career to the Red Cross. Barton was both an administrator and a hands-on relief work participant. She was involved in helping victims of a Michigan forest fire, Ohio and Mississippi River floods, a drought in Texas, an Illinois tornado, and much more.

Though Barton was intensely committed to the Red Cross, infighting within the organization led her to resign in 1904. A year later, the U.S. government assumed control of the national organization. Barton moved to Glen Echo, Maryland, and wrote *A Story of the Red Cross*, published in 1905, and *Story of My Childhood*, published in 1907. Never married, Barton died of pneumonia on April 12, 1912.

Becquerel, Antoine-Henri
(1852–1908)
French
Physicist

Henri Becquerel followed in the footsteps of his father, Alexandre-Edmond Becquerel, and grandfather, Antoine-César Becquerel, who were both famous physicists. Henri established himself in his father's prominent posts at the Museum of Natural History and the National Conservatory of Arts and Crafts, and he held several other prestigious positions before he made his landmark discovery of radioactivity in 1896. For this work he earned the 1903 Nobel Prize in physics.

Henri Becquerel's discovery of radioactivity in 1896 earned him the 1903 Nobel Prize in physics. *(AIP Emilio Segrè Visual Archives, William G. Myers Collection)*

Becquerel was born on December 15, 1852, in Paris, France. Eminent scholars in his father and grandfather's circles surrounded him from the very beginning, introducing him at a young age to the world of science, and physics in particular. He began his formal education at the Lycée Louis-le-Grand, from which he graduated to the École Polytechnique in 1872. Upon his departure from the Polytechnique in 1874, he married Lucie-Zoé-Marie Jamin, daughter of physicist J.-C. Jamin. She died four years later, only weeks after giving birth to their son, Jean, who in turn followed in his father's footsteps to become a physicist. In 1890, two years after he earned his doctorate from the Faculty of Sciences of Paris, Becquerel married his second wife, the daughter of E. Lorieux, an inspector general of mines.

Becquerel earned positions from the institutions where he studied soon after his graduations. In 1876 the Polytechnique appointed him as a *répétiteur*, from which he rose to the position of full professor two decades later in 1895. In 1877 the Administration of Bridges and Highways appointed him as an *ingénieur* after he studied from 1874 to 1877 at the School of Bridges and Highways; he later ascended to the position of *ingénieur de première classe* for the administration. In 1878 the Museum of Natural History appointed him to his father's former position of *aide-naturaliste*. Henri's father Edmond died in 1891, and the subsequent year his father's chairs of physics at the museum and at the National Conservatory of Arts and Crafts succeeded to Henri.

In 1895 the German physicist WILHELM CONRAD RÖNTGEN discovered X rays. At that time Becquerel was studying fluorescence, as his father and grandfather had, and so he experimented with the substance he was studying then, potassium uranyl sulfate. Hypothesizing a connection between X rays and luminescence, he placed this uranium salt, known to luminesce, on top of photographic plates wrapped in black paper, then exposed this setup to sunlight. On February 24, 1896, he reported his results to the Academy of Sciences: the luminescence of the uranium salt exposed the plates through the black paper, thus suggesting the existence of X rays. Becquerel repeated this same procedure twice, but he stored these setups in dark drawers since the sun did not shine on February 26 or 27. By happenstance, he developed these photographic plates on March 1 and found them fully exposed, suggesting some type of radiation other than X rays, as the lack of exposure to sunlight meant that luminescence could not have been triggered.

The rays that exposed the photographic plates were actually called Becquerel's rays until his doctoral student, Marie Curie, named this phenomenon radiation. Her research for her dissertation proved so fundamental to the understanding of radiation that she, along with her husband, Pierre Curie, shared the 1903 Nobel Prize for physics with Becquerel. Their joint findings shifted the direction of physics in the 20th century. Becquerel died on August 25, 1908, in Le Croisic, in Brittany, France.

Benz, Karl Friedrich
(1844–1929)
German
Engineer, Transportation Industry

Karl Benz built the first vehicle successfully powered by an internal-combustion engine, patenting his design in 1886 and introducing the automobile later that year. His design drew on advances made by other scientists, and he continued to improve upon his own engine. His company eventually merged with the company founded by GOTTLIEB WILHELM DAIMLER, his rival automobile manufacturer. Clearly, Benz played a pivotal role in the transportation revolution that transformed world culture, allowing for personal long-distance travel.

Karl Friedrich Benz was born on November 25, 1844, in Karlsruhe, Germany. His father was a railway mechanic and engine driver. He attended the Karlsruhe Grammar School and then studied mechanical engineering at the Karlsruhe Lyzeum and Polytechnikum, graduating in 1864. He worked a series of factory jobs (draftsman, designer, and works manager) at Maschinenbau Gesellschaft, Karlsruhe, then for two years at Schweizer & Cie, Mannheim, then with the brothers Benskiser at Pforzheim in 1868.

In 1871, Benz established his own business—a small machine tool works—with August Ritter in Mannheim. The next year, Benz married Bertha Ringer, and together the couple had five children. The economic downturn in the middle of the decade required Benz to identify a product that would distinguish itself in a tight marketplace. He focused his efforts on the internal-combustion engine, though in 1877, N. A. Otto filed a patent on the four-cycle engine, forcing Benz to work on a two-cycle engine.

On New Year's Eve 1879, Benz turned out his first engine, which ran for long periods without experiencing problems (as many early engines did). Its success was due partly to its cylinder system, which was refreshed with air on each stroke. Benz continued to work out technical improvements while simultaneously searching for financial backing. It wasn't until 1882 that Benz secured such backing, when he founded the Gasmotorenfabrik Mannheim. Within a year, he left that company over differences with the other partners, but he found more backing elsewhere to found Benz & Co., Rheinische Gasmotorenfabrik, Mannheim.

In 1885, Benz starting driving his motor vehicle around Mannheim, reaching speeds of three miles per hour. The next year saw the revocation of Otto's four-cycle patent, and Benz jumped at the opportunity to patent his own four-cycle design. His patent was granted on January 29, 1886. On July 3, 1886, he introduced the world's first vehicle to be propelled by an internal-combustion engine—a three-wheeled carriage with a specially mounted engine.

Benz's first engine was equipped with a battery-buzzer electric ignition; it was water-cooled (when the water boiled off, the cylinder jacket simply had to be refilled); and it was capable of reaching speeds of three miles per hour, running 0.8 horsepower at 250 revolutions per minute. It featured a huge flywheel mounted horizontally in the rear. Unlike Gottlieb Daimler, Benz's fiercest rival, who simply mounted an engine on a carriage, Benz designed a completely integrated automobile. However, when he exhibited it at the Paris Fair of 1889, it drew scant popular interest.

Benz continued to make advances in his design and features, adding a fourth wheel in the 1890s and equipping the new car with kingpin steering. By 1899, the Benz company had manufactured 2,000 automobiles. In 1903, Benz left his company's board, only to return briefly in 1904. Finally, in 1905, he retired to Ladenburg.

The inflation resulting from reparations in the wake of World War I devastated the German economy and imperiled the fledgling automobile market. The number of automobile manufacturers had plummeted from 77 to 30. In 1925, the combined production of the two primary manufacturers, Daimler and Benz, numbered only 3,666 passenger cars, less even than in 1911. In seeking refinancing, the bank stipulated that Benz join forces with Daimler, in part as a means of ousting the 45 percent of a shareholder, the speculator Jakob Schapiro. In June 1926, the Daimler and Benz companies entered an "agreement of mutual interest" and merged their businesses.

Benz sat on the board of the newly formed amalgamation from 1926. He died on April 4, 1929, in Ladenburg (near Mannheim), Germany. The house he lived in his retirement was bought first by the town of Ladenburg and later by the Daimler-Benz foundation, a nonprofit philanthropic organization, as its headquarters.

Berliner, Emile
(1851–1929)
German/American
Inventor, Communications Industry

Emile Berliner made two of the most important inventions that established the modern era of mass media and telecommunications: the telephone microphone, which transformed Alexander Graham Bell's primitive contraption into a usable device; and the gramophone, an innovation that was an improvement over THOMAS EDISON's phonograph since it enabled the mass production of recordings. Berliner established companies in Britain, Germany, and the United States for the production and marketing of his gramophones and records; these companies eventually transformed into major global corporate powers (EMI, DGG, and RCA.) Berliner demonstrated not only scientific ingenuity but also business acumen, as he trademarked one of the most recognizable corporate symbols ever— Nipper, the dog that recognizes "His Master's Voice" over a gramophone.

Berliner was born on May 20, 1851, in Hannover, Germany. He grew up in a family of 11 children. He left school at the age of 14 to work as a printer's devil, then as a dry-goods clerk at age 15. At the age of 19, Berliner embarked on a stormy, two-week crossing of the Atlantic aboard the *Hammonia*, arriving in Washington, D.C., in 1870.

In New York, Berliner studied physics part-time at Cooper Institute (now Cooper Union) while working as a bottle washer in a chemistry laboratory for six dollars a week. By 1875, chemist Dr. Fahlberg (who discovered saccharine) had promoted him to lab assistant. The next year, he witnessed Alexander Graham Bell demonstrating the telephone at the U.S. Centennial Exposition; as with many other burgeoning scientists and inventors, Berliner endeavored to improve upon Bell's primitive sound-transmission system.

Berliner hit upon inspiration while sending a telegram, when the telegraph operator commented on a quirk in telegraphy—the harder he pressed on the keys, the better the transmission. Berliner recognized that an electrical contact responded to pressure, prompting him to experiment with the pressure of the voice's sound waves, thereby inventing the carbon microphone transmitter. In 1877, Thomas A. Watson (acting as agent for the Bell Telephone Company) negotiated with Berliner to buy the rights to his microphone for $50,000. Berliner's invention improved the Bell telephone immensely, and the principle behind his microphone remains behind the technology to this day.

That year, Thomas Edison invented the tinfoil phonograph, capable of replaying a recording a single time. Bell improved upon Edison's design with his graphophone, substituting a wax cylinder that could be replayed numerous times but had to be individually recorded. In 1887, Berliner provided the breakthrough innovation by switching from a cylinder to a flat disk as a recording medium that allowed for mass production, thus making the gramophone (patent number 372,786) and its sound recordings available to the mainstream public.

In 1889, the German company Kaemmer & Reinhardt first produced the gramophone commercially as a toy, playing discs pressed in the United States (such as Berliner record number 26, "Twinkle, Twinkle, Little Star") and shipped to Germany by Berliner himself. The gramophone was not produced in the United States until 1894, and even then it did not catch on, so Berliner went overseas to introduce his technology to new markets. He founded The Gramophone Company, Ltd., in England (now EMI—Electric and Musical Industries, Ltd.) to produce both gramophones and disks (or records) to play on them. He also founded the Berliner Grammophon Gesellschaft (now Deutsche Grammophon Gesellschaft—DGG), which pressed its first disks on June 11, 1898.

On September 15, 1897, the Gramophone Company purchased Francis Barraud's painting, "His Master's Voice," which depicted a dog named Nipper recognizing the sound of his owner's voice through an Edison-Bell phonograph (later repainted as a gramophone.) On May 26, 1900, Berliner registered the Nipper trademark for the American incarnation of his gramophone enterprise, the Berliner Gramophone Company, which struggled against Edison and Bell on the legal front of patent disputes as well as the popular front, each vying to establish its technology as the popular standard. One of Berliner's tactics in this battle was to persuade popular artists, such as Enrico Caruso and Dame Nellie Melba, to record on disks for playback on the gramophone. His company was renamed the Victor Talking Machine Company in celebration of the victory over Edison-Bell, first in the patent battle and consequently in the struggle to establish primacy as the standard-bearer in the marketplace.

Berliner made a series of other contributions to society, though none of them equaled the significance of his telephone or gramophone inventions. In 1908, he invented a radial aircraft engine, and in 1919, he invented a helicopter (actually an autogyro) with his son, Henry Adler Berliner. Berliner's contributions were not only scientific inventions, however; for example, in 1907, he organized the first conference on milk pasteurization and quality, in Washington, D.C., and in 1911, he established the Esther Berliner fellowship in his mother's name, to support women's scientific research.

In 1929, the Radio Corporation of America (RCA) acquired Victor and went on to become one of the major forces in the media establishment in the United States (and internationally). Berliner died soon thereafter, on August 3, 1929.

⊠ Bessemer, Sir Henry
(1813–1898)
English
Engineer, Steel and Iron Industries

Sir Henry Bessemer invented a steel production process that fueled the Industrial Revolution, as decarburized (or carbon-free) steel was used in the rails that extended train tracks throughout Europe and America, as well as in the ships that increased transatlantic commerce. The Bessemer process, which blasts air into molten pig iron to burn off the carbon that weakens the integrity of wrought iron, continues to form the basis of steel production.

Henry Bessemer was born on January 19, 1813, at Charlton (near Hitchin), in Hertfordshire, England. His father was Anthony Bessemer, an inventor and engineer. Bessemer left school early to continue his education by learning metallurgy at his father's type foundry, which produced gold chains for jewelry. At the age of 17, he moved to London, where he supported artistic endeavors by following in his father's footsteps as an inventor.

Bessemer gained no financial rewards from his first innovation, but he gained experience in the cutthroat tactics of the business world: he invented a deed stamp that prevented the reuse of old stamps, prompting the Stamp Office at Somerset House to offer him a comfortable salary as superintendent of stamps—until Bessemer suggested the dating of stamps, rendering his services moot, and prompting the government to retract its job offer. He sold his next invention—a method for compressing soft plumago dust (or native graphite) to form hard "lead" pencils—for 200 pounds to a friend who subsequently made a fortune on it.

He learned from his mistakes and ensured the success of his next invention. When his sister asked him to emboss lettering on the cover of her portfolio of flower paintings, he realized that the gold paint he used could be produced much more cheaply by using bronze powder, so he invented a machine to do so (the first prototype failed, the second succeeded). Wary of patenting his process, he chose instead to keep it secret. He had married Anne Allan in 1837, and he employed her three brothers to run a factory established in the St. Pancras section of London. Bessemer became a wealthy man from this invention.

During the Crimean War of the early 1850s, he turned his fertile mind to the invention of new technology to replace the inaccurate targeting of cannons. He realized that if he carved spiral grooves into long thin projectiles, they would spin in flight, thus improving accuracy. He patented his new artillery shell in November 1854; the British War Department ignored his innovation, but Napoleon III of France wooed Bessemer by financing lavish visits to Paris.

However, Bessemer's artillery proved too powerful for existing gun barrels, made from iron too weak to withstand the increased pressure generated by the new shells. So Bessemer applied himself to the fortification of gun barrels by strengthening their wrought iron. Inspiration struck him as he was lying sick in bed: blow air into molten pig iron to burn off carbon, which weakens the integrity of iron.

Bessemer conducted trials at Baxter House in St. Pancras, blasting 15–20 psi (pounds per square inch) of air through six horizontal pipes into the bottom of an egg-shaped caldron containing seven hundredweight of pig iron made molten by a four-foot cylindrical furnace. After 10 minutes, the molten iron sent up sparks and white flames like an erupting volcano, marking the success of the Bessemer Converter, which he patented in October 1855.

Bessemer established the Bessemer Steel Company in Sheffield, to produce steel by the Bessemer process, and to license to other manufacturers this process that reduced steel production costs significantly. After sustaining losses in 1858 and 1859, Bessemer Steel Company introduced a new Tilting Converter, which increased production capacity to 30 tons of high-grade steel in half-an-hour.

The Bessemer process spread worldwide, reaching France in 1858, Germany in 1862, and Austria in 1863. Andrew Carnegie observed the Bessemer process on a European tour and employed the process in the United States first in 1864, though under the patent of WILLIAM KELLY, who had earlier and independently developed a similar process. Some commentators have attributed the Bessemer process as the single most significant contribution to the Industrial Revolution. The Bessemer process remains the basis of steel production, though many innovations have improved the process.

The Royal Society inducted Bessemer into its fellowship in 1877, and two years later, Bessemer was knighted. Throughout his career, he filed more than 110 patents. Bessemer died on March 15, 1898, in London.

Best, Charles Herbert
(1899–1978)
American/Canadian
Medical Researcher

Charles Best collaborated with FREDERICK G. BANTING in identifying insulin as a diabetes-stabilizing substance. This discovery radically transformed the lives of diabetics, extending their life expectancy and improving their quality of life significantly. Best also conducted important research on choline, identifying its ability to reduce fat accumulation in the liver, as well as promoting the use of heparin as an anticoagulant, and the use of histaminase as an antiallergen.

Charles Herbert Best was born on February 27, 1899, in West Pembroke, Maine, a border town straddling New Brunswick, Canada. Both his mother, Luella Fisher, and his father, the doctor Herbert Huestes Best, who practiced on both sides of the border, originally hailed from Nova Scotia. World War I interrupted Best's liberal arts study at the University of Toronto, as he served as a sergeant in the Canadian Tank Corps (qualifying him for Canadian citizenship). Upon his return to the university, Best financed his premed course of study by playing baseball professionally.

In 1921, J. J. R. Macleod, Best's physiology professor, sponsored Frederick Banting's experiments on pancreatic extracts as diabetes regulators. To determine which of two students would serve as Banting's assistant, Macleod flipped a coin—Best won the coin toss. On May 17, 1921, the same day Best finished the examinations for his bachelor of arts degree in physiology and biochemistry (which he earned with honors), he commenced on experiments as Banting's assistant. This line of research resonated with Best, as he had lost his paternal aunt (who lived with his family) to diabetes only three years earlier.

Following up on the 30-year-old discovery by Oscar von Mering and Joseph Minkowski that dogs deprived of their pancreases developed diabetes, Banting conceived of an experiment to induce atrophy in the pancreas of a dog, from which to prepare an extract for injection into a pancreas-deficient, diabetic dog. Banting hypothesized that secretions from the pancreatic islets of Langerhans, devoid of the digestive enzyme trypsin that destroyed the stabilizing effects of the pancreas on blood-sugar levels, would counteract diabetes. After overcoming obstacles, the pair of researchers removed one dog's atrophied pancreas, chopped it up, ground it in a cold mortar, mixed it with saltwater, filtered it through cheesecloth, then injected this extract into the diabetic dog. The dog's blood-sugar level dropped from 0.2 to 0.12, representing a qualified success.

In September, Macleod returned from his summer vacation to suggest ways to improve the experiment; he also hired biochemist J. B. Collip to purify the extract chemically. Best and Banting published their initial results in an article entitled "The Internal Secretion of the Pancreas" in the February 1922 edition of the *Journal of Laboratory and Clinical Medicine*. The month before, Best and Banting had injected Collip's purified extract into themselves to test for side effects; finding none, they injected this insulin (Latinate for "island," after the islets of Langerhans) into Leonard Thompson, a 14-year-old diabetic degenerating toward certain death. The boy's health improved, and he lived another 13 years, dying in 1935 not of diabetes but of pneumonia contracted after a motorcycle accident.

Best continued to work toward his master of arts degree, which he received in 1922. At the same time, recognition of the profound significance of Best and Banting's discovery showered upon the scientists, as they essentially extended the life span of diabetics while also allowing them to live more normal lives. In 1923, while Best was delivering an address to Harvard medical students, word arrived of that year's Nobel Prize winners in physiology or medicine: the Royal Swedish Academy of Sciences granted the prestigious award to Banting and Macleod. Soon thereafter, a telegram arrived from Banting, who expressed fury that Best had been overlooked and promised to share not only the credit but also his prize money equally with Best. Macleod, who also considered the accomplishment collaborative, shared his prize money with Collip. In 1924, Best married Margaret Mahon, a historian and botanist, and the couple eventually had two sons.

In 1925, the University of Toronto granted Best his medical degree, as well as the Ellen Mickle Fellowship, awarded to the graduate with the highest standing in the medical course. The next year, he traveled to England for two years of postgraduate research under Sir Henry Dale, earning his doctorate from the University of London in 1928. While there, Best discovered the antiallergic enzyme histaminase.

In 1927, Best returned to the University of Toronto to take up the directorship of the department of physiological hygiene. Two years later, in 1929, when Macleod retired, Best assumed the chairmanship of the physiology department as well. Over the next decade, Best discovered the "lipotropic" function of choline, or this component of lecithin's ability to prevent fat accumulation in the liver. He also investigated the anticoagulating effects of heparin and organized Canadian efforts to supply dried blood serum to Allied wounded in World War II. In 1941, the Canadian Navy appointed him director of its medical research unit.

That year, when Banting died in a plane crash in Newfoundland, Best assumed the directorship of the department of medical research that the University of Toronto named after this pair of scientists in the wake of their landmark discovery of insulin, a post he retained for the remainder of his career. From 1948 through 1949, he presided over the American Diabetes Foundation; thereafter, he remained honorary president of this organization as well as the International Diabetes Foundation. In 1953, the International Union of Physiological Sciences named him its first president. That same year, the University of Toronto named a new medical research building the Best Institute.

Best retired in 1965, and in the next years, a group of his friends purchased and donated his parents' Maine house to the American Diabetes Association for use as a museum. Best died at Toronto General Hospital on March 31, 1978, several days after he ruptured an abdominal blood vessel upon hearing of his son's fatal heart attack.

Binet, Alfred
(1857–1911)
French
Psychologist

Alfred Binet instituted the first intelligence tests, though his original conception differs radically from how his ideas were applied by others. Binet considered intelligence too complex to be captured by a single number, as it would be impossible to take into account many unquantifiable factors. Binet, in collaboration with Théodore Simon, devised the Binet-Simon Intelligence Test specifically to identify students who developed *en retard*, or "late," which translated as "retarded," a term that has taken on a negative connotation absent from the original French word. This kind of bastardization exemplifies the misapplication of Binet's work, which survives now in the Stanford-Binet Intelligence Test. Binet specifically opposed Wilhelm Wundt's notion of an intelligence quotient, though Binet's name is associated with IQ. More correctly, Binet can be seen as a precursor to JEAN PIAGET's developmental psychology.

Binet was born on July 11, 1857, in Nice, France. His parents divorced when he was young, and his mother, who was an artist, raised her only child. After he graduated from the Lycée Louis-le-Grand in Paris, his father tried to persuade him to follow in the family tradition by becoming a physician, so Binet studied medicine briefly before switching to law for his degree.

However, law did not hold Binet's interest either, and he began independently studying the psychological works of Bain, Sully, and especially John Stuart Mill's theory of associationism (or environmentalism—the belief that environment dictates psychology) at the Bibliothèque Nationale. In 1880, he started publishing papers in his adopted field, bringing him to the attention of the neurologist Jean Martin Charcot, director of the Salpêtrière Hospital in Paris, who met Binet in 1883 and invited him to work at the hospital. There, Binet investigated hypnosis, hysteria, and abnormal psychology, as well as the more dubious fields of phrenology (the study of skulls to discern character traits) and physiognomy (the belief in a direct correlation between animal-appearance and psychology). However, Binet soon became disillusioned with Charcot's lack of scientific integrity.

In 1884, Binet married Laure Balbiani, daughter of the embryologist E. G. Balbiani, who invited his new son-in-law to work in his laboratory at the Collège de France. There, Binet conducted research for his doctoral dissertation, which he wrote on various aspects of insects, such as their behavior, physiology, histology, and anatomy. Also important to Binet's later research was the birth of his two daughters, Madeline and Alice, as he later based his psychological theories on his play with them.

Recognition of the significance of Binet's research came early in his career, as the French Academy of Moral and Political Sciences named him a *lauréat*, granting him a 1,000-franc prize in 1887 (a significant sum at that time). In 1891, he met Dr. Henri Beaunis, who offered him a position at the Sorbonne's laboratory of physiological psychology despite their disagreement over the validity hypnosis—perhaps because the independently wealthy Binet did not require a salary. By 1894, Binet became codirector of the lab, ascending to the director's position in 1895. That year, he and Beaunis cofounded *L'Année Psychologique*, the first French psychological journal, which he edited from 1897 until his death. He also sat on the board of the American journal, *Psychological Review*.

Binet's most significant research commenced when he took on Théodore Simon as a doctoral student. In the fall of 1904, the French government appointed Binet to a ministerial commission in conjunction with a new law requiring universal education for all French children, a regulation that raised the question of how to identify students who developed *en retard*, or later than most students. While conducting casual "research" on his two daughters, he realized the correlation between attention span and the progression of intellectual stages, and he devised sets of tasks appropriate to progressive developmental stages that could predict slow intellectual maturation.

In 1905, Binet and Simon established a pedagogical laboratory, testing about 50 "normal" children of different ages, as well as about 45 "subnormal" children. He gave each group about 30 simple tasks to perform: if three-quarters of the children of the same age successfully completed a task, it was considered age-appropriate. Binet and Simon developed a "Test of Intelligence," whereby they would introduce students to tasks appropriate to one year younger than their age. If the students passed this test, Binet and Simon would administer the tasks appropriate to their own age to see if their intellectual development was on par with their peers' age. If students could not perform the tasks appropriate to two years younger than their own age, they were considered *en retard*, in need of further surveillance and evaluation.

Binet and Simon first published what became known as the Binet-Simon Intelligence Test in 1905 and followed up with revisions in 1908 and 1911. Their test was adopted almost universally, except, interestingly enough, in France, where Binet's work was largely ignored. Unfortunately, the Binet-Simon test was also almost universally misapplied, as many failed to heed Binet's caveats: that test scores were to be used in practical matters only; that there was no proposed theory of intellect underlying the test; and that the test was meant to discover mild retardation in children, not rate differences in "normal" children. The validity of the test became bastardized when evaluators disregarded these distinctions.

Binet published prolifically throughout his career, and wrote several important books in his later period: *L'étude expérimentale de l'intelligence (The Experimental Study of Intelligence)*, published in 1903; *Les enfants anormaux (Abnormal Children)*, published in 1907; and *Les idées sur les enfants (Modern Ideas about Children)*, published in 1909. Binet died relatively young, on October 18, 1911, in Paris. In 1916, the Stanford-Binet Intelligence Test was instituted, and it has become a standard test. However, the fact that Binet was not alive to oversee the application of his ideas to practice resulted in their gross misinterpretation.

⊠ **Boulton, Matthew**
(1728–1809)
English
Inventor, Energy and Power Industries

Matthew Boulton and his partner, JAMES WATT, developed the steam engine for industrial applications, most significantly for coin stamping. While Watt provided the engineering expertise necessary to redesign the steam engine, moving the site of steam condensation outside the piston to vastly improve efficiency, Boulton provided the entrepreneurial drive and unflagging enthusiasm to promote the new technology. Together, Boulton and Watt helped spur the Industrial Revolution, as their steam engines provided the power that drove many factories. Over the quarter-century that they held a patent for the steam engine, they built and sold more than 500 of them, placing them throughout the industrializing world.

Boulton was born on September 3, 1728, in Birmingham, England. He received little formal education, leaving school at the age of 14 to join his father's silver-stamping business. By 1750, he had become a full partner in his father's business, inheriting it in 1759. At first, he continued to manufacture buttons and buckles, but he soon expanded operations.

Inspired by the ingenious American inventor BENJAMIN FRANKLIN's visit to England in 1758, Boulton joined with the scientist Erasmus Darwin and his own doctor, William Small, to found the Lunar Society of Birmingham, which met on the bright nights of the full moon (for safer travel, and hence the name) to discuss scientific developments. Boulton concurrently developed his manufacturing interests, which always went hand in hand with his scientific interests. In 1761, he established the "Soho Manufactory," a "community" of factories loosely linked to perform successive steps in the manufacturing of different products, such as flatware and snuffboxes or other trinkets, all derived from common sheets of metal.

Boulton's interest in steam engines was peaked in 1764, in response to Darwin's suggestion of building a steam carriage. He further developed this interest upon meeting James Watt, a Scottish engineer, who shared his enthusiasm for steam engines. While conducting research at a Glasgow University laboratory in 1769, Watt studied a model of THOMAS NEWCOMEN's steam engine and endeavored to improve upon the design. That same year, he joined with John Roebuck to patent his own steam engine design.

By 1773, Roebuck had gone bankrupt while still in debt to Boulton, who agreed to forgive loan repayments in exchange for the rights to Roebuck's share of Watt's steam engine patent. When Roebuck agreed, Watt and Boulton joined forces to extend the patent another 25 years, as of 1775. Watt further improved his design, patenting a double-action rotative steam engine in 1882 and the Watt engine in 1888. Without Boulton's entrepreneurial skill, however, Watt's steam engines would not have effected as significant an influence as they did—it was Boulton who placed the engines in the Cornish tin mines, driving the lapping machines that drained them for excavation.

The Royal Society recognized the influence wrought by Boulton on the world of science by inducting him into its fellowship in 1885, despite the fact that he never published a single scientific paper. The next year, Boulton applied the steam engine to his own silver-stamping business by designing steam-driven coinage machines that he patented in 1790. Before he even received this patent, however, the director of the United East India Company, Robert Wissett, contracted him to mint coinage for the colony of Bencoolen on the island of Sumatra. Boulton and Watt proceeded to mint coinage on contract to several other colonies under the auspices of the East India Company, including Bombay and Calcutta. In 1797, the Royal Mint followed suit by commissioning Boulton to reform its copper currency throughout the realm. Boulton steam engines continued to drive the machines in the Royal Mint for almost a century, remaining in operation until 1882.

By the time that Boulton passed his share of Boulton, Watt & Company along to his son, Matthew Robinson Boulton, in 1800, it had manufactured more than 500 steam engines. These engines drove factories, such as Sir RICHARD ARKWRIGHT's textile mills, that figured significantly in the development of the Industrial Revolution. In his retirement, Boulton contributed to the social structure of his hometown, Birmingham, establishing a theater in 1807 in addition to founding the General Hospital and Birmingham Assay Office. Boulton died in Birmingham on August 18, 1809.

Brindley, James
(1716–1772)
English
Engineer

James Brindley was one of the first canal-builders in England, launching the "canal era" that saw the proliferation of these man-made waterways across the country to fuel the Industrial Revolution by providing efficient transportation routes for sending raw materials to manufacturing centers and finished products back to the land of the consumer. Brindley was the first to employ tunnels and aqueducts in English canals. In all, he was responsible for the design and construction of some 365 miles of canals.

Brindley was born in 1716 in Newstead, near Buxton, in Derbyshire, England. He was the eldest child born to Susannah and James Brindley, a cottager. He almost certainly received little formal education, though stories of his illiteracy may be apocryphal, based less on hard evidence than on the fact that he performed all calculations in his head and hence left no record of his work. In fact, he had the unorthodox method of solving technical challenges by "sleeping on" the problem, often for several days, until the solution appeared to him—this technique (as well as his involvement in multiple projects simultaneously) earned him the nickname of "the Schemer."

In 1726, Brindley's father purchased a portion of Lowe Hill Farm in Leek. There, Brindley worked as a farm laborer until 1733 when, at the age of 17, he apprenticed to Abraham Bennett, a mill- and wheelwright near Macclesfield. Upon Bennett's death in 1742, Brindley established his own business as a millwright in Leek. He ran flint mills for Josiah Wedgwood's pottery and also concurrently involved himself in multiple other projects: he designed and constructed an engine to drain the Clifton coal pits; he held a share in the Golden Hill colliery; and he established a partnership with his brother, John, in the Longport Pottery.

In 1758, Brindley patented a different design for THOMAS NEWCOMEN's steam engine, though the actual engine itself proved a failure. That same year, Earl Gower contracted Brindley to survey a canal between the Trent and Mersey Rivers, a project that never made it off the ground. However, Gower recommended Brindley for another canal project, which secured Brindley's reputation as an ingenious engineer.

On July 1, 1759, Francis Egerton, the duke of Bridgewater, hired Brindley to design and build the Worsley Canal to link his coal mines to the manufacturing center of Manchester. It took Brindley six days to survey the route, and six years to complete the 10-mile canal. Three features especially distinguished the "Duke's Canal": a subterranean channel; the Barton aqueduct, which carried the water over the River Irwell; and the draining of the coal mines to supply water to the canal (though these last two innovations were subsequently attributed to John Gilbert, the duke's land agent).

In 1760, a son, John Bennett, was born to Brindley. On December 8, 1765, Brindley married Anne Henshall, and together the couple had two daughters—Susannah and Anne—in addition to the son he previously sired. Brindley returned to the Trent and Mersey Canal that he had surveyed for Earl Gower but did not build due to lack of funding; by 1865, Josiah Wedgwood had realized the intelligence of sending his fragile pottery to market via waterways as opposed to by horse and cart, so he backed the project.

Called the Grand Trunk Canal, it ultimately measured 93 miles, transversing Cheshire and Staffordshire; it featured 213 bridges, 160 aqueducts, 76 locks, and five tunnels, including the famous one-and-a-half-mile Harecastle Tunnel. While the Grand Trunk Canal was in construction, Brindley worked concurrently on several other projects: he was the supervising engineer of the Staffordshire Canal as well as of the Worcestershire Canal, both constructed in 1766; and the next year, he assisted in planning for the Birmingham and the Coventry Canals.

While conducting a survey for the Caldon Canal, a branch of the Trent and Mersey connector, Brindley got caught in a rainstorm and slept in his soaked clothing, resulting in a fatal cold. He died on September 27 (or 30), 1772, at Turnhurst, in Staffordshire, England, before the completion of the Grand Trunk Canal. Brindley's brother-in-law, Hugh Henshall, continued work on the canal, and it finally opened in 1777. Others of his canals were improved after his death, mostly by straightening them out to allow for more efficient transportation.

Brunel, Isambard Kingdom
(1806–1859)
English
Engineer

Isambard Kingdom Brunel started his career overseeing the Thames Tunnel project that his father, the French engineer MARC ISAMBARD BRUNEL, initiated with his invention of the tunneling shield. He spent the rest of his career performing incredible feats of engineering, made more incredible by their sheer number. He engineered the Great Western Railway, introducing a wider track gauge to allow faster service. He also engineered bridges and docks and even designed prefabricated hospitals for use in the Crimean War. Perhaps his greatest contributions, though, were a series of ships he designed and constructed—the *Great Western*, the *Great Britain*, and the *Great Eastern*—each of which was the largest ship of its kind when it was launched. Due to the diversity of projects he worked on, he was considered one of the greatest engineers of the 19th century.

Brunel was born on April 9, 1806, in Portsmouth, England, to Sophia Kingdom. He attended English private schools until he was 14 years old, when his father sent him to the College of Caen in Normandy, France. Brunel then attended the Collège Henri Quatre in Paris.

In 1823, he commenced work on the Thames Tunnel for his father. He proved such a distinguished engineer that when William Armstrong, the managing engineer, resigned due to ill health on August 7, 1826, Brunel the younger took over as resident engineer of the project at the age of 20. The tunnel flooded twice within two years: the first time, Brunel personally submerged in a diving bell to fill the breach with clay and gravel bags; the second time, the floodwater killed six men and knocked him unconscious. He spent a year recuperating, and the project spent the next seven years in limbo due to lack of financing.

During his year off, Brunel designed a suspension bridge to span the Avon River Gorge at Clifton, submitting his plans to a competition; however, the Scottish engineer THOMAS TELFORD awarded himself the design prize in 1829. Brunel's revised design won a subsequent competition, leading to his appointment as the engineer of the Clifton Suspension Bridge, though construction was not completed until after his death. However, his accomplishments thus far in his career qualified him for induction into the prestigious Royal Society on June 10, 1830. In 1831, Brunel designed the Monkwearmouth Docks in Bristol, the first in a series of commissions to design docks for other harbors, including Brentford, Briton Ferry, Milford Haven, and Plymouth. Then, on March 7, 1833, he was appointed engineer of the Great Western Railway (GWR), which featured some of his most innovative engineering feats: the Box Tunnel outside Bath, the Maidenhead Bridge, the viaducts at Hanwell and Chippenham, and the Temple Meads Station in Bristol. The GWR also advanced the "battle of the gauges," in which Brunel challenged the standard four-foot, eight-inch-wide track by introducing the wider gauge of seven feet, which remained in use until 1892. In all, Brunel was responsible for more than 1,000 miles of railway in Britain.

Brunel married Mary Horsley on July 5, 1836, whom he had met in the company of composer Felix Mendelssohn. His next projects moved him from land to sea, as he designed a series of three ships, each of which was the largest seafaring vessel in the world upon its launch. His first, the PS *Great Western* (named after the Great Western Railway Company, which owned it and financed its construction), measured 236 feet long, while the longest existing steamship was a mere 208 feet. The paddle-driven timber steamship was first launched on July 19, 1837, but embarked on its maiden voyage on April 8, 1838, reaching New York a mere 15 days later. Over the next eight years, it made 60 transatlantic crossings.

Later that same year, on June 4, the first section of the GWR opened, and on June 30, 1841, the Great Western Railroad between London and Bristol was finally completed. The next year saw the opening of the first section of the Bristol & Exeter Railway, also engineered by Brunel, on July 1. On March 25, 1843, two decades after Brunel started working on it, the Thames Tunnel opened.

Brunel also continued designing and constructing huge ships. After four years in the making, the iron-hulled SS *Great Britain* was launched on July 19, 1843, and remained in service for some three decades. Capable of carrying 250 passengers, 130 crew, and 1,200 tons of cargo, the behemoth was the first propeller-driven ship to cross the Atlantic Ocean, making its maiden voyage in 1845. Also that year, Brunel completed the Hungerford suspension bridge. A decade later, in 1855, Brunel designed a prefabricated hospital, which was shipped in parts to Crimea.

Brunel's final nautical achievement, the double-hulled, quintuple-decked PSS *Great Eastern*, was four times as big as the *Great Britain*. Brunel designed it to be made of plates and rivets of standardized size, but its construction was fraught with other problems, as its coal consumption doubled Brunel's estimate, and its launching, which cost 120,000 pounds more than expected, had to happen sideways. Though it was capable of carrying 4,000 passengers, it never made the Australian crossing that the Eastern Steam Navigation Company intended it for. However, it was used in 1865 to lay cable across the Atlantic.

Brunel's final achievement, the Royal Albert bridges—two 455-foot spans with a central pier built on rock 80 feet above high-water mark that crossed the river Tamar at Saltash near Plymouth—opened in 1859. However, the year before, Brunel had collapsed from a seizure of paralysis upon learning of an explosion on the *Great Eastern* at its launching on January 31, 1858. Brunel died on September 15, 1859, in London, and was buried in Kensal Green cemetery on September 20, 1859.

Brunel, Sir Marc Isambard
(1769–1849)
French/English
Engineer

Sir Marc Isambard Brunel made his mark first by establishing a mass-production facility for manufacturing pulley blocks at Portsmouth Dock. However, he is best remembered for his invention of the tunneling shield, a large device providing support of the subaqueous strata during the incremental digging of tunnels. He and his son, ISAMBARD KINGDOM BRUNEL, oversaw the initial implementation of this shield in the 20-year Thames Tunnel project.

Brunel was born on April 25, 1769, in the hamlet of Hacqueville, in the Normandy region of France. His mother, Marie Victoria Lefevre, was the second of four wives married to Jean Charles Brunel, a wealthy farmer who sent Brunel to seminary in Rouen in preparation for entering the clergy to quell his intention of becoming an engineer. However, the Superior recognized Brunel's irrepressible inclination toward mechanics, mathematics, and drawing, and directed him toward naval training.

From 1786 through 1792, Brunel served as an officer in the French navy. Upon his discharge, he reentered a society torn apart by the French Revolution; as a monarchist, he fled to the United States, where he became a successful architect and civil engineer, eventually ascending to the post of chief engineer of New York City. Another measure of his success was his winning of the competition to design the city plan for Washington, D.C., which was not implemented because it proved economically unfeasible.

He returned to Europe in 1799, landing in England in March. There, he reconnected with Sophia Kingdom, an English woman he had met in France before immigrating to America; the couple married on November 1, 1799. On April 9, 1806, their only son, ISAMBARD KINGDOM BRUNEL, was born. In the meanwhile, Brunel had established his reputation by founding one of the first completely mechanized mass-production efforts, making pulley blocks for the British military.

The British Admiralty needed 100,000 pulley blocks a year. Sir Samuel Bentham, inspector-general of naval works, compared his own plans with those submitted by Brunel, judged the young French engineer's design superior to his own, and so awarded Brunel the contract. Brunel set up a manufacturing facility at the Portsmouth dockyard, equipping it with 43 machines, driven by 30-horsepower engines and built by Henry Maudsley. This automated system decreased the necessary manpower from more than 100 skilled workers to 10 unskilled workers, and the final product ended up cheaper, yet of higher and more consistent quality than those made by hand.

An innovative inventor, Brunel was both unskilled and unlucky in business dealings. In 1814, the sawmills he built at Battersea, London, were all but destroyed by fire. The next year, as he was set to launch an army-boot factory, the Battle of Waterloo ended the Napoleonic Wars, folding his market; he ended up bankrupt. He and his wife were imprisoned for debt in 1821 and were released only upon the intervention of the duke of Wellington, who compelled Parliament to grant 5,000 pounds to free Brunel and his wife from debtors' prison.

Brunel earned his lasting legacy working on the Thames Tunnel. Previous attempts to excavate a tunnel beneath the river in 1801 and 1807 proved unsuccessful when they encountered quicksand. The key to Brunel's success was his 1818 patenting of his tunneling shield, consisting of three cast-iron tiers of 12 separate frames constituting 36 cells, each with a worker removing six-inch-wide planks one at a time, digging out the width of a brick, and then replacing the plank four-and-a-half inches further in. When all the planks in all 36 cells were dug, screw jacks pushed the entire shield forward, and then masons bricked another single layer.

On July 20, 1824, Brunel was appointed engineer of the Thames Tunnel, and digging began in 1825. However, working conditions were dangerous, as the roof leaked raw sewage (the Thames had become London's garbage repository). On May 18, 1827, the roof sprung a major leak, flooding the tunnel. After it was drained, it flooded yet again on January 12, 1828, killing six men. By August, with financial backing dried up, the project was put on hold, and the tunnel was bricked over.

When excavation recommenced seven years later, a larger shield (some 400 feet) was installed. However, the going was slow, as the work was illuminated only by candlelight, and hydrogen sulfide and inflammable methane explosions plagued the project: progress averaged less than one inch per day in 1837. By the early 1840s, as completion of the tunnel drew near, recognition was showered on Brunel, despite the fact that his son, Isambard, had taken over as resident engineer one year into the project, as Brunel the elder's invention was the key to success. In 1841, the Royal Society inducted him into its fellowship, and that same year, he was knighted

Work on the Thames Tunnel, a horseshoe 23 feet high, 37 feet wide, and 1,506 feet long, was completed in 1842. At its lowest point, the tunnel was only 14 feet beneath the riverbed. On March 25, 1843, the tunnel was opened to the public. Brunel died on December 12, 1849, in London, before the tunnel was transformed from a pedestrian walkway into a conduit for the London Underground; in 1865, the first trains traveled from Wapping to Rotherhythe.

⊠ Bunsen, Robert Wilhelm Eberhard von
(1811–1899)
German
Chemist

Although Robert Bunsen is best remembered for the ubiquitous piece of laboratory equipment named after him, the Bunsen burner, he did not actually invent it (he introduced a crucial design improvement), nor was this his most significant contribution to science. More significant was his introduction (in collaboration with Gustav Kirchoff) of the spectroscopic method of chemical analysis. He also discovered cacodyl as well as the elements cesium and rubidium.

Robert Wilhelm Eberhard von Bunsen was born on March 31, 1811, in Göttingen, Germany, the youngest of four sons. His father was a librarian and professor of linguistics at the University of Göttingen, thus exposing Bunsen to an academic environment from an early age. He attended school in Holzminden, graduating to his father's university to study chemistry. His keen intelligence made itself apparent, as he earned his doctorate at the age of 19.

A long period of travel throughout his homeland and Europe (financed in part by the German government) to conduct scientific stud-

Best known for his design improvements of the Bunsen burner, Robert Bunsen also introduced the spectroscopic method of chemical analysis. *(Smith Collection, Rare Book & Manuscript Library, University of Pennsylvania)*

ies ensued, as he visited laboratories in Berlin, Giessen, and Bonn, Paris, and Vienna (where he remained from 1830 through 1833). The University of Göttingen appointed him as a lecturer upon his return. While studying arsenic's insolubility, he discovered an antidote against the poison (which remains the best-known anodyne): the use of iron oxide as a precipitating agent.

In 1836, the University of Kassel hired Bunsen; then two years later, the University of Marsburg hired him. There, he distilled arsenic with potassium acetate to arrive at a cacodyl (also known as alkarsine or "Cadet's liquid"), a malodorous compound that he named after the Greek term for "stinking" (*kakodhs*). However, his arsenic studies blinded him in one eye when the compound exploded, rocketing glass shrapnel into his eye.

After recovering, Bunsen collaborated with Lyon Layfair to devise a means to improve the efficiency of blast furnace exhaust (which lost 50 to 80 percent of the heat produced) by recycling gases and other by-products, such as ammonia. In 1841, he invented the Bunsen battery, replacing the costly platinum electrode of Grove's battery with a carbon electrode. The next year, he was elected a member of the Chemical Society of London. In 1846, he devised Bunsen's theory of geyser action by heating water at the bottom and middle of a tube to the boiling point, thereby generating an ejaculatory explosion.

In 1852, the University of Heidelberg appointed Bunsen to the chair vacated by Leopold Gmelin. There, he performed electrolytic studies using chromic acid instead of nitric acid to generate pure metals of chromium, magnesium, aluminum, manganese, sodium, barium, calcium, and lithium. He innovated a means of measuring the specific heat of metals in order to determine their precise atomic weights by developing an ice calorimeter that gauged not the mass but rather the volume of the melted ice. Also in 1852, he commenced a collaboration with Sir Henry Roscoe lasting the remainder of that decade, in which they determined the exponential equivalency between light radiated by the sun per minute and the chemical energy generated by combining equal volumes of hydrogen and chlorine gases to form hydrochloric acid. The next year, he was inducted into the Académie des Sciences, and in 1858, he was named a foreign fellow of the Royal Society of London.

In 1859, Bunsen commenced collaboration with Gustav Kirchoff, a Prussian physicist, and together they innovated spectroscopy, or the study of light emitted by burning chemicals by passing it through a prism, thereby facilitating the chemical analysis of its constituent parts. This process required a nonluminous flame burning at a high temperature, so Bunsen redesigned a burner (by premixing the gas and air *before* combustion) originally designed by the University of Heidelberg technician Peter Desaga, but thereafter referred to as the Bunsen burner. Using the Bunsen-Kirchoff spectroscope, which they pieced together out of a prism, a cigar box, and two discarded telescopes, they discovered the elements cesium (from the Latin for "blue sky," *caesium*, after its blue spectral lines) and rubidium (from the Latin for "darkest red color," *rubidus*, after the violet spectral lines, in 1861.

Bunsen received numerous honors recognizing his accomplishments: the 1860 Copley Medal of the Royal Society, the 1877 Davy Medal (with Kirchoff), and the 1898 Albert Medal. He retired in 1889, and a decade later, on August 16, 1899, in Heidelberg, he died after three days of peaceful sleep.

C

Caldicott, Helen
(1938–)
Australian
Physician

A pediatrician who has worked extensively with children who have cystic fibrosis, Helen Caldicott is best known as an antinuclear activist. She has used her knowledge of the medical effects of radiation and her impassioned rhetoric as an activist to inspire people around the world to lobby against nuclear armament and nuclear energy. In the early 1970s, she organized opposition to the nuclear tests France was conducting in the South Pacific, which were in violation of the International Atmospheric Test Ban Treaty of 1962. Her speeches publicizing the violation and educating the public about the effects of radiation, particularly on children, resulted in the French government ending these tests. She also was responsible for reviving the U.S.-based organization Physicians for Social Responsibility and leading it during a period of rapid expansion in the early 1980s.

The daughter of Philip Broinowski, a factory manager, and Mary Mona Enyd Coffey Broinowski, an interior designer, Caldicott was born in Melbourne, Australia, on August 7, 1938. She attended public schools with the exception of four years spent at Fintona Girls School, a private secondary school in Adelaide. As an adolescent, she read Nevil Shute's novel *On the Beach* and was significantly affected by its depiction of nuclear holocaust. She entered the University of Adelaide medical school when she was just 17, graduating in 1961 with a B.S. in surgery and an M.B. in medicine (the equivalent of an American M.D.). The following year, she married William Caldicott, a pediatric radiologist, and the couple had three children—Philip, Penny, and William Jr.

After completing a three-year fellowship in nutrition at Harvard Medical School in Boston, Caldicott returned to Adelaide and accepted a position in the renal unit of Queen Elizabeth Hospital. There, she completed her residency and a two-year internship in pediatrics. She also established a clinic for the treatment of cystic fibrosis.

It was after working with children in the cystic fibrosis clinic and having children of her own that Caldicott began to organize others and speak out against nuclear proliferation. Her first challenge concerned France's illegal testing over Mururoa, a French colony in the South Pacific. Caldicott learned that in 1972, following five years of testing by France, there were higher than normal radiation levels in drinking water and

Helen Caldicott has used her knowledge of the medical effects of radiation and her impassioned rhetoric as an activist to inspire people around the world to lobby against nuclear armament and nuclear energy.
(Courtesy Helen Caldicott)

rain in Australia. She set about to educate the public on the effects of radiation. Her speeches in opposition to the testing inspired a mass popular movement that resulted in the Australian government taking legal action against France through the International Court of Justice and in France putting an end to its atmospheric testing.

Caldicott also led a struggle against the commercial uranium industry. In 1975, the Australian Council of Trade Unions passed a resolution banning the mining, transport, and sale of uranium, and the government implemented an export ban. But these measures prevailed only until 1982, when international pressure forced the ban to be lifted.

In the years following 1975, Caldicott and her family spent an increasing amount of time in the United States. She held appointments at the Children's Hospital Medical Center in Boston and became an instructor in pediatrics at Harvard Medical School. In 1978, she became involved with Physicians for Social Responsibility, a group whose membership grew rapidly following the near meltdown of Pennsylvania's Three Mile Island nuclear reactor on March 28, 1979. In 1980, Caldicott stopped practicing medicine in order to devote all her time to leading the organization. Her work involved lots of travel and public speaking to raise awareness among the general population. The organization made a documentary film, *Eight Minutes to Midnight*, which was often part of Caldicott's presentation and was nominated for an Academy Award in 1982. She also wrote *Nuclear Madness: What You Can Do!* with coauthors Nancy Herrington and Nahum Stiskin.

Eventually Physicians for Social Responsibility lobbied for a more mainstream platform than what Caldicott espoused, and she resigned as president in 1983. She went on to help found the Medical Campaign against Nuclear War, the Women's Party for Survival, and the Women's Action for Nuclear Disarmament, among other organizations. She has written two other books: *Missile Envy: The Arms Race and Nuclear War*, which came out in 1984, and *If You Love This Planet: A Plan to Heal the Earth*, which was published in 1992. She also ran for Parliament in Australia in 1990, losing by a very small margin.

Her many awards and honors include the Humanist of the Year Award from the American Association of Humanistic Psychology in 1982, the International Year of Peace Award from the Australian government in 1986, and a nomination for the Nobel Peace Prize in 1985. In the late 1990s, she was living in Canberra, Australia, with her husband.

Campbell-Swinton, Alan Archibald
(1863–1930)
Scottish
Engineer, Communications Industry

Alan Archibald Campbell-Swinton provided the theoretical basis for electronic television some two decades before the technology existed to implement his theories. He readily acknowledged the impossibility of realizing his "Distant Electric Vision" theory in practice, yet this did not discourage him from anticipating the invention of such technology. In the 1930s, electronic television supplanted mechanical television as the standard for broadcast in Britain and subsequently became the standard worldwide. Campbell-Swinton can thus be considered one of the fathers of television, an idea that spawned in many different parts of the world simultaneously and now affects all the world by disseminating information and entertainment.

Campbell-Swinton was born on October 18, 1863, at Kimmerghame, in Berwickshire, Scotland. A mere two years after Alexander Graham Bell invented the telephone, 15-year-old Campbell-Swinton had managed to wire a telephone connection between two houses, an early indication of his engineering skills. After graduating from the Cargilfield Trinity School, he attended Fettes College in Edinburgh from 1878 through 1881, then spent a postgraduate year touring France.

In 1882, Campbell-Swinton commenced an engineering apprenticeship at Sir William George Armstrong's works at Elswick in Newcastle. During his five-year tenure there, he solved a problem with insulating electrical wires onboard ships by encasing them in lead. In 1887, he left Newcastle bound for London, where he established his own electrical contracting business. He aided GUGLIELMO MARCONI in gaining an audience for his telegraphic invention by sending a letter of introduction to the chief engineer of the British Post Office, William H. Preece, who invited Marconi to demonstrate his instrument in 1896.

Also in 1896, soon after WILHELM CONRAD RÖNTGEN announced his discovery of X rays, Campbell-Swinton was one of the first scientists to conduct research into the possibility of using radiography for medical applications. He experimented extensively with CROOKES tubes, the same apparatus Röntgen had used to discover X rays, reporting no ill effects on his eyes after many hours of exposure. However, the hazardous effects of X rays and radiation generally were later established, prompting scientists to take precautionary measures while working with X rays and radiation.

Before the discovery of X rays, Crookes tubes were used to transmit and receive cathode rays. Campbell-Swinton realized a possible application for these functions—the transmission and reception of images. As early as 1903, he commenced preliminary experiments that eventually led to the development of television transmission, though he initially employed not a Crookes tube but a Braun oscilloscope tube, which similarly transmitted and received cathode rays. The next year, he dissolved his company to work independently as an engineering consultant while simultaneously continuing his work on television. His developing theory incorporated the scanning of images by means of electron beams, as well as the synchronization of the transmitter with the receiver.

In its June 18, 1908, edition, the prominent British science journal *Nature* published a letter from Campbell-Swinton under the headline "Distant Electric Vision" (a precursor to the term "television.") In it, he explained how a NIPKOW scanning disk could capture magnetic deflections off an image and transmit them line-by-line from one set of cathode-ray tubes to a receiver containing another set, which could display this image on a phosphorous-coated screen. However, he also acknowledged the impossibility of employing his theory in practice within the limitations of electronics technology of that day.

Three years later, in his vice presidential address to the Röntgen Society in London, Campbell-Swinton described an all-electronic television system, replacing the mechanical Nipkow disk with a electron gun that neutralized the charge to create a varying current. However, 20 years passed before the technology caught up with his theory, allowing for the implementation of electronic television. In the meanwhile, Campbell-Swinton worked as an engineer with the W. T. Henley Telegraph Works Company, PARSONS' Marine Steam Turbine Company, and Crompton Parkinson, Ltd., where he became a director.

Campbell-Swinton also spent the end of his career contributing to his field by serving numerous professional societies. He served as manager of the Royal Institution from 1912 through 1915; as president of the Radio Society of Great Britain from 1913 through 1921; as vice president and chairman of the council of the Royal Society of Arts from 1917 through 1919 and 1920 to 1921; as the vice president of the Institution of Electrical Engineers from 1921 through 1925; and as a member of the council of the Royal Society in 1927 and 1929.

Campbell-Swinton died on February 30, 1930, in London. Two years later, in 1932, EMI engineers W. F. Tedham and J. D. McGee realized an all-electronic television system based on Campbell-Swinton's theory in experiments kept secret even from their bosses at Electric and Musical Industries, Ltd., who had ordered them not to pursue such a line of investigation. The Marconi Company assisted in further development, and by early 1937, this electronic system had replaced JOHN LOGIE BAIRD's mechanical system as the standard of television transmission and reception on the British Broadcasting Company.

Carlson, Chester
(1906–1968)
American
Inventor, Xerography

The German philosopher Walter Benjamin placed great significance on the effect of mechanical reproducibility on art, claiming its ability to emancipate the masses. Chester Carlson, the inventor of the electrostatic photocopying process, would probably be surprised by such a grand claim, as his invention was ignored at first. Two decades after the first successful photocopy in 1938, the Xerox Company introduced its Model 914, a machine that revolutionized the modern office by allowing for quick and clear reproduction of print and image documents. Now, the ability to copy documents is taken for granted, as it has become an indispensable aspect of business, education, and personal life.

Chester Floyd Carlson was born on February 8, 1906, in Seattle, Washington, to Olaf Adolph Carlson, an itinerant barber who suffered from arthritis, and his wife, both of whom had contracted tuberculosis. The family moved to Mexico in search of healthy climes, then eventually settled in San Bernardino, California. To help support the family, Chester worked odd jobs—washing windows before school, sweeping offices after school—from the age of 12, and by 14 he was the family's sole means of support. In high school, he published a chemical magazine for extra income. When he was 17, his mother died, leaving him to care for his father alone.

Carlson worked his way through college, graduating from a junior college in Riverside, California, to the California Institute of Technology. Despite his jobs as a janitor and printer's devil, he was $1,400 in debt by the time he graduated with a bachelor of science degree in physics in 1930, at the onset of the depression. After applying to 82 companies for a job in his field (and receiving only two replies, neither containing a job offer), Carlson settled for a $35-per-week job as a research engineer at the Bell Labs in New York City. He attended law school at night, and when Bell laid him off, he shifted to the legal field for work, securing a position in the patent department at the P. R. Mallory Company, an electronics firm that manufactured batteries, which eventually promoted him to manager of the department.

Carlson's work here sowed the seeds of his invention, as patent applications required numerous copies of the drawings and descriptions, which had to be done photographically or by hand, both of which were expensive and time-consuming. Carlson pored over scientific tracts in the New York Public Library in search of a solution to his copy problem; after ruling out photographic copying, he happened upon photoconductivity, or electrostatics, a field recently established by the Hungarian physicist Paul Selenyi.

Carlson then conceived the idea of employing electrostatics to copy. He conducted experiments in his Jackson Heights apartment in Queens until his wife, Linda, complained about the noxious sulfur fumes, forcing him to move his lab to an apartment owned by her mother behind a beauty salon in the Astoria section of Queens. He hired Otto Kornei, an out-of-work German physicist, as his assistant.

On October 22, 1938, Carlson and Kornei conducted their first successful experiment in electrostatic copying, transferring writing from a microscopic slide to a zinc plate covered in sulfur. On the slide, he wrote the date and place ("10-22-38, Astoria") in India ink and rubbed the sulfur on the plate with a handkerchief to activate its electrostatic charge, then placed the slide on the plate and exposed them to a short flash of intense light. He then removed the slide and covered the plate in lycodium powder (derived from the waxy spores of clubmoss); in one big breath, he blew off the powder to reveal a copy of the message he had written! He preserved this image by heating wax paper to seal the powdered letters.

Carlson patented his invention in 1940 while simultaneously trying to interest companies in developing and marketing it—between 1939 and 1944, he pitched his photoconductive copying process to some 20 corporations, including IBM (which twice expressed some interest), Eastman Kodak, General Electric, and RCA, but he had no takers. Finally, in 1944, while on business from P. R. Mallory (a position he retained throughout his experimentation), he mentioned his reproduction process offhandedly on a visit to the Battelle Memorial Institute of Columbus, Ohio, and officials of this nonprofit technological research organization expressed interest in the process. Realizing the expense necessary to bring such a mechanism to market, Battelle devised a royalty-sharing plan giving Carlson 25 percent of profits until he reimbursed Battelle for its research investment (which amounted to $17,000 by the time Carlson paid it off) within five years, at which point his cut would rise to 40 percent.

During the remainder of the war, Battelle assigned a lone researcher—physicist and former printer Roland M. Schaffert—to develop the technology. He (along with a small group of assistants added after the war) introduced a new plate, covered in selenium instead of sulfur, as well as a corona wire for applying the electric charge and transferring the powder from the plate to paper. The group also developed new inks, including ammonium chloride salt (to clarify the blurry images) and plastic materials to melt iron particles in the ink into the paper.

On January 2, 1947, Battelle licensed the technology to the Haloid Company of Rochester, New York. The first public demonstration of the electrophotographic copying process took place on October 22, 1948, exactly a decade after Carlson's first successful experiment in Astoria. The process took a new name—xerography—after a suggestion by an Ohio State

University classics professor, based on the Greek words for "dry" (*xeros*) and "writing" (*graphos*). Haloid introduced its first photocopier, the Xerox Model A, in 1949, though it was not until 1959 that Haloid Xerox (as the company was renamed the year before) introduced a truly viable machine, Model 914 (named after the dimensions of legal-sized paper, which the copier could handle). *Fortune* magazine later called it "the most successful product ever marketed in America." By 1961, the Xerox Company (as it had truncated its name) earned almost $60 million annually in revenue, with revenues growing to more than $500 million by 1965.

By the 1960s, Carlson had married Dorris Hudgins. He never worked directly for Xerox but maintained his independence as a consultant. Of course, he received 40 percent of the profits, but instead of amassing personal wealth, he devoted himself to philanthropy and promoting world peace. He donated more than three-quarters of the $200 million he earned from his invention to charitable organizations, and he made sure to gift a block of Xerox stock to Kornei, who had helped develop the invention that radically transformed the business world (as well as the academic and publishing worlds).

On September 19, 1968, while walking down 57th Street in New York City from a conference to a movie theater, Carlson collapsed and died.

⊠ Carothers, Wallace Hume
(1896–1937)
American
Chemist, Textile Industry

Wallace Carothers conducted research at the DuPont Company that led to the discovery of two synthetic materials that radically transformed society through their versatility and utility: neoprene, a rubber analog, and nylon, a

Wallace Hume Carothers is shown here in the early 1930s, demonstrating the elastic properties of neoprene. *(Hagley Museum and Library)*

silk-like fiber. The advent of nylon freed manufacturers from dependence on natural fibers for the production of textiles; this development became increasingly significant during shortages of natural fibers, such as those experienced during World War II, when trade blockades made it impossible to access essential natural materials. The continued improvement of nylon has resulted in the tailoring of specific characteristics, such as insulation and water repulsion.

Wallace Hume Carothers was born on April 27, 1896, in Burlington, Iowa, the eldest of four children. His mother was Mary Evalina McMullin, and his father, Ira Hume Carothers, taught at the Capital City Commercial College in Des Moines, Iowa, eventually ascending to the vice presidency there. Carothers attended public school in Des Moines, graduating from North High School in 1914. He then matriculated to Capital City Commercial College and completed the accounting curriculum in a mere year to graduate in 1915.

Tarkio College in Missouri granted Carothers teaching assistantships in English and commercial studies, thus underwriting his study of the sciences—particularly chemistry and physics. Carothers completed the entire chemistry curriculum by the time that Arthur M. Pardee, the head of Tarkio's chemistry department, departed for World War I (Carothers did not qualify for military service due to his health), so the college appointed the undergraduate student to head the department. Carothers graduated with a bachelor of science degree in 1920.

Carothers completed a master's degree in chemistry within the next year at the University of Illinois. Pardee, who had moved to the University of South Dakota upon his return from the war, persuaded his former student (and replacement) to teach chemistry there for a year, after which Carothers returned to the University of Illinois for doctoral study. He won the Carr fellowship, the highest award granted by the department of chemistry, to finance the writing of his dissertation under Roger Adams on aldehydes in reactions catalyzed by platinum. He earned his Ph.D. in 1924, and the university retained him to teach in the chemistry department. That same year, he also published his first paper, "The Double Bond," in the *Journal of the American Chemical Society*, applying physicist Irving Langmuir's atomic concepts to organic chemistry.

In 1926, Harvard University hired Carothers as an organic chemistry instructor. After only three semesters there, Carothers was wooed by Charles Stine, research director for the DuPont Company, to join his fundamental research program as a director at the new Experimental Station in Wilmington, Delaware. What enticed Carothers away from one of the most prestigious institutions in the country was the prospect of conducting pure research supported by a team of postdoctoral associates using state-of-the-art equipment; furthermore, he would have no teaching duties, which had distracted him from his research.

Carothers and his team focused their research on acetylene, and by 1931, they had combined vinylacetylene with a chlorine compound to create a rubber analog, which DuPont marketed as "neoprene." However, Carothers had already conceived of the line of research that would earn him lasting fame. In a letter to Cornell University's John R. Johnson dated February 14, 1928, Carothers outlined his theory of linear polymerization, or the stringing together of monomers into long chains, called macromolecules. He eventually published a series of 31 papers which presented this theory in its entirety and laid the foundation for polymer chemistry, a field that has had a profound impact not only on the world of science but also on mainstream society, which benefits immensely from these man-made fibers.

The disintegration of trade relations between the United States and Japan, its main source of silk, spurred Carothers and his colleagues to investigate the synthesis of an artificial fiber similar to silk. Toward this end, they developed the condensation reaction method of polymerization; Carothers honed this technique to distill the by-product of water, thereby preventing it from dripping back into the concoction and weakening it. He prepared a number of polyesters by this method, some of which seemed commercially unpromising, and some of which he stored in vials for future analysis.

Carothers married Helen Everett Sweetman, an employee in DuPont's patent division, on February 21, 1936. A little less than a year later, in January 1937, Carothers's favorite sister, Isobel, who was a singer in the popular radio trio of Clara, Lu, and Em, died unexpectedly. Carothers, who was manic-depressive, fell into deep melancholia over the loss. Even the pregnancy of his wife could not buoy his spirits; he became obsessed with famous chemists who had committed suicide, and Julian Hill, a DuPont colleague, noticed a ration of cyanide in his briefcase. On April 29, 1937, Carothers administered himself a lethal dose of the poison and died in Philadelphia. His daughter, Jane, was born on November 27, 1937, half a year after his family had interred his ashes in Glendale Cemetery in Des Moines.

Carothers left behind him a legacy of 62 technical papers and more than 50 patents, enough to establish him as a significant contributor to science. However, his most significant contribution was only discovered posthumously, when DuPont tested the polyamide samples that Carothers had left in his lab. One, marked "Tiber 66" (after its dual-stranded structure containing six carbons on each side, created by the combination of adipic acid with hexamethylene-diamine), turned out to be a strong and versatile fiber, which DuPont marketed as "nylon" starting in 1938. Since then, DuPont has earned about $4 billion a year on this material that now sports innumerable diverse applications.

Carson, Rachel Louise
(1907–1964)
American
Marine Biologist, Ecologist

Rachel Carson combined a professional knowledge of science (she trained as a marine biologist) with poetic writing skill to create best-selling books about the sea. Her impact on history, however, resulted from a book she did not really want to write: a warning that unless human exploitation of the environment was curbed, much of nature might be destroyed. Carson's book *Silent Spring* introduced the idea of ecology to the American public and almost single-handedly spawned the environmental movement.

Rachel Carson was born in Springdale, Pennsylvania, on May 27, 1907. Her father, Robert, sold insurance and real estate. Her family never had much money, but the 65 acres of land around their home near the Allegheny Mountains was rich in natural beauty, which her mother, Maria, taught her to love. "I can remember no time when I wasn't interested in the out-of-doors and the whole world of nature," Carson once said.

When Carson entered Pennsylvania College for Women (later Chatham College) on a scholarship, she planned to become a writer. A biology class with an inspired teacher made her change her major to zoology, however. She graduated magna cum laude in 1929 and obtained another scholarship, to do graduate work at Johns Hopkins University. She also began summer work at the Marine Biological Laboratory in Woods Hole, Massachusetts. Carson had always loved reading about the sea, and Woods Hole gave her a long-awaited chance not only to see the ocean but also to work in it. She obtained a master's degree in zoology from Johns Hopkins in 1932.

Carson taught part-time at Johns Hopkins and at the University of Maryland for several years. Then in 1935 her father died suddenly and her mother moved in with her. A year later her older sister also died, making orphans of Carson's two young nieces, Virginia and Marjorie. Carson and her mother adopted the girls.

Needing a full-time job to support her family, Carson applied to the U.S. Bureau of Fisheries. In August 1936 she was hired as a junior aquatic biologist—one of the first two women employed there for anything except clerical work. Elmer Higgins, Carson's supervisor, recognized her writing ability and steered most of her work in that direction. He rejected one of her radio scripts, however, saying it was too literary for his purposes. He suggested that she make it into an article for *Atlantic Monthly*, and it appeared as "Undersea" in the magazine's September 1937 issue.

An editor at Simon & Schuster asked Carson to expand her article into a book. The result,

Under the Sea Wind, appeared in November 1941. Critics liked it, but the United States entered World War II a month later, and book buyers found themselves with little interest to spare for poetic descriptions of nature. The book sold poorly.

Carson continued her writing for the U.S. Fish and Wildlife Service, created in 1940 when the Bureau of Fisheries and the Biological Survey merged. She became editor in chief of the agency's publications division in 1947. In 1948, she began work on another book, drawing on information about oceanography that the government had obtained during the war. That book, *The Sea Around Us,* described the physical nature of the oceans. It was more scientific and less poetic than Carson's first book. Published in 1951, it became an immediate best-seller, remaining on the *New York Times* list of top-selling books for a year and a half. It also received many awards, including the National Book Award and the John Burroughs Medal.

Suddenly Rachel Carson found herself famous and, for the first time, relatively free of money worries. In June 1952, she quit her U.S. Fish and Wildlife Service job to write full-time. A year later she built a home on the Maine coast, surrounded by "salt smell and the sound of water, and the softness of fog." She shared it with her mother; her niece, Marjorie; and Marjorie's baby son, Roger. When Marjorie died in 1957, Carson adopted Roger and raised him.

Carson's third book, *The Edge of the Sea,* described seashore life. Published in 1955, it sold almost as well as *The Sea Around Us* and garnered its own share of awards, including the Achievement Award of the American Association of University Women. The book that gave Rachel Carson a place in history, however, was yet to be written. It grew out of an urgent letter that a friend, Olga Owens Huckins, sent to her in 1957 after a plane sprayed clouds of the pesticide dichlorodiphenyltrichloroethane (DDT) over the bird sanctuary that Huckins and her husband owned near Duxbury, Massachusetts. Government officials told Huckins that the spray was a "harmless shower" that would kill only mosquitoes, but the morning after the plane passed over, Huckins found seven dead songbirds. "All of these birds died horribly," she wrote. "Their bills were gaping open, and their splayed claws were drawn up to their breasts in agony." Huckins asked Carson's help in alerting the public to the dangers of pesticides.

Silent Spring appeared in 1962. It took its title from the "fable" at the book's beginning, which pictured a spring that was silent because pesticides had destroyed singing birds and much other wildlife. The health of the human beings in her scenario was imperiled as well. This was the only fiction in the book.

Carson's book did more than condemn pesticides. These toxic chemicals, she said, were just one example of humans' greed, misunderstanding,

Rachel Carson's book *Silent Spring* is often cited as the inspiration for the creation of the Environmental Protection Agency. *(Beinecke Rare Book & Manuscript Library, Yale University)*

and exploitation of nature. "The 'control of nature is a phrase conceived in arrogance born of the . . . [belief] . . . that nature exists for the convenience of man." People failed to understand that all elements of nature, including human beings, are interconnected, and that damage to one meant damage to all. Carson used the word *ecology*, from a Greek word meaning "household," to describe this relatedness.

The reporter Adela Rogers St. Johns wrote that *Silent Spring* "caused more uproar . . . than any book by a woman author since *Uncle Tom's Cabin* started a great war." The powerful pesticide industry claimed that if Carson's supposed demand to ban all pesticides—a demand she never actually made—were followed, the country would plunge into a new Dark Age because pest insects would devour its food supplies, and insect carriers such as mosquitoes would spread disease everywhere. The media portrayed Carson as an emotional female with no scientific background, ignoring her M.S. degree and years as a U.S. Fish and Wildlife Service biologist. Many scientists took Carson's side, however. President John F. Kennedy appointed a special panel of his Science Advisory Committee to study the issue, and the panel's 1963 report supported most of Carson's conclusions.

Rachel Carson died of breast cancer on April 14, 1964. The trend she started, however, did not. It resulted in the banning of DDT in the United States and the creation of the Environmental Protection Agency (EPA). Most importantly, it reshaped the way that the American public viewed nature.

Chain, Sir Ernst Boris
(1906–1979)
German/English
Biochemist

Sir Ernst Chain followed up on ALEXANDER FLEMING's 1928 discovery of penicillin by isolating its antibacterial agent, thus proving its effectiveness as one of the first antibiotics. For this work, Chain shared the 1945 Nobel Prize in physiology or medicine with his colleague Sir HOWARD WALTER FLOREY, as well as with Fleming. Chain spent the rest of his career promoting the development of penicillin as a kind of wonder drug—for example, he elucidated the chemical structure of crystalline penicillin, identifying four distinct types, and he later discovered penicillinase, an enzyme that destroys penicillin. The discovery of penicillin had a profound effect on medicine in the 20th century, enabling doctors to combat bacterial infections simply and effectively.

Ernst Boris Chain was born on June 19, 1906, in Berlin, Germany. His mother was Margarete Eisner, a Berliner, and his father, Dr. Michael Chain, was a Russian immigrant and chemical engineer who established a prosperous chemical manufacturing plant. Visits to the laboratory at his father's factory inspired Chain's early interest in chemistry, but when Chain was 14, in 1919, he lost his father. He attended secondary school at Berlin's Luisengymnasium. Despite the fact that his family lost its fortune in the postwar inflation, forcing his mother to transform their home into a guest house, they still afforded to send him to Friedrich-Wilhelm University, where he earned degrees in chemistry and physiology in 1930.

For the next three years, Chain held concurrent positions conducting enzyme research at Berlin's Charité Hospital and at the Kaiser Wilhelm Institute for Physical Chemistry and Electrochemistry. In January 1933, Adolf Hitler ascended to chancellor of Germany, prompting Chain, a Jew, to emigrate (his mother and sister chose not to and later died in concentration camps). He landed in England in April 1933 and worked briefly at University College Hospital Medical School before securing a position at Cambridge University's School of Biochemistry investigating phospholipids under Sir Frederick Gowland Hopkins.

In 1935, when Howard Florey, newly appointed head of Oxford University's Sir William Dunn School of Pathology, sought nominations for an accomplished biochemist to join his laboratory, Hopkins recommended Chain. Chain became a demonstrator and lecturer in chemical pathology there the next year. While investigating the protein lysozyme, his literature search yielded Alexander Fleming's overlooked paper on the penicillin mold, a study that interested Chain and, in turn, Florey.

Funded by a Rockefeller Foundation grant, Chain and Florey (joined by Norman Heatley) set out to identify and isolate the active ingredient in *Penicillium notatum,* which Fleming had failed to do. Whereas Fleming cooked up a broth from the penicillin mold, Chain and his colleagues took the further step of freeze-drying this broth to obtain a powder, which they administered to laboratory mice. Successful trials prompted the team to experiment on humans, which required them to prepare huge amounts of penicillin broth (125 gallons of broth yielded one tablet's worth of powdered extract). In 1941, they injected eight bacterial infection patients with penicillin; two died of unrelated complications, while the other six recovered.

Inspired by their discovery of penicillin's antibacterial agent, Chain and his colleagues devised ways to maximize production of this wonder drug for use in treating the wounded in World War II, which they commenced in 1943. After the war, Chain advocated the unrestricted use of penicillin, even behind the Iron Curtain in the Communist bloc.

In 1945, Chain shared the Nobel Prize in physiology or medicine with Florey and Fleming for their roles in the discovery of the antibacterial action of penicillin. In 1948, he married fellow biochemist Dr. Anne Beloff, and together the couple eventually had three children: two sons, Benjamin and Daniel, and one daughter, Judith.

By that year, he grew discouraged with Oxford's lack of support for advancing penicillin research, so when the Istituto Superiore di Sanità in Rome offered him the position of scientific director of its newly established International Research Center for Chemical Microbiology, Chain jumped at the opportunity. Over the next dozen years, Chain distinguished the institute as a premiere international research facility. In 1958, he and his colleagues isolated the basic penicillin molecule, leading to the production of hundreds of new strains of penicillin.

In 1961, Chain returned to England as a biochemistry professor at Imperial College in London. During this decade, he discovered penicillinase, an enzyme that destroys penicillin. He retired to emeritus status in 1973, though he continued to conduct research at the Wolfson Laboratories as a senior research fellow until 1976.

Besides the Nobel Prize, Chain also received the 1946 Berzelius Medal and the 1954 Paul Ehrlich Centenary Prize. In 1947, the French government appointed him as a commander in the Légion d'Honneur, and in 1969, Queen Elizabeth II knighted him. He died of heart failure in Ireland on August 14, 1979.

Cooke, Sir William Fothergill
(1806–1879)
English
Inventor, Communications Industry

Sir William Fothergill Cooke collaborated with Sir CHARLES WHEATSTONE to introduce the first working electric telegraph in England. Wheatstone developed the scientific workings of the five-wire telegraphic system, while Cooke promoted its introduction to the business community and the general populace. Samuel Morse introduced his single-wire system in the United States that same year, and he is generally accorded credit for telegraphy, less for his primacy than for the fact that his Morse Code became the standard language of telegraphy.

Strictly speaking, Cooke and Wheatstone deserve credit for introducing telegraphy, as they beat Morse to the punch.

William Fothergill Cooke was born in 1806 at Ealing in London, England. He was the son of a surgeon and followed in his father's footsteps to study medicine. He commenced his higher education at the University of Durham, then continued at the University of Edinburgh before enlisting in the Madras Army in 1826. After serving for five years, he recommenced his studies at the University of Paris in 1831.

By 1836, Cooke was an officer in the Indian army. On furlough, he attended a telegraphic demonstration by Professor Monke at the University of Heidelberg on March 6, 1836, and he immediately determined to forestall his medical studies to devote himself to developing the electric telegraph for practical use—specifically in alarm systems and railway signaling. By January 1837, Cooke exhibited his three-needle telegraph in London. He realized that the limitations of his scientific knowledge would prevent him from further developing the telegraph, so he consulted Michael Faraday and Dr. Peter Mark Roget (of thesaurus fame), who recommended that he meet with Charles Wheatstone.

On February 27, 1837, Cooke visited Wheatstone's Conduit Street home to discuss telegraphy with the eminent scientist, who immediately recognized the pitfalls of Cooke's telegraph. Cooke suggested a partnership, with Wheatstone contributing the scientific expertise and Cooke promoting the project for practical application, an arrangement the scientist reluctantly agreed to in May, solidifying this agreement by filing jointly for a patent of their five-needle telegraph on June 10. They signed a deed of partnership on November 19, 1837.

The five-needle telegraph functioned on an established principle, with the completion of a battery-line current controlled by a make-and-break switch; the flow of current activated an iron needle, which rested vertically when no current flowed but moved left or right when prompted by the electromagnetic charge. By deflecting any two of the five needles to one of the 20 letters arranged in a diamond formation, an unskilled operator could send a message letter-by-letter (improvising the six missing letters of the alphabet.)

On July 25, 1837, Cooke and Wheatstone demonstrated their telegraphic system to the directors of the London and Birmingham Railway. They had buried their five wires, covered in cotton, in iron pipes beside a one-and-a-half mile stretch of tracks between Euston, where Wheatstone sent the signal, to Camden Town, where Cooke received it. Although Wheatstone and Cooke were elated, the directors were indifferent, failing to recognize the potential of such an innovation. That same year, Samuel Finley Breese Morse introduced his own telegraphic system in the United States.

Within a year, the insulation had deteriorated around the five wires, victim of water damage. Cooke and Wheatstone solved this problem by insulating the wires in glass and raising the line on iron posts; the partners paid for this improvement out of their own pockets, recouping their investment by charging a shilling for public viewing of the system after it was in place. The directors of the Great Western Railway recognized the potential of Cooke and Wheatstone's telegraphic system, arranging a trial between Paddington and West Drayton. The success of this trial led to the implementation of a telegraphic line on the Great Western Railway in 1839.

The partnership between Cooke and Wheatstone was volatile and came to a head in 1841, when they disagreed over claims of credit for their invention. The renowned engineer MARC ISAMBARD BRUNEL intervened as a mediator, determining equal credit to each for the telegraph, attributing the scientific innovation to Wheatstone and the public introduction of the technology to Cooke. Thereafter, Cooke promoted the installation of telegraphic lines, paying Wheatstone a royalty on the miles of line laid. For his part, Wheatstone continued to improve the technology, developing a two-wire, one-needle system used to announce the news of the birth of Queen Victoria's second son in 1844 on the first public telegraph line opened between London and Gosport. On September 2, 1845, Cooke bought out Wheatstone's interest in the joint patent for 120,000 pounds, and on his own he formed the Electric Telegraph Company.

The fate of telegraphy was sealed on New Year's Day of 1845, when it proved instrumental in the capture of John Tawell as he fled his mistress's murder from the Slough railway station in Salt Hill, boarding a train for London. Beside the tracks ran a telegraphic line that carried an instant message describing the suspect, who was apprehended by police upon his arrival in London. This news captured the public interest in telegraphy, indelibly establishing its importance.

The Royal Society granted Cooke its Albert medal in 1867. The next year, Wheatstone was knighted, with Cooke receiving such honors a year after that, perhaps signifying the popular attribution of primary credit to the scientist for developing the technology, with secondary credit attributed to the entrepreneur who introduced the system to the populace. Cooke died a decade after his knighting, on June 25, 1879, at Farnham, in Surrey, England.

Cousteau, Jacques-Yves
(1910–1997)
French
Oceanographer

Jacques Cousteau popularized the study of the world beneath the water's surface with his books, films, and television series. Though not a scientific expert, Cousteau gained his appreciation of the seas through experience, and he experimented with cinematographic media to try to re-create a hands-on experience for his audience. Later in his career, he used his vast influence to raise environmental awareness in hopes of curbing the destruction wrought on the natural environment by its human inhabitants.

Cousteau was born on June 11, 1910, in Saint-André-de-Cubzac, France. His mother was Elizabeth Duranthon and his father was Daniel Cousteau, a legal adviser. Cousteau suffered from chronic enteritis, a painful intestinal condition, for his first seven years. A 1936 car accident mangled his left arm, but he opted against amputation in favor of rehabilitation, during which time he experienced a spiritual connection with the sea. On July 12, 1937, he married Simone Melchior. Jean-Michel was born in March 1938, and Phillipe was born in December 1939. Though both sons worked extensively with their father, Phillipe was slated to inherit his father's role before he died in a plane crash on June 28, 1979, near Lisbon, Portugal.

Cousteau attended the École Navale, graduating second in his class in 1933. After a military stint in Shanghai, China, Cousteau returned to the aviation academy and graduated in 1936. He served as a second lieutenant and a gunnery officer in the French navy before World War II. During the war he used his oceanographic experimentation as cover for his participation in the resistance movement, earning a Croix de Guerre with a palm after the war for his contributions to the resistance.

In 1942 Cousteau produced an 18-minute underwater film entitled *Sixty Feet Down*, commencing his career as an underwater filmmaker. As a pioneer in this field, he had to pave his own way. For example, he invented the Aqua-Lung with Emile Gagnan, which they patented in 1943, increasing the mobility and facility of underwater diving. In 1947 Cousteau set the world free diving record at 300 feet. On July 19, 1950, he purchased the *Calypso*, an old U.S. minesweeper that he converted into an oceanic laboratory and film studio. Thus outfitted,

Cousteau produced a steady stream of award-winning films from the *Calypso*. In 1955 he filmed a version of his 1953 book, *The Silent World*, which sold more than five million copies worldwide. The 90-minute film won the prestigious Palme D'Or at the 1956 Cannes International Film Festival and a 1957 Oscar from the American Academy of Motion Picture Arts and Sciences.

Over the next three decades, Cousteau produced a string of successful television programs and series, including ABC's *The Undersea World of Jacques Cousteau*, which ran from 1968 through 1976, PBS's *Cousteau Odyssey*, which commenced in 1977, and TBS's the *Cousteau Amazon*, which ran in 1984. More than 40 Emmy nominations were bestowed on Cousteau's television programs, often for their informational content. The Cousteau Society was founded in the 1970s as a nonprofit peace and environmental awareness initiative.

Several awards honored Cousteau's lifetime achievements, starting with the National Geographic Society's Gold Medal in 1961 and its Centennial Award in 1988. In 1985, the French government awarded him with the Grand Croix dans l'Ordre National du Mérite, and the U.S. government awarded him the Presidential Medal of Freedom. Though scientific purists may balk at Cousteau's lack of training and qualifications, few scientists have done more to raise worldwide awareness of scientific issues than Cousteau. Jacques Cousteau died at the age of 87 on June 25, 1997.

Crick, Francis
(1916–)
English
Molecular Biologist

Francis Crick, along with his research partner James Watson, revolutionized the scientific understanding of genetic transfer with their 1953 discovery of the structure of deoxyribonucleic acid (DNA). Although he expresses pride for his and Watson's persistence and ambition in tackling a seemingly impossible task, he also admits that the time was ripe for this discovery—if they had not made the discovery, some other scientist would have. Their discovery helped establish the field of molecular biology, and Crick's subsequent research on genetic coding advanced the field significantly. He shared the 1962 Nobel Prize in physiology or medicine with Watson and Maurice Wilkins.

Francis Harry Compton Crick was born on June 8, 1916, in Northampton, England, the older of two sons born to Anne Elizabeth Wilkins, a teacher, and Harry Crick, a shoe and boot factory manager. After studying locally at Northampton Grammar School until 14 years of age, Crick won a scholarship to Mill Hill School in North London, where his family moved. At age 18, he matriculated at University College, whose curriculum he completed in three years to graduate in 1937 with a second-class honors degree in physics (as well as a minor in mathematics).

University College retained Crick as a graduate student, studying the viscosity of water at high temperatures under Edward Neville de Costa Andrade. In 1940, he married Ruth Doreen Dodd, and together the couple had one son, Michael, who was born during an air raid on London on November 25, 1940. Also that year, Crick commenced work at the British Admiralty Research Laboratory, designing nautical mines to destroy German merchant ships in World War II. There he met Odile Speed, an art student serving as a naval officer translating captured German documents. Crick divorced his first wife in 1947 and married Speed in 1949. The couple eventually had two daughters together, Gabrielle and Jacqueline.

After the war, Crick chose his professional direction by the "gossip test" method, reasoning that scientific discovery follows human curiosity, which is expressed through "gossip." He and his colleagues thought and talked incessantly about the perpetuation of life, and Crick specifically wondered about the biological mechanism of life transferal, the motivating question in the formation of the field of molecular biology. Crick combined his background in physics with his fledgling interest in biology by studying the physical properties of cytoplasm in cultured chick fibroplast cells under Arthur Hughes as doctoral research at the Strangeways Laboratory in Cambridge, starting in 1947.

In 1949, Crick transferred to the Cavendish Laboratories under a Medical Research Council Unit studentship. After studying protein structure with Max Perutz and John Kendrew for two years, Crick shifted his attention to DNA upon the arrival of James Watson, a precocious 23-year-old visiting fellow with the Phage Group (short for bacteriophages, or bacterial viruses) from the United States. The two office mates quickly recognized their symbiosis of interests (Crick's structuralism and X-ray diffraction expertise, Watson's biological knowledge) in considering the question of genetic transfer. Watson identified the "intellectual chaos" surrounding this vital question, with no coherent theory organizing research efforts, and the pair boldly endeavored to solve the seemingly impenetrable mystery of genetic replication by educating themselves in the myriad disciplines involved in the equation.

Crick and Watson's discovery of DNA's structure proceeded from information that was largely available at the time—they simply assimilated this information into a coherent theory. OSWALD THEODORE AVERY had demonstrated in 1944 that purified DNA (not protein, as was previously assumed) was the primary carrier of genetic information. Erwin Chargaff had determined that the bases in DNA combined in one-to-one ratios, specifically pairing adenine with thymine and guanine with cytosine. And ROSALIND ELSIE FRANKLIN had taken X-ray crystallographic photographs of DNA that revealed its helical (or spiral) structure, though she did not know that her adversarial colleague Maurice Wilkins showed Watson this photo.

The key to Crick and Watson's solution was their application of the model-building method first employed by Linus Pauling in determining the three-dimensional alpha helix of protein. After much trial and error, Crick and Watson finally built a model on March 7, 1953, that conformed to all known physical rules. They published the Watson-Crick model (the former won a coin toss to decide the name order) in a paper simply entitled "A Structure for Deoxyribonucleic Acid" in the April 25, 1953, edition of the British scientific journal *Nature*, accompanied by a drawing of the double helix by Crick's wife, Odile.

Besides asserting DNA's structure, Crick and Watson suggested the possibility that DNA contains the mechanism for genetic replication. This hypothesis soon gained confirmation, representing what most commentators consider the most significant scientific discovery of the 20th century—their discovery solidified the field of molecular biology, opening the door to a deeper understanding of the transmission of the genetic code from one generation to the next.

In the wake of his announcement, Crick finished his doctoral dissertation, entitled "X-ray diffraction: polypeptides and proteins," to earn his Ph.D. in 1954. Crick remained at the Cavendish Laboratories, advancing the understanding of genetic coding through DNA in collaborative research with Vernon Ingram. In 1957, he commenced a collaboration with South African geneticist and molecular biologist Sydney Brenner, with whom he worked on genetic coding. At that year's symposium of the Society for Experimental Biology, Crick presented his landmark paper, "On Protein Synthesis," which "permanently altered the logic of biology," according to Horace Judson, author of *The Eighth Day of Creation*. In this paper, Crick advanced both the "sequence hypothesis," which held that the sequence of bases in DNA amounts to genetic coding, and the "central dogma," which asserted the irreversibility of the transfer of genetic information from nucleic acid into protein.

For his work on the structure of DNA, Crick shared the 1962 Nobel Prize in physiology or medicine with Watson and Wilkins. Throughout the rest of the 1960s, he focused his research on histones, or proteins associated with chromosomes. Then, in 1976, he took a sabbatical from the Cavendish Laboratories to the Salk Institute for Biological Sciences in La Jolla, California, which subsequently appointed him to the Kieckhefer endowed professorship.

At the Salk Institute, Crick shifted his research from molecular biology to neurobiology, focusing on consciousness, or what he calls "awareness." However, he admits that he "has yet to produce any theory that is both novel and also explains many disconnected experimental facts in a convincing way," as his 1953 discovery of DNA's structure did.

Crookes, Sir William
(1832–1919)
English
Physicist, Chemist

Sir William Crookes is best remembered for the practical application of his scientific genius, as he discovered a new element (thallium) and invented two important scientific instruments—the radiometer and the Crookes tube. The latter innovation acted as a key precursor to the discovery of X rays and the electron. Crookes himself participated in research on radioactivity, inventing the spinthariscope for counting alpha particles emitted by radium. However, the significance of this invention is overshadowed by his establishment of the most accurate technique for determining atomic weight, a model in scientific precision.

Crookes was born on June 17, 1832, in London, England. He was the oldest son of the tailor Joseph Crookes and his second wife, Mary Scott. Crookes received his early education at the Chippenham grammar school until 1848. Then, at the age of 15, he enrolled in London's Royal College of Chemistry to study under August Wilhelm von Hoffman, working as his assistant from 1850 through 1854. He published his first scientific paper, on the selenocyanides of the element selenium, in 1851.

In 1854, the Radcliffe Observatory in Oxford named Crookes superintendent of its Meterological Department, and the next year,

Sir William Crookes invented the Crookes tube, a key precursor to the discovery of X rays and the electron. *(AIP Emilio Segrè Visual Archives, W. F. Meggers Collection)*

Chester College of Science appointed him as a lecturer in chemistry. He returned to London in 1856 and married Ellen Humphrey; together, the couple had four children—three sons and a daughter. In 1859, he became the founding editor of the *Chemical News*, which distinguished itself from scientific journals by its informal tone; Crookes remained its editor until 1906.

In 1861, while performing spectroscopic analyses (a technique recently introduced by R. W. E. BUNSEN and G. R. Kirchoff) of seleniferous deposits from a sulfuric acid factory, Crookes noticed an inexplicable green line in the spectrum of selenium. After isolating the metallic substance corresponding to this line, Crookes announced his discovery of the new element thallium. The Royal Society recognized the significance of this discovery by inducting him into its fellowship in 1863 (despite a controversy over propriety with C. A. Lamy, a French chemist who displayed the same metal at the 1862 International Exhibition in London).

In 1871, Crookes published *Select Methods in Chemical Analysis*, and two years later, he published a paper announcing his final determination of the atomic weight of thallium. This 1873 paper established a benchmark in the methodology of atomic weight calculation and was universally lauded for its accuracy and clarity. In order to make this determination, Crookes had to employ a vacuum balance, which inspired his next line of research.

Crookes published "Attraction and Repulsion resulting from Radiation" in 1874, a paper that laid the groundwork for his invention of the Crookes radiometer, announced in a paper the next year. The action of the apparatus resembled that of a paddle-wheel riverboat, except that Crookes colored his four paddles black on one side to absorb radiant energy, and silver on the other to reflect it, thus creating a directional draft of energy that set the blades in circular motion when exposed to radiant sunlight. The practical application of this instrument outweighed Crookes's flawed explanation of its theoretical basis.

Crookes next turned his attention to study of rarefied gases in vacuum tubes. He discussed his investigations in his 1878 Bakerian Lecture and his 1879 British Association Lecture, explaining in the latter the apparatus he invented to carry out these experiments—the Crookes tube. This instrument consisted of an evacuated glass tube containing a negative electrode and a positive anode; when Crookes introduced a small electrical charge between the two, he observed an inexplicable ray emitting from the negative electrode, which he called a "molecular ray" (but subsequently became known as a "cathode ray," after Michael Faraday's term for the negative electrode). Furthermore, Crookes observed a light-free zone at the point of discharge on the negative electrode that became known as a Crookes dark space.

Recognition for his significant contributions to the practice of scientific inquiry showered on Crookes, as he was knighted in 1897 and inducted into the Order of Merit in 1910. He served as president of the prestigious Royal Society from 1913 through 1915 and received its Royal, Davy, and Copley Medals. Crookes died on April 4, 1919, in London.

D

Daily, Gretchen
(1964–)
American
Biologist

Gretchen Daily practices "population biology," a discipline she helped create in collaboration with Paul and Anne Ehrlich by applying scientific research to the issue of overpopulation. Overpopulation is not a future problem but rather a present reality, the three pointed out, requiring a radical shift in political and economic policy as well as personal behaviors. Daily uses an interdisciplinary approach, fusing the voices of science, law, business, and government, to promote the notion of sustainability. Interestingly, Daily and the Ehrlichs profess "equity," or equal consideration for all humans as well as for other species and the environment, as the key to the survival of a life-sustaining world.

Gretchen C. Daily was born on October 19, 1964, in Washington, D.C. Her mother, Suzanne R. Daily, was an antique dealer, and her father, Charles D. Daily, was an ophthalmologist. Daily attended high school in West Germany, then returned to the United States for her postsecondary education, which she conducted exclusively at Stanford University. She received her bachelor of science degree in 1986, then earned her master of science degree a year later.

Daily continued at Stanford, conducting doctoral work in biological sciences under Paul Ehrlich, the renowned author of *The Population Bomb*. The two commenced a partnership that continued throughout the 1990s, focusing their research on the issue of populations, both animal and human. Their coauthored papers focused on detailed microcosms (such as a 1988 paper on the feeding habits of red-naped sapsuckers) as well as generalized macrocosms (such as a 1990 paper on the effects of rapid climate change on the world food situation, published in the prestigious *Proceedings of the Royal Society*). By the time Daily received her Ph.D. in 1992, she had already developed into a distinguished and respected scholar, earning the Frances Lou Kallman Award for Excellence in Science and Graduate Study upon graduation.

Daily joined the University of California at Berkeley's Energy and Resources Group while continuing joint studies with Ehrlich. In 1994, she received a concurrent appointment at Stanford's Center for Conservation Biology as a Pew Fellow in Conservation and the Environment, a position created expressly for her. That year, she collaborated with Ehrlich and his wife, Anne Ehrlich, in writing an influential paper, "Optimum Human Population Size," published in the July issue of the academic journal *Population and Environment*.

In this paper, Daily and the Ehrlichs extrapolated the ideal number of human inhabitants based on their own commonsense criteria for the impact the earth can sustain. They set a relatively high standard of living, about comparable to that enjoyed by the Swedes or French, while also allowing for the preservation of natural ecosystems, such as stretches of wilderness populated by diverse animal species, necessary for the survival of an inhabitable planet. Based on these variables, they calculated an "optimum human population" of between one-and-a-half and two billion people, which corresponded to the population of 1900 through 1930. In other words, the current human population already exceeds the earth's ability to sustain a decent standard of living, and we are presently in a state of acute overpopulation.

The Ehrlichs and Daily expanded this paper into a book, *The Stork and the Plow: The Equity Solution to the Human Dilemma*, published in 1995. They based their argument on the notion of "equity." For example, instead of "throwing condoms at women" as a means of family planning, the equity solution promotes sexual equality in the household and in society as a better way to lower birthrates. Having no more than two children is the most fundamental way to reduce overpopulation. Other critical factors include recycling and reduced consumption, and adopting a political stance in favor of sustainability.

Published in 1997, Daily's second book, *Nature's Services: Societal Dependence on Natural Ecosystems*, resulted from her Pew fellowship: she acted as editor, soliciting chapter contributions from 18 other Pew fellows discussing the issues of biodiversity and ecosystem conservation. From 1997 through 1998, Daily served on the subcommittee of the Presidential Committee of Advisors of Science and Technology, and in 1999, she served as a fellow in the ALDO LEOPOLD Leadership Program.

Daily remains on the Stanford faculty as the Bing Interdisciplinary Research Scientist in the department of biological sciences. She also continues to conduct field research at sites in Costa Rica and Mexico, studying bird, butterfly, and insect populations and their response to earth changes brought about by human overpopulation. She is married to Gideon W. Yoffe.

Daimler, Gottlieb Wilhelm
(1834–1900)
German
Engineer, Transportation Industry

Gottlieb Daimler, together with his partner Wilhelm Maybach, was one of the first to build a working motor vehicle, though KARL FRIEDRICH BENZ is generally attributed with making the first successful automobile (actually a three-wheeler). Daimler and Benz worked less than a hundred miles from one another, and developed their motor vehicles at almost exactly the same time, but the two purportedly never met. About a quarter-century after Daimler's death, his company joined forces with Benz's. Both men are considered seminal figures in establishing the automobile industry, a development that revolutionized personal transportation.

Gottlieb Wilhelm Daimler was born on March 17, 1834, in Schorndorf, near Stuttgart, Germany. He was the second of four sons born to Wilhelmine Friederike Fensterer, a dyer's daughter. His father, Johannes Daimler, was a master baker who also ran a wine parlor. Daimler attended the local Latin school and then, starting in 1848, the Royal trade school in Stuttgart, where he trained as a gunsmith's apprentice, constructing his journeyman's piece in 1852—an elaborately adorned double-barreled pistol.

In 1853, the industrialist Ferdinand Steinbeis secured Daimler a job at a steam engineering works in Graffenstaden, near Strasbourg. In 1857, Daimler continued his technical education, studying mechanical engineering at the Stuttgart Polytechnic, and upon completing his course of study in 1859, he returned to his job in Graffenstaden. In 1861, he traveled to England, where he worked with engineer Joseph Whitworth in Manchester. He also visited France, where he saw a gas engine designed by J. J. E. Lenoir in Paris, inspiring him to continue his own design of a small engine that was inexpensive to run, an idea he had conceived as early as 1859.

Hired as technical manager in 1863, Daimler reorganized the mechanical engineering section of the Brudersaus Maschinenfabrik in Reutlingen. There, he met Wilhelm Maybach, and the pair remained partners throughout the rest of Daimler's career. In 1867, Daimler married Emma Kurtz. In 1869, the Maschinenbau Gesellschaft in Karlsruhe appointed him as its director, a position he retained until 1872, when the Gasmotoren-Fabrik Deutz hired him as its technical director and Maybach as head of its design office. There, the pair worked with NIKOLAUS AUGUST OTTO to perfect the Otto oil engine, a four-stroke internal combustion motor.

While in Russia on business for Deutz in 1881, Daimler conceived of the notion of "independent driving" in reaction against "overcrowded trains" and "limitations imposed by the railways." However, Deutz would not benefit from this line of innovation, as Daimler left the company the next year, taking Maybach with him. The pair installed themselves in Cannstatt (near Stuttgart), intent on developing their own internal combustion engine, but with a new twist—they intended to mount their motors in vehicles. By 1883, they had patented a design, and two years later, their second engine, nicknamed the "grandfather clock," received the German Imperial patent number 34926. Daimler and Maybach mounted this engine on a bicycle to create what they called a "riding car"—better known as a motorcycle. By 1887, Daimler and Maybach had established a factory in Seelberg to build two-cylinder V-engines.

About a hundred miles away, Karl Benz was concurrently introducing his own engine-powered vehicle, a three-wheeled contraption that is generally considered the first successful automobile. By 1889, both Benz and Daimler had developed four-wheeled motor vehicles, which they each displayed at the Paris Exhibition. Neither raised particular public interest, though two entrepreneurs—R. Panhard and E. Levassor—contracted with Daimler (and not Benz) to sell his product in France. Also in that year, Daimler's first wife, Emma, died; he subsequently married Lina Hartmann in 1893.

The Daimler-Motoren Gesellschaft (DMG) was founded in 1890, with Daimler himself working as a deputy member of the supervisory board. However, Daimler and Maybach left the company within four years to focus their attention on exhibiting their engine around the country and entering their cars in road races, such as the first international car race from Paris to Rouen in 1894, which a Daimler car won. Daimler rejoined DMG as chairman of the supervisory board in 1895.

In 1896, a wealthy Austrian living on the French Riviera named Emil Jellinek (who also happened to be a speed freak) influenced Daimler and Maybach to design a faster engine, capable of reaching the unheard-of speed of 40 kilometers-per-hour. Daimler and Maybach reluctantly complied (worried about safety, but reassured by Jellinek's cash payment for four such cars), and Jellinek drove the car to 42 kilometers-per-hour. Jellinek made further demands—a longer wheelbase, wider track, lower center of gravity, and a 35-horsepower engine—and secured these requests by ordering the entire production run of 36 cars.

Jellinek also requested that the new model be named after his daughter, Mercedes; the name was officially registered on June 6, 1902. However, Daimler did not live to see the dubbing of perhaps the most recognizable name in the automobile industry, as he died on March 6, 1900, in Cannstatt. A little over a quarter-century later, in 1926, the Daimler and Benz corporate entities joined forces.

Dalton, John
(1766–1844)
English
Physicist, Chemist

John Dalton changed the course of the human understanding of physical makeup with his atomic theory of matter, which states that minute, indestructible particles, called atoms, comprise all elements. Dalton arrived at this theory by considering the properties of gases. Dalton's law, or the law of partial pressures, states that the total pressure of mixed gases amounts to the sum of the pressure of each individual gas. Dalton also contributed to the understanding of the aurora borealis, the origin of the trade winds, the barometer, the thermometer, the hygrometer, the dew point, rainfall, and cloud formation. His limited educational background freed him from academic prejudice, and he carefully guarded his freedom of thought against undue influence by accepted theories. Dalton trusted his own observations and experience to guide his scientific research.

Dalton was born on September 6, 1766, in Eaglesfield, in Cumberland, England. His parents were Mary Greenup Dalton and Joseph Dalton, a Quaker weaver. The younger Dalton inherited his family's modest farm in 1834 when his older brother, Jonathan, died. Dalton devoted himself exclusively to his work and his religion, never marrying.

By the age of 12 Dalton had acquired enough education to commence teaching in Cumberland's Quaker School. At the age of 14 he moved to Kendal, where he taught with his brother at the Quaker School for the next 12 years. In 1793 he moved to Manchester, where he taught mathematics and natural philosophy at the New College in Manchester, established by the Presbyterians as an alternative to Cambridge and Oxford, which required oaths to the Church of England.

Dalton published his first book, *Meterological Observations and Essays*, in 1793, based on his journal of meteorological observations started in 1787. The next year he published *Extraordinary Facts Relating to the Vision of Colours*, an issue of personal concern as he was color-blind. In 1800 the Manchester Literary and Philosophical Society appointed Dalton as secretary, and he read most of his papers there throughout his lifetime.

Best known for his atomic theory of matter, John Dalton also contributed to the understanding of the aurora borealis, the barometer, the thermometer, and other scientific principles. *(E. F. Smith Collection, Rare Book & Manuscript Library, University of Pennsylvania)*

The society appointed him president in 1817, a position he maintained until his death.

Dalton presented four important papers to the Society in 1801. "On the Constitution of Mixed Gases" expounded his law of partial pressures; "On the Force of Steam" discussed the dew point and represented the founding of exact hygrometry; "On Evaporation" proposed that the quantity of water evaporated was proportional to the vapor pressure; and "On the Expansion of Gases by Heat" stated that all heated gases expand equally. This last principle is known as Charles's Law, as Jacques Charles discovered the effect in 1787, though Dalton published his statement first.

In 1802 Dalton presented his paper "On the Absorption of Gases by Water," to which he appended the first table of atomic weights. In a December 1803 lecture to the Royal Institution, Dalton explicated for the first time his atomic theory, which held that all atoms of a particular element are alike, having the same atomic weight. Dalton published this theory in his 1808 text, *A New System of Chemical Philosophy*, which he revised in 1810 and 1827.

Dalton received a Gold Medal from the Royal Society in 1826, but it was not until after his death that Stanislao Cannizzaro, in 1858, reasserted Amedeo Avogadro's theories from a half-century earlier that confirmed Dalton's atomic theory undisputedly. Even without this final confirmation, Dalton's influence was immense, as evidenced by the 40,000 people who attended his Manchester funeral after his death on July 27, 1844.

Daniell, John Frederic
(1790–1845)
English
Chemist, Meteorologist

John Frederic Daniell invented a voltaic cell, now named the Daniell cell after him, that maintained its charge much longer than the existing electric cells at the time. This invention bolstered the telegraph industry through its infancy, allowing for sustained transmissions with its constant current. Earlier, he had invented the dew-point hygrometer to measure atmospheric humidity. In testament to his unprecedented intelligence, the Royal Society inducted him into its ranks at a very early age, and King's College created a professorship in chemistry for Daniell, its first such chair, despite the fact that he had never attained a postsecondary education.

Daniell was born on March 12, 1790, in London, England. His father was a lawyer. Daniell's relatives gave him his first job at their sugar refinery and resin factory, where he first encountered the chemical process. Chemistry lectures by William T. Brande inspired him to pursue his own chemical investigations. The excellence of this independent research brought him to the notice of the prestigious Royal Society, which inducted him into its fellowship in 1814, when he was a mere 23 years old.

Daniell conducted meteorological research in addition to his chemical studies, and in 1820, he invented a device to measure the humidity of the atmosphere, a dew-point hygrometer. Three years later, he published *Meteorological Essays*, a collection of his papers on Earth's atmosphere, the trade winds, and instructions for constructing meteorological instrumentation. What distinguished this text was Daniell's use of physical laws to explain atmospheric phenomena, as well as his meticulous exactitude in meteorological observations and measurements. On a practical note, his suggestion that the moisture of hothouses required monitoring led to a transformation in the management of hothouses. He revised this text to include a discussion on radiation for its second edition, which came out in 1827.

In 1831, King's College in London appointed Daniell as its first professor of chemistry on the strength of his research and writings, and despite the fact that he lacked academic credentials. In the mid-1830s, he turned his attention to electric cells in response to the demand for more consistent and longer lasting power sources for the burgeoning telegraph industry. At the time, telegraphy depended on the voltaic cell, invented by ALESSANDRO VOLTA in 1797, which lost its potential once the energy was drawn, due to hydrogen bubbles gathering on the copper plate and creating resistance to the free flow of the circuit. Voltaic cells thus had an extremely brief shelf life, forcing telegram messages to remain exceedingly brief lest the energy supply fail mid-message.

In 1836, Daniell devised a new type of cell consisting of a negative zinc amalgam electrode immersed in a dilute solution of sulfuric acid contained in a porous pot, surrounded by a solution of copper sulfate contained in copper with a positive copper electrode immersed in it. The porous pot allows hydrogen ions to pass through to the copper sulfate, but it prevents the mixing of the two electrolytes. This cell, now known as the Daniell cell, sustained a constant current over long periods of time and thus served as a perfect energy source for telegraphy. British and American telegraph companies employed the Daniell cell exclusively, though other constant current cells were developed thereafter (namely, Grove's nitric acid depolarized cell and Sand batteries). Interestingly, telegraph operators measured the cell's power by the degree of pain it induced upon contact with their nerves.

In 1839, Daniell attempted to fuse metals by means of a 70-cell battery. However, this arrangement generated such a powerful electric arc that the ultraviolet rays damaged Daniell's vision, as well as harming the eyes of other observers, who walked away from the experiment with an artificial sunburn. He followed up on these investigations more carefully to demonstrate that a metal's ion, not its oxide, carries the electric charge in electrolysis of metal-salt solutions.

Daniell dedicated his 1839 book, *Introduction to the Study of Chemical Philosophy*, to the eminent chemist, Michael Faraday, who was a close friend. The Royal Society granted Daniell its Rumford Medal in 1832, and then after he had invented his eponymous cell, the society presented him its Copley Medal in 1837. He died on March 13, 1845, in London.

Darby, Abraham
(1678–1717)
English
Engineer, Steel and Iron Industries

Abraham Darby contributed two significant innovations to the iron- and brass-working industries: first, he patented a method of sand-casting that replaced individual production with mass production; and second, he introduced coke—a residue from coal that has been distilled to remove its gases—as a fuel for firing foundries. Ironically, Darby's coke-firing method did not catch on until after his death when his son, Abraham Darby Jr., expanded the influence of the family iron-smelting business. His grandson, Abraham Darby III, built the world's first iron bridge at the seat of the family's ironworks dynasty.

Darby was born at Wren's Nest, near Dudley, in Worcestershire, England, in 1678. His father was a Quaker farmer, and Darby honored this religion's belief in social simplicity throughout his life. He apprenticed to a malt-kiln maker near Birmingham and trained in engineering, before establishing his own malt-mill-making business at Baptist Mills in Bristol in 1698. Two years later, he married, but his first son, Abraham Darby II, was not born until 1711.

In about 1704, Darby visited the low countries of Holland to recruit Catholic brass founders, skilled in the making of brass battery, harnessing water to power hammers in shaping cold brass plate into hollowware. Four years later, in 1708, Darby patented a new technique for sand-casting iron (without loam or clay) for iron pots and ironware. This innovation introduced mass production into the metalworking industry, replacing the handmade process of producing one item at a time.

Also in 1708, Darby founded the Bristol Iron Company, and soon thereafter he established works at Coalbrookdale on the Severn River, a perfect location, as it provided ample supplies of iron ore as well as low-sulfur coal. This Clod-type coal, mined deep in the ground, proved to be the key to Darby's most significant innovation—the use of coke instead of charcoal to fire the iron. Within a year, Darby had devised a technique for smelting iron with coke. Three years later, horses delivered 400 loads of coal a week to Baptist Mills, where the brass and iron works used 250 tons of coal in 1712.

Abraham Darby the elder died at Madley Court, in Worcestershire on March 8, 1717, before his coke-smelting process caught on. Coke was traditionally used in the smelting of copper and lead, but ironworkers remained true to charcoal, despite its softness and increasing scarcity. As well, Darby tended to conduct business exclusively with other Quakers, thus limiting the influence of his coke-smelting process in the ironworking industry at large.

Abraham Darby II, proved to be a much more astute business person than his father. He started working at Coalbrookdale in 1728 and moved into management within two years. He eventually took over the company, transforming it into a highly successful enterprise. Perhaps the major factor fueling his success was the advent of THOMAS NEWCOMEN's steam engine, which improved the market for iron, in that it powered six-ton, 10-foot-long mine-pumping machines that required the strong iron produced by coke-smelting. By 1758, the younger Darby's reorganized company had cast more than a hundred cylinders for such mine-pumping machines. Darby also used Newcomen steam engines to blast his coke-burning furnaces, increasing efficiency over water-powered blasting.

Abraham Darby Jr. died in 1763, before he could realize plans to build the world's first iron bridge. Five years later, his son, Abraham Darby III, took over the family business at the age of 18 and resurrected his father's plans to span the Seven at Coalbrookdale from Medley parish to Broseley parish with an iron bridge. It was not until 1777, however, that work began on the project. By 1779, after two years of solid work casting and constructing the bridge, Darby III completed the first iron bridge in the world. The Society of Arts presented him with its gold medal in honor of this achievement.

Unfortunately, Darby III did not inherit his father's business acumen, and he ran the family company into debt, in part by purchasing Hadeley May house in 1781. However, he did contribute other legacies to his family's name, establishing Coalport as a shipping facility linking Coalbrookdale to the world before selling it to Richard Reynolds in 1788, a year before his death in 1789.

Darby III also continued the family lineage by fathering Abraham Darby IV, who also entered the family business. It was under the fourth generation of Darbys that the Coalbrookdale works produced the iron used in the first railway locomotive, designed with a high-pressure boiler by RICHARD TREVITHICK.

Darwin, Charles Robert
(1809–1882)
English
Naturalist

Charles Darwin's name is synonymous with his documentation of evolution and his theory of evolution, known as Darwinism. His assertion of this theory rocked the world, as it allowed no room for divine intervention in the process of evolution. Theologians balked at the notion, since it explained the historical progression of species in purely scientific terms, suggesting that nature controlled its own progression without intervention by a Creator. However, Darwin's methods were exacting, making it difficult to punch holes in his theory. Darwinism continued to hold sway as an explanation of evolution long after his death.

Darwin was born on February 9 (or 12), 1809, at The Mount, in Shrewsbury, England. His father, Robert Waring Darwin, was a physician and the son of the famous physician Erasmus Darwin. His mother, Susannah Wedgwood Darwin, the daughter of the potter Josiah Wedgwood I, died prematurely. Darwin was the couple's fifth child and their second son.

Darwin's elder sisters commenced his education, until he entered the Shrewsbury School in 1818, studying under Dr. Samuel Butler. Darwin moved on to Edinburgh University in 1825 to study medicine, which he found utterly boring. His father made a last stab at salvaging his young son's education and future by enrolling Darwin at Cambridge in 1827 for clerical study. Darwin was not inclined toward academics and barely passed his exams to receive his degree in 1831.

The true commencement of Darwin's education happened when he signed on as the unpaid naturalist for HMS *Beagle*. Both his father and Captain FitzRoy required convincing of the worthiness of this project, but Darwin's uncle Josiah Wedgwood supported the 22-year-old Darwin's decision. On December 27, 1831, the *Beagle* set sail on its five-year voyage, forever changing the history of science. The most significant location the ship visited was the Galápagos Islands off the western coast of South America, as these islands isolated themselves from one another and from the mainland, allowing species to evolve in separation. Species that existed on one island did not exist on another, and certain species, for example tortoises, developed distinctly on the different islands.

Though Darwin published the *Journal* of the journey in 1839, he waited another two decades to publish his broad theory of evolution. In the intervening time Darwin married his first cousin Emma Wedgwood on January 29, 1839, and together they had 10 children, seven of whom survived past childhood. A letter to Darwin from Alfred Russel Wallace in Malaya in 1858 expressed a similar hypothesis concerning evolution and spurred Darwin into action. The two men presented a joint paper to the Linnean Society in London that year, announcing their coinciding theories. The following year, Darwin published his immensely influential text, *On the Origin of Species by Means of Natural Selection*. In this book he set forth the three main principles of Darwinism—variation, heredity, and natural selection.

Darwin's later work, *The Descent of Man, and Selection in Relation to Sex*, published in 1871, followed up on his theory of evolution, but he also made contributions outside of the realm of evolution. In 1868 he published *Variation in Animals and Plants Under Domestication*. His other work included the study of barnacles, atolls, and earthworms, among other things. Darwin died on April 19, 1882, at Down House in Kent, England.

Davy, Sir Humphry
(1778–1829)
English
Chemist

Sir Humphry Davy is best known for his discovery of several chemical elements, including sodium and potassium. He established himself with a study on the effects of nitrous oxide. Later in his career, he turned his attention toward very practical concerns, inventing the miner's safety lamp. For his achievements he was knighted on April 8, 1812, and further honored with the title of baronet in 1818.

Davy was born on December 17, 1778, in Penzance, England. His father, Robert Davy, a wood-carver, speculated unsuccessfully on farming and tin mining, and he died in 1794. After his death, Davy's mother, Grace Millett, managed a milliner's shop until 1799, when she inherited a small estate. Davy married Jane Apreece, a widow, on April 11, 1812.

Davy's formal education was quite limited. He attended grammar school in Penzance before transferring to school in Truro in 1793. In 1795, he apprenticed to a surgeon and apothecary, intending to enter the field of medicine. He proved quite adept at chemical science, and in 1798 he was appointed as the chemical superintendent at the Pneumatic Institute in Clifton. One of his responsibilities included experimenting with gases to understand their effects. Davy worked with nitrous oxide, which came to be known as laughing gas due to the way it released inhibitions in those who inhaled it. Since his name was associated with such a pleasant substance, Davy quickly established a positive reputation. In 1801, the newly organized Royal Institution of Great Britain in London appointed him as a lecturer. His lectures, which became very popular, sometimes included demonstrations of the scientific principles discussed, such as the effects of nitrous oxide. Davy gained friendships with the socially elite, including the poets Samuel Coleridge and William Wordsworth. In 1802 he was promoted to the position of professor of chemistry.

Davy subsequently turned his attention to the effects of electricity on chemicals, a discipline known as electrochemistry. In 1806, Davy reported on some of his findings in the paper entitled "On Some Chemical Agencies of Electricity," which won the 1807 Napoleon Prize from the Institut de France despite the fact that England and France were warring at the time. That year Davy used electrolysis to discover sodium and potassium. He followed this with the discoveries of boron, hydrogen telluride, and hydrogen phosphate. In 1812, he published the first part of the text *Elements of Chemical Philosophy*, though he never managed to complete another section of the book. In 1813, Davy published a companion piece, *Elements of Agricultural Chemistry*. In 1815, he put his experimental expertise to use by inventing a safe lamp for mining.

Besides his publications and lectures, Davy earned recognition through awards and memberships. The Royal Society elected him a fellow in 1803, appointing him secretary in 1807, and promoting him to president from 1820 to 1827. In 1805, he won the Society's Copley Medal. In 1826, Davy suffered from a stroke, and he never fully recovered. On May 29, 1829, he died in Geneva, Switzerland.

De Laval, Carl Gustav Patrik
(1845–1913)
Swedish
Inventor, Food Technology

Gustav De Laval invented the continuously operating cream separator, a perfection of an existing separation mechanism. More significant perhaps was his improvement of the steam turbine, an advancement that helped fuel the Industrial Revolution. Later in his career, he invented the vacuum milking machine, automating the cow-milking process. De Laval's scientific interests were wide-ranging, and his inventions covered a broad range of applications. His work ethic matched his creative impulse, and he performed a staggering amount of inventive work in his career. In this sense, he was a kind of Swedish THOMAS EDISON.

Carl Gustav Patrik De Laval was born on May 9, 1845, at Orsa, in the Dalecarlia province of Sweden. His family had emigrated from France to Sweden in the seventeenth century. De Laval revealed his ingenuity as a child, and he studied mechanics at the Stockholm Institute of Technology and chemistry at the University of Uppsala, where he earned his doctorate.

Antonin Prandl, a German inventor, was the first to mechanize the cream-separation process when he invented a centrifuge that used the force of rotation to split the cream from the milk in 1864. This machine did not improve efficiency immensely, though, as the cream still had to be skimmed off the milk by hand, thus requiring the farmer to stop the machine every time it finished separation. In 1877, De Laval worked from Prandl's basic design, introducing turbines that allowed for continuous operation—the centrifuge could be emptied and refilled while the machine kept running. This innovation proved the key to improving dairy farmers' efficiency, such that De Laval is now recognized as the inventor of the high-speed centrifugal cream separator.

De Laval patented his invention in 1878 and simultaneously collaborated with Oscar Lamm, who owned an engineering firm, to establish a company for making and marketing the continuous-operation cream separators. The company proved so successful that it had to change status in 1883 to a limited company, now named AB Separator, with Lamm as chairman of the board and managing director. Clemens von Bechtolsheim, a German inventor, subsequently improved upon De Laval's separator design by adding alfa-plates that divided the milk into thin layers, a process that increased the efficiency of the mechanism. This innovation was so successful that the company's name was eventually changed to Alfa Laval AB in the 20th century.

De Laval's work on the separator required him to experiment with turbines, which he used to drive the machine. In 1887, he invented a small, high-speed turbine with a single row of blades that could reach the speed of 42,000 revolutions per minute. However, this model did not prove practical for commercial applications, so over the next decade, De Laval experimented with many different turbine designs, building a variety of different kinds of turbines. He employed more than a hundred engineers to construct his different designs, allowing for trial-and-error discovery that would have been impossible on his own.

In 1890, De Laval invented the steam turbine, despite the fact that he lacked knowledge of the properties of steam. He compensated for this shortcoming by simply designing the turbine to the properties of steam that he observed. For example, he designed a convergent-divergent exit nozzle, which was shaped so that the steam could travel in different directions. He also made innovations to the vane-wheel and the turbine disc, which sat on a flexible shaft and functioned properly well above the critical whirling speed.

In 1897, depression set in, a condition that plagued De Laval the rest of his life. Despite this, he continued to produce inventions of scientific and practical merit. In 1913, he invented the vacuum milking-machine, which used mechanized suction instead of hand-action to extract milk from cow teats, thus drastically decreasing the time necessary to milk cows and thus increasing the efficiency of the milking process.

De Laval interested himself in a wide array of scientific pursuits, ranging from aerodynamics to electric lighting to electrometallurgy. In testament to his prodigious output, he received 92 Swedish patents, and founded 37 companies for the marketing of his inventions. His diaries, in which he recorded his scientific and technical ideas, numbered in the thousands and are now housed in the Stockholm Technical Museum. De Laval died on February 2, 1913, in Stockholm.

De Vries, Hugo
(1848–1935)
Dutch
Botanist, Genetics Researcher, Physiologist

At the beginning of the 20th century, Hugo De Vries recovered from obscurity the laws of heredity that JOHANN GREGOR MENDEL had formulated some 34 years earlier. Independently Karl

Correns and Erich Tschermak von Seysenegg rediscovered Mendel's work simultaneously. De Vries also advanced the study of plant physiology by identifying such processes as plasmolysis, or osmosis in plant cells.

De Vries was born on February 16, 1848, in Haarlem, Netherlands. His parents were Maria Everardina Reuvens, who hailed from a family of scholars, and Gerrit De Vries, who was a representative of the Provincial State of North Holland, a member of the Council of State, and the minister of justice under William III. De Vries studied medicine at the University of Leiden from 1866 until 1870, when he moved to the University of Heidelberg, where he studied under Wilhelm Hofmeister. Reading works by Julius von Sachs and CHARLES ROBERT DARWIN influenced him tremendously, and in 1871, when he moved on to the University of Würzburg, he had the opportunity to study under Sachs, with whom he maintained a longstanding professional relationship.

Later that year, De Vries accepted a position teaching natural history at the First High School in Amsterdam while continuing to do research in Sachs's laboratory in the summers. In 1875, he landed a position with the Prussian Ministry of Agriculture in Würzburg, writing monographs on red clover, potato, and sugar beets, as well as on the processes of osmosis in plant cells. This post lasted two years, until he moved to the University of Halle as a privatdocent, lecturing on the physiology of cultivated plants. De Vries resigned this position in favor of a lectureship in plant physiology at the University of Amsterdam later in the year of 1877. His appointment represented the first academic position in the field of plant physiology in the Netherlands, and De Vries stayed on at the University of Amsterdam for the remainder of his career. In 1878, the university promoted him to the position of assistant professor of botany, and in 1881, he ascended to the status of full professor.

In 1886, De Vries commenced his research in plant genetics when he noticed that some species of evening primrose differed from others and sought to explain this distinction. He reported his early findings in the 1889 book *Intracellular Pangenesis*. He continued to work with plant breeding until 1900, when he formulated the laws of heredity that restated Mendel's work from 1866, though De Vries discovered Mendel's papers only after he had formulated his own version of the same ideas. De Vries developed the theory of mutation, which held that there existed mutation periods that represented the process of evolution. De Vries described progressive mutants as productive characteristic transformations, whereas retrogressive mutants represented changes that did not benefit the continuation of the species. De Vries published this work in 1901 through 1903 in the book *The Mutation Theory*, which appeared in 1910 and 1911 in an English translation.

In 1896, the University of Amsterdam named De Vries a senior professor of botany, and in 1904, he served as a visiting lecturer at the University of California at Berkeley. In 1912, he visited the Rice Institute in Houston, Texas, to participate in its opening ceremonies. De Vries retired in 1918, though he continued to work in the fields that made him famous, plant genetics and plant physiology. De Vries died on May 21, 1935, in Lunteren, Netherlands.

Diesel, Rudolf
(1858–1913)
French/German
*Mechanical Engineer,
Transportation Industry*

Rudolf Diesel invented the engine that bears his name, a design that ignites a variety of fuels, which spontaneously combust when introduced into the cylinder to meet the intense heat generated by air pressure. The high compression ratio of the diesel engine allowed for the burning of low-grade fuels, which was incredibly efficient and economical, but also highly pollutant, as these less-refined fuels burned with impurities, such as nitric oxide, and produced copious soot. Diesel died before the applications for his innovation became more universal: submarines in World War I used diesel engines almost exclusively, and the diesel locomotive overtook the steam engine as the primary source of power in the railroad industry after World War II.

Rudolf Christian Karl Diesel was born on March 18, 1858, in Paris of Bavarian parents. After the Battle of Sedan during the Franco-Prussian War in 1870, Diesel's family was expelled to England, where they lived in poverty. Diesel went to live with an uncle in his father's native town of Augsburg, Germany, where he attended secondary school. In 1875, he enrolled in the Technische Hochschule in Munich, where he studied thermodynamics under Carl von Linde.

Upon his graduation in 1880, Diesel moved back to Paris, where Linde had secured him a job in his building-refrigeration plant. Within a year, Diesel was promoted to plant manager. In 1885, Diesel set up his own laboratory to design and test an expansion engine fueled by ammonia, as an alternative to the huge, expensive, and inefficient steam engine that powered most industrial applications at the time. Although this particular design proved unsuccessful, it paved the way for his later innovation of the engine named after him.

In 1890, Linde's company transferred Diesel to Berlin. At that same time, Diesel conceived of a new engine design, which he patented in 1892. The next year, he published a paper entitled "*Theorie und Konstruktion eines rationellen Wäremotors*" ("Theory and Construction of a Rational Heat Motor") describing his innovation: theoretically, the four-stroke engine could burn any fuel, ignited not by a spark but by the intense temperature (about 1,000 degrees Fahrenheit) attained by compressing air to a high pressure (about 500 pounds-per-square-inch) before the fuel was sprayed into the cylinder, thus expanding the gases (thereby avoiding the sudden pressure increase inherent in the internal combustion engine).

Diesel obtained financial backing from the Augsburg Maschinenfabrik, and from Baron Friedrich von Krupp of Essen, and set about building a prototype—a single 10-foot iron cylinder with a flywheel at its base. He conducted his first trial in Augsburg on August 10, 1893, which was a success, but it took several more years of improvements before his engine was commercially viable.

Part of the problem for these delays stemmed from his use of coal dust as fuel. A by-product of the Saar coal mines in the Ruhr valley, coal dust was not only cheap and plentiful, but also it dovetailed with one of the philosophical underpinnings of Diesel's engine: adaptability, so that individual craftsmen and artisans could use whatever fuels were locally available to power the engine, thus counteracting the de facto monopoly large corporations enjoyed due to the overbearing expense of operating expensive, inefficient engines in manufacturing. However, controlling the rate of introducing the coal powder into the cylinder proved difficult, and after one of his coal-dust-burning prototypes exploded, nearly killing its inventor, Diesel abandoned coal dust in favor of refined mineral oil and later heavy petroleum oils as fuel.

In 1897, an independent trial of the 25-horsepower engine, conducted by Professor M. Schröter, confirmed its incredible mechanical efficiency (75.6 percent efficient, in theory). Diesel displayed the engine at the Munich Exhibition of 1898, where the brewer Adolphus Busch witnessed it in action, prompting him to install the first commercial engine built on the Diesel patent in his St. Louis, Missouri, brewery. Duly impressed, Busch purchased the U.S. and Canadian licenses for manufacture and sales.

Diesel licensed his design worldwide. The industrialist and inventor ALFRED BERNHARD NOBEL, for example, manufactured diesel engines in his St. Petersburg plant and became a millionaire on royalties. He established his own factory in Augsburg in 1899.

Diesel died mysteriously at sea during an overnight crossing of the English Channel. It was generally presumed that he fell overboard from the mail steamer *Dresden* on September 29 or 30, 1913, while it was traveling between Antwerp and Harwich, but some historians believe he may have committed suicide. Throughout his life, Diesel had restricted licensing to abide strictly to his 1893 patent stipulating combustion at nearly constant pressure, which required operation at low speeds, thereby limiting the application to large engines (and hence preventing the innovation of smaller, more utile engines).

After Diesel's death, however, more utilitarian applications were innovated, and the benefits of the diesel engine design were more fully exploited. Diesel engines, such as the Busch-Sulzer, and the Nelesco, built by the New London Ship and Engine Company of Groton, Connecticut, served as the primary technology powering submarines serving in World War I and continued to serve an important role in the nautical industry. Diesel technology subsequently became extremely important in the development of the railroad industry and, to a lesser extent, the automobile industry.

Domagk, Gerhard
(1895–1964)
German
Bacteriologist

Gerhard Domagk responded to his wartime experience of powerlessness as a physician to treat the wounded for bacterial infections by conducting research on antibacterial agents. In 1932, he discovered an unlikely candidate as the "magic bullet"—a red leather dye—that fought bacterial infection without poisoning the infected. This discovery transformed the medical field, which had previously been handcuffed against treating bacterial infection.

Gerhard Johannes Paul Domagk was born on October 30, 1895, in Lagow, Brandenburg, which was then part of Germany and is now in Poland. His mother was Martha Reimer, and his father, Paul Domagk, was assistant headmaster of a school in Sommerfeld. Domagk attended this school, which specialized in science instruction, until he turned 14, whereupon he transferred to school in Silesia. In 1914, he commenced study as a medical student at the University of Kiel, but within months World War I broke out, so he enlisted in the German army, which assigned him to the fighting in Flanders.

In December 1914, the army transferred him to the eastern front, where he was wounded in battle. He joined the Sanitary Service, serving as a medical officer throughout the remainder of the war. Working in the cholera hospitals of Russia (among other assignments), he experienced the powerlessness of medical treatment of bacterial infections.

In 1918, Domagk returned to the University of Kiel to complete his medical studies, and in 1921, he passed his state medical examinations to earn his medical degree. He conducted laboratory research for the next three years, first on creatin and creatinin under Max Bürger, and then on metabolism under Professors Hoppe-Seyler and Emmerich. In 1924, the University of Greifswald appointed him university lecturer in pathological anatomy; the next year, he moved to the University of Münster, which appointed him to the same position. Also in 1925, he married Gertrude Strube, and together, the couple eventually had four children—three sons and one daughter.

From 1927 through 1929, Domagk took a leave of absence to conduct research at

I. G. Farbenindustrie, in Wuppertal. In 1929, the company established a new research institute for pathological anatomy and bacteriology and appointed Domagk its director of research in experimental pathology and bacteriology. Motivated by his wartime inability to treat bacterial infections, Domagk focused his investigations on finding antibacterial agents, first in vitro, or in test tubes, and then in vivo, or in living organisms, such as mice and rabbits.

In 1932, after testing thousands of potential antibacterial agents, Domagk discovered a red coal-tar dye used on leather, Protonsil Rubrum, which exhibited effects against bacteria in test tubes and proved nontoxic to mice. He conducted an experiment whereby he injected 26 mice with a hemolytic streptococcal bacteria culture, then injected 12 mice with a single dose of Protonsil Rubrum an hour and a half later. The 14 control mice, which did not receive the potential antibacterial agent, all died within four days, as expected. The 12 treated mice, on the other hand, all survived!

For reasons unknown, Domagk waited three years before publishing his findings. During this period, his daughter contracted a streptococcal infection at his laboratory when she was accidentally pricked with a needle; after all traditional treatments failed to respond, Domagk injected her with a dose of Protonsil Rubrum, and she recovered! He finally published his results in 1935 in the German journal *Deutsche medezinische Wocherschrift*, in an article entitled *"Ein Beitrag zur Chemotherapie der bakteriellen Infektionen."*

Subsequent independent research confirmed his findings, extending them to identify the sulfonamide group as the active ingredient in Protonsil Rubrum. Ensuing studies discovered that this treatment does not in fact kill bacteria but rather prevents the bacteria from reproducing by blocking metabolism. Sulfanilamide derivatives proved effective against pneumonia, meningitis, blood poisoning, and gonorrhea. Domagk's discovery thus transformed the medical field, empowering doctors to treat bacterial infections.

In recognition of the significance of this discovery, the Royal Swedish Academy of Sciences awarded Domagk the 1939 Nobel Prize in physiology or medicine, but Adolf Hitler had forbidden German citizens from receiving the prize. In fact, the Nazi government arrested Domagk when he informed it of this honor. Not until 1947, after World War II (during which his mother died of hunger in a refugee camp in 1945) had ended, did Domagk receive the medal, though the prize money had reverted to the Nobel Foundation.

Domagk followed up on his antibacterial studies with tubercular chemotherapy research, discovering the antitubercular compounds Conteben and Tibione (which fight tuberculosis effectively despite their toxicity). He retired in 1958. Besides the Nobel Prize, he received numerous other honors: he was knighted in the Order of Merit in 1952 and was awarded the Grand Cross of the Civil Order of Health of Spain in 1955; the University of Frankfurt granted him its Paul Ehrlich Gold Medal and Prize in 1956; the Royal Society of London and the British Academy of Science inducted him into their fellowships in 1959; and the Japanese government bestowed on him its Order of Merit of the Rising Sun in 1960. Domagk died of a heart attack on April 24, 1964.

Drake, Edwin Laurentine
(1819–1880)
American
Entrepreneur, Energy and Power Industries

"Colonel" Edwin Laurentine Drake established the petroleum industry in one fell swoop with his 1859 discovery of underground oil in Titusville, Pennsylvania. Prior to his discovery, petroleum saw few practical uses; afterward, its utility was recognized, and it became an important commodity for energy production. Drake proved inept in his business dealings, and he lost all his money to bad investments, dying in poverty.

Drake was born on March 29, 1819, in Greenville, New York. When he was still young, his family moved to Castleton, Vermont, where he attended the local schools. He worked on the family farm until the age of 19, when he left for Buffalo, New York, where he secured a job as a night clerk on a lake steamer. Over the next dozen years, he held a variety of other jobs, most of them related to the transportation industry in one way or another.

In 1850, Drake landed a position as a conductor on the New York & New Haven Railroad. When his wife died in 1854, he moved into a hotel in New Haven, Connecticut, where he crossed paths with James Townshend, one of the founders of the Pennsylvania Rock Oil Company, an outfit that harvested ground-seepage oil for use in medicinal preparations. When Drake resigned from his conductor's job due to ill health in December 1857, Townshend (as shrewd a businessman as Drake was an incompetent one) recognized an opportunity. Drake retained rights to free passage on the rail lines, so Townshend contracted him to travel to northwest Pennsylvania to investigate oil deposits at its property in Oil Creek as a representative of the company in which he now held stock.

Rail tracks connected New Haven to Erie, Pennsylvania, but Drake completed the rest of his journey to Titusville in a stage coach that traveled over smooth roads the first 15 miles to Waterford, but the remaining 45 miles of mud-filled roads took two days to traverse. Townshend's letter of introduction addressed his envoy as "Colonel" Edwin Laurentine Drake (who had never served in the military) to impress the local populace. In Titusville, Drake met George Bissell, another principal in the Pennsylvania Rock Oil Company.

Through his extensive travels, Drake had observed the drilling of artesian water wells and salt wells in New York and Pennsylvania. He applied this notion of drilling to the collection of oil, an idea that excited Bissell, as he had read research by Benjamin Silliman identifying petroleum as a potentially valuable commodity. Bissell established the Seneca Oil Company, with Drake installed as president, to search for oil by drilling. Drake leased land and set up a drill, powered by an old steam engine, in 1858; by then, he had moved his family to Titusville (he had remarried in 1857).

On August 27, 1859, after three-and-a-half months of drilling, Drake struck oil at the depth of 69.5 feet. This date effectively marks the beginning of the petroleum age. Drake's drill, constructed by a local blacksmith with salt-well experience, pumped only 40 barrels of oil a day at the height of its productivity (decreasing to 15 barrels a day later). Unfortunately, Drake neglected to patent the key innovation of his drill—a pipe encasing the drill in a protective layer—and others soon used his drilling technique to discover and extract huge quantities of oil, thereby establishing the petroleum industry.

In 1860, the town of Titusville appointed Drake as justice of the peace, a position with a salary of $3,000 a year. He worked concurrently as an oil commission merchant for a New York company. In 1863, he sold his Titusville property and moved to New York City, where he promptly lost his savings to bad investments. At this point, he was suffering from a disease that paralyzed his legs. His wife began taking in boarders and sewing to earn enough money to keep the family afloat.

Hearing of his plight, some of Drake's Titusville friends raised $5,000 to move the family back to town. In 1876, the State of Pennsylvania granted Drake an annuity of $1,500 a year to support his family. Drake, who established the industry that proved to be extremely lucrative, died a pauper on November 8, 1880, in Bethlehem, Pennsylvania.

Dunlop, John Boyd
(1840–1921)
Scottish
Inventor, Transportation Industry

John Boyd Dunlop is credited with inventing the pneumatic tire, or the air-filled rubber tire. In actual fact, Robert William Thomson had invented a similar tire in the mid-19th century, but his design failed to inspire its practical use. Dunlop, unaware of Thomson's patent, introduced a better design, which happened to coincide with the explosion of the personal transportation revolution, as both bicycles and automobiles came into popular use at the very time of Dunlop's invention. However, travel by these modes remained exceedingly uncomfortable until the pneumatic tire absorbed the shock of traveling over bumpy roads. Dunlop started a company (named after him) for producing pneumatic tires, thereby establishing an industry that thrived throughout the 20th century and proved key to the development of the era of improved transportation.

Dunlop was born on February 5, 1840, in Drehorn, in North Ayrshire, Scotland. He hailed from a family of farmers, but he decided not to return to the land, instead studying veterinary medicine at Edinburgh University (he undoubtedly witnessed the importance of vets while growing up on a farm). He qualified to practice at the young age of 19 and established his veterinary practice in Edinburgh, where he remained for almost a decade before moving his practice to Ireland, near Belfast, in 1867. There, he built up an extensive clientele and a successful practice in veterinary surgery.

Dunlop's profession required wide-ranging travel over cobbled city streets or rough country roads on wheels made of the only materials available at the time—iron, wood, and solid rubber—each of which made for an exceedingly uncomfortable ride, he found. Handy with rubber, he began experimenting with ways to fashion a wheel that would reduce the shock of bumps. He requisitioned his nine-year-old son's tricycle for use in his investigations, mounting his experimental wheels on its axles—legend has it that Dunlop did so to help his son win a race, but he may have simply used the nearest available vehicle.

In October 1887, Dunlop happened upon the optimal solution—the pneumatic tire, which consisted of a canvas jacket (or linen tape, according to one source) fitted with rubber treading and flaps attached with rubber cement to a rim; enclosed under this outer layer was a rubber inner tube that could be inflated with air by using a football pump. The air acted as a shock absorber, smoothing the ride. A subsequent design improvement attached the tire to the rim of the wheel by means of a wire. A little over a year later, in December 1888, he patented his invention.

Little did Dunlop realize that his invention had already been patented—by Robert William Thomson in 1845. However, Thomson's innovation went practically unnoticed. What distinguished Thomson's from Dunlop's invention was less design (though Dunlop's was an improvement) than timing: Dunlop's introduction came as the personal transportation revolution was gaining momentum, as the automobile had recently been invented, and the bicycle was gaining popularity as a form of transportation as well as recreation.

At a bicycle race, the performance of Dunlop's pneumatic tire impressed W. H. Du Cros, inspiring him to form a partnership with Dunlop. Together, they founded a company for manufacturing pneumatic tires, naming it after the product's inventor—the Dunlop Rubber Company. The company had to pitch a legal battle against Thomson, which it won, in order to market what is now commonly acknowledged as Dunlop's invention. The Dunlop Tire Company continues to exist, merged with the Goodyear Tire Company (named after the inventor of rubber vulcanization, CHARLES GOODYEAR).

Within a decade of the invention of the pneumatic tire, it had replaced the solid rubber tire in most applications, including both bicycles and cars. By then, however, Dunlop was ready to leave the business that he had founded. In 1896, he sold Du Clos not only his patent but also his share of the business for 3 million pounds, a pittance compared to the profitability of the tire industry. At about that same time, André and Edouard Michelin had translated Dunlop's invention for use on automobiles and founded the company that bears their name.

After selling the business, Dunlop retired to Dublin, Ireland. He spent the remainder of his life there in relative obscurity, running a local drapery business. He died in Dublin on October 23, 1921.

E

Eastman, George
(1854–1932)
American
Inventor, Photography

George Eastman introduced photography to the mainstream population by simplifying the development process, first introducing the "dry" process and later marketing stripping film and roll film. He also made photography more accessible to amateurs by making simple and inexpensive cameras. His Kodak box camera was the first to introduce the idea of professional development: after shooting a roll of film, the entire camera would be sent back to the plant for development, which allowed photographers unskilled at development to take pictures. The advent of the one-dollar camera made the Kodak brand ubiquitous; for the first several decades of the century, the Kodak name was synonymous with camera.

Eastman was born on July 12, 1854, in Waterville, New York. Due to his family's poverty, Eastman quit school at the age of 14 to work as a messenger boy while studying accounting at night. He managed to save $3,000 to fund his experimentation in photography, which he began in 1877 after he read an entry on the process in a British almanac.

Although many photographic processes existed in his day, Eastman realized that a "dry" process would eventually replace the "wet" process (which required immediate exposure and development) on the commercial market. Most importantly, this simplification would start to open up the market to mainstream use of photography. Although dry processes existed before Eastman introduced his, none of them were practical: a colloidan-albumen-plate method required six times the exposure of a wet plate, for example.

In 1879, Eastman was granted patent number 226,503, entitled "Method and Apparatus for Coating Plates," covering his photographic emulsion coating machine. The next year, he rented a loft in Rochester, New York, to mass-produce dry plates, thus launching his business. All the while, he continued to experiment with the photographic process, in search of improvements. In 1884, he patented flexible film, and in 1886, he introduced the Eastman-Walker roller slide. This type of stripping film consisted of a paper base attached to layers of soluble gelatin, collodion, and sensitized gelatin emulsion, which formed negatives that could be cut up and then developed directly onto glass plates coated with glycerine that dissolved under hot water to allow for the stripping of the paper. The image was then ready for printing on a gelatin sheet.

Although this innovation simplified the development process, it did not capture the mainstream market, so Eastman continued to innovate, coming up with the "Detective Camera" in 1888. This camera (subsequently known by its more popular name, the Kodak box camera) came loaded with film enough for 100 exposures and cost $25, which included a shoulder strap and case. When all the exposures were used up, the camera was sent back to the Rochester plant for development and reloading of new film for $10. This model did capture the popular imagination, and it sold more than 100,000 units by 1895.

In 1889, Eastman introduced roll film made from transparent nitrocellulose, otherwise known as celluloid, a flammable film invented by JOHN WESLEY HYATT in 1868 that was later replaced by nonflammable cellulose acetate. Eastman sold a considerable amount of this film before a patent, first filed in 1887 by Rev. Hannibal Goodwin, was granted in September 1898, only two years before the reverend's death. A protracted legal battle ensued, pitting the owners of Goodwin's patent against Eastman's company, which he had founded in 1892. The suit was settled in March 1914, with Eastman paying $5 million in back royalties to the owners of Goodwin's patent.

Also in 1892, Eastman introduced spool film that was loadable in daylight, thus avoiding the necessity of transporting a full darkroom to remote locations, which had prevented photographers from taking landscape pictures and other outdoor shots in the past. In 1900, Eastman marketed a pocket-sized camera for one dollar. This inexpensive camera truly introduced photography to the masses, earning Eastman a fortune.

Eastman used his fortune philanthropically, donating more than $100 million to charity by the time he died in 1932. He committed suicide on March 14, 1932, after suffering extensively from personal problems. In his suicide note, he wrote, "My work is done. Why wait?"

Edison, Thomas
(1847–1931)
American
Inventor, Energy and Power Industries

Thomas Edison held 1,069 patents on his inventions, which included such important innovations as the incandescent electric lamp and the phonograph. He combined scientific acumen with an uncanny entrepreneurial sense for identifying the technological advances that would prove profitable. He happened upon his only discovery in the realm of pure science accidentally when he noted what became known as the Edison effect, the emission of electrons from a heated cathode. Edison did not recognize the importance of this discovery, though subsequent scientists used the effect as the basis for the electron tube.

Edison was born on February 11, 1847, in Milan, Ohio. He was the youngest of seven children born to Samuel Ogden Edison Jr. and Nancy Elliot. Edison married Mary Stillwell on December 25, 1871, and the couple had three children together—Marion Estell, Thomas Alva, and William Leslie. Edison's first wife died on August 8, 1884; he then married Mina Miller on February 24, 1886. Together they had three children—Charles, Madeleine, and Theodore.

Edison's formal education was limited to home schooling, though he had set up his own chemical laboratory by the age of 10. At age 12 he started developing his entrepreneurial skills by riding trains to sell newspapers. He even set up a makeshift laboratory in one of the trains he rode regularly. He compensated for his lack of education by reading voraciously, and he later characterized his reading of the first two volumes of Michael Faraday's *Experimental Researches in Electricity* in one sitting in 1868 as a seminal event in his scientific life.

Edison became a telegraph operator in 1863, traveling from city to city until in 1868 he reached Boston, where he worked for Western Union Telegraph Company. There he patented his first invention, the electric vote recorder, though the technology proved too efficient for legislators who would rather waste their time in argumentation. In 1869, he moved to New York City, where he landed a $300-a-month job as general manager at Law's Gold Indicator Company. That year he invented a paper tape ticker for standardizing the reporting of stock prices, and he joined Franklin L. Pope and James N. Ashland to establish Pope, Edison and Company, a firm devoted to technological innovation. In 1870 Gold and Stock Telegraphic Company bought out the company, at the same time purchasing the rights to Edison's tape ticker for $40,000. Edison invested this windfall in a scheme that industrialized the invention process, implementing a factory line approach at a plant in Newark, New Jersey. Edison outgrew this facility in 1876 and moved to Menlo Park, New Jersey.

There Edison invented the phonograph in 1877, though he did not immediately recognize its commercial value. He began work on the incandescent light bulb in 1878, and after experimenting with more than 6,000 substances, he discovered a carbon filament from bamboo fiber that lasted more than 1,000 hours in a vacuum. Edison then devoted his efforts to supplying electricity for widespread use, operating the first power station from Pearl Street in New York City as of September 4, 1882. He again outgrew his facilities and moved them to West Orange, New Jersey, in 1887. There he invented the kinetograph, a precursor to modern motion-picture technology.

Edison received numerous honors for his work, both during his lifetime and after his death. He won the 1892 Albert Medal of the British Society of the Arts. In 1927, the National Academy of Sciences elected him a member, and in 1928, the U.S. Congress awarded him a gold medal for his life's work. In 1960, he was inducted into the Hall of Fame of Great Americans. Edison died on October 18, 1931, in West Orange, New Jersey. His prolific output combined with the social significance of

Thomas Edison invented the phonograph in 1877 and the incandescent light bulb in 1878. *(Smith Collection, Rare Book & Manuscript Library, University of Pennsylvania)*

his inventions earned him an unquestionable place among the ranks of the most important scientists of the 20th century.

Einstein, Albert
(1879–1955)
German/Swiss/American
Physicist

Albert Einstein is perhaps the most important scientist of the 20th century. In one year, 1905, he published four papers that radically revised the scientific understanding of the world. He explained the equivalence of mass and energy and developed both the photon theory of light and the special theory of relativity. He expanded upon this latter theory 11 years later with the general theory of relativity. Once he had established his status as an extremely influential scientist, he used his voice to speak out for Zionism and pacifism, and against Hitler's Nazi regime.

Einstein was born on March 14, 1879, in Ulm, Germany. His mother was Pauline Koch and his father was Hermann Einstein, an electrical engineer. Einstein married twice. When his first marriage ended in divorce, he married his cousin, Elsa, a widow with two daughters.

The collapse of his father's electrochemical business, combined with Einstein's revulsion at all things German, led the family to immigrate to Switzerland, where Einstein gained citizenship in 1901. In 1896, he entered the Swiss Federal Institute of Technology in Zurich to study physics and mathematics. After failing to secure a teaching position after graduation, Einstein settled into a job with the Swiss Patent Office in Bern as a technical expert, third class, for seven years, starting in 1902. He loved this arrangement, as the light duties allowed him to concentrate his free time on science. This period was the most fertile in his career, as it spawned not only a doctorate, which he earned in 1905 from the University of Zurich with the dissertation "A New Determination of Molecular Dimensions," but also his four papers that same year, which shaped scientific history.

All four papers were published in the journal *Annals of Physics*. The first paper, "On the Motion of Small Particles Suspended in a Stationary Liquid According to the Molecular Kinetic Theory of Heat," concerned Brownian motion, which is the random movement of microscopic particles suspended in liquids or gases as a result of the impact of molecules in the surrounding fluid. The second paper, "On a Heuristic Point of View about the Creation and Conversion of Light," described electromagnetic radiation as a flow of quanta, or discrete particles, now known as photons. The third paper, "On the Electrodynamics of Moving Bodies," contained the special theory of relativity, which stated that the speed of light is constant for bodies moving uniformly, relative to one another. It rejected the notions of absolute space and absolute time, and it asserted the notion of time dilation: that time slows down for a moving body. The final paper, "Does the Inertia of a Body Depend on Its Energy Content?" contained Einstein's famous equation, $E = mc^2$.

A series of university positions followed, commencing in 1908 with a spot at the University of Bern. In 1909, Einstein became an associate professor of physics at the University of Zurich. Two years later he taught at the German University in Prague, and the year after that he returned to the Federal Institute of Technology in Zurich as a professor. Finally in 1914 he gained some stability as a member of the Prussian Academy of Sciences in Berlin and the director of the Kaiser Wilhelm Institute for Physics.

In 1916, Einstein followed up on his special theory of relativity with the publication of his paper "The Foundation of the General Theory of Relativity." In it he extended his special theory of relativity to apply to all situations. When his prediction that a ray of light from a distant star passing near the Sun would appear to be bent slightly, in the direction of the Sun, was observed to be correct during a solar eclipse in 1919, Einstein gained international fame. In 1921, he won the Nobel Prize in physics.

Einstein left Nazi Germany for a position at the new Institute for Advanced Study at Princeton in 1933. He became a U.S. citizen in 1940 and in fact convinced President Franklin Roosevelt of the possibility of the Nazi regime's developing a weapon of mass destruction using nuclear power. Einstein continued to work for peace until an attack due to an aortic aneurysm led to his death on April 18, 1955, in Princeton, New Jersey.

⊗ **Einthoven, Willem**
(1860–1927)
Dutch
Physiologist

Next to WILHELM CONRAD RÖNTGEN's discovery of X rays, Willem Einthoven's invention of the string galvanometer, which measured the heart's electrical charges, and his development of the electrocardiogram (ECG), which recorded the galvanometer's readings for diagnostic purposes, represent the most significant advancement in diagnostic tools for the practice of clinical medicine. The ECG enabled physicians to diagnose heart abnormalities and diseases, the first step toward curing them.

Einthoven was born on May 21, 1860, in Semarang on the island of Java, then part of the Dutch East Indies and now part of Indonesia. His father, Jacob Einthoven, was an army medical officer and then Semarang parish doctor who died in 1866. His mother, Louise M. M. C. de Vogel, daughter of the director of finance for the Indies, moved her three daughters and three sons (Willem was the third born and the eldest son) back to her homeland of Holland, settling in Utrecht. Einthoven passed the *Hogere Burgerschool* to graduate from secondary school in 1878, whereupon he matriculated at the University of Utrecht as a medical student.

Einthoven served as an assistant to the opthalmologist H. Snellen Sr. at the Gasthuis voor Ooglidders eye hospital while earning his *candidaat* diploma (the equivalent of a bachelor of science degree) under the anatomist W. Koster. A broken wrist resulting from a sporting accident (he was an avid gymnast, fencer, and rower) influenced his first published paper, "*Quelques remarques sur le mécanisme de l'articulation du coude,*" or "Some remarks on the elbow joint," which gained some notoriety. He continued his doctoral work, studying under the physicist C. H. D. Buys Ballot. He wrote his doctoral dissertation, which was published in 1885 as "*Stereoscopie door kleurverschil,*" or "Stereoscopy by means of color variation," under the physiologist F. C. Donders.

When he received his Ph.D. in medicine in 1885, the University of Leiden offered him A. Heynius's chair of physiology, a post he filled upon passing his final state medical examinations to qualify as a general practitioner on February 24, 1886. His inaugural address was titled "*Der leer der specifieke energieen,*" or "The

Willem Einthoven developed the first electrocardiogram, enabling physicians to diagnose heart diseases and abnormalities. (*AIP Emilio Segrè Visual Archives, E. Scott Barr Collection*)

theory of specific energies." Later in 1886, he married his cousin, Frédérique Jeanne Louise de Vogel, and together the couple had four children: Augusta, born in 1887; Louise, born in 1889; Willem, born in 1893, who became an electro-technical engineer who developed the vacuum string galvanometer for use in wireless communications; and Johanna, born in 1897, who became a physician.

In 1887, the English physiologist Augustus D. Waller used a capillary electrometer, an instrument invented four years earlier by Gabriel Lippmann by encapsulating mercury in a glass tube, to record the electrical current generated by each heartbeat. However, the instrument's sensitivity (capable of reading changes of a millivolt) was counteracted by its inertia, which required mathematical correction of its photographically recorded readings. Dissatisfied with this method, which essentially only registered the heart's electrical activity without registering reliable measurements, Einthoven sought to devise a better method of measuring a beating heart's electrical output.

In 1901, Einthoven designed the string galvanometer, then spent the next three years refining its precision construction—a fine quartz wire suspended between magnetic poles. The sensitive instrument measured the strength of an electrical current by the distance it displaced the wire, at a right angle to the magnetic direction of the force. He recorded the deflection photographically, allowing for precise readings of the degree of displacement. He placed the electrode leads not in the heart, but rather in one of three standardized placements: right to left hand, right hand to left foot, and left hand to left foot. He connected these leads to a recording instrument, resulting in a curved reading known as an electrocardiogram.

Einthoven published his findings in *Die Konstruktion des Sitengalvanometers* in 1909. Over the next four years, Einthoven established standard ECG readings, whereby a one-centimeter deflection of the recording stylus on standardized paper corresponded to a one-millivolt electrical charge, thus allowing for comparison between readings. He also identified five different electrical waves (labeled P, Q, R, S, and T waves) corresponding to different types of heartbeats, allowing for the distinction between normal and abnormal heartbeats, and hence the diagnosis of heart diseases.

For his invention of the string galvanometer and his development of the electrocardiogram, Einthoven won the 1924 Nobel Prize in physiology or medicine. Although he received the prize solo, he was not alone in the advancement of electrocardiography—the English physicians Sir William Ogler and James Herrick devised ECG diagnosis techniques several years after (and independent of) Einthoven, and from 1910 through 1935, Sir Thomas Lewis and the American F. N. Wilson correlated ECG curves with clinical data.

Einthoven went on to modify his invention for use as a long-distance radio telegraph receiver, and as an instrument to measure electrical charges in nerves. After a protracted and painful illness, Einthoven died in Leiden on September 28, 1927.

⊗ Elion, Gertrude Belle ("Trudy")
(1918–1999)
American
Chemist, Medical Researcher

Although Nobel Prizes in science are usually awarded for basic research, the 1988 prize in physiology or medicine went to three people in applied science—drug developers. "Rarely has scientific experimentation been so intimately linked to the reduction of human suffering," the 1988 *Nobel Prize Annual* said of their work. One of the researchers was Gertrude Elion.

Gertrude, whom everyone called Trudy, was born on January 23, 1918, to immigrant parents in New York City. Her father, Robert, was a dentist.

The family moved to the Bronx, then a suburb, in 1924. Trudy spent much of her childhood reading, especially about "people who discovered things."

In 1933, the year Trudy graduated from high school at age 15, her beloved grandfather died in pain of stomach cancer, and she determined to find a cure for this terrible disease. There was no money to send her to college, however, because her family had lost its savings in the 1929 stock market crash. Trudy enrolled at New York City's Hunter College, which offered free tuition to qualified women. She graduated from Hunter with a B.A. in chemistry and highest honors in 1937.

Elion failed to win a scholarship to graduate school, so she set out to find a job—not easy for anyone during the Depression, let alone for a woman chemist. One interviewer turned her down because he feared she would be a "distracting influence" on male workers. She took several short-term jobs and also attended New York University for a year, beginning in 1939, to take courses for her master's degree. She then did her degree research on evenings and weekends while teaching high school and finally completed the degree in 1941.

World War II removed many men from workplaces, making employers more willing to hire women, and in 1944 it finally opened the doors of a research laboratory to Gertrude Elion. Burroughs Wellcome, a New York drug company, hired her as an assistant to a researcher, George Hitchings. Most drugs in those days were developed by trial and error, but Hitchings had a different approach. His lab looked systematically for differences between the ways that normal body cells and undesirable cells such as cancer cells, bacteria, and viruses used key chemicals as they grew and reproduced. The researchers then tried to find or make chemicals that interfered with these processes in undesirable cells but not in normal ones.

Elion at first worked mostly as a chemist, synthesizing compounds that closely resembled the building blocks of nucleic acids. The nucleic acids, deoxyribonucleic acid (DNA) and ribonucleic acid (RNA), carry inherited information and are essential for cell reproduction. Scientists had theorized around 1940 that an antibiotic called sulfanilamide killed bacteria, by "tricking" them into taking it up instead of a nutrient that the bacteria needed, thus starving them to death, and Hitchings thought that a similar "antimetabolite" therapy might work against cancer. If a cancer cell took up compounds similar but not identical to parts of nucleic acids, he reasoned, the chemicals would prevent the cell from reproducing and eventually kill it, much as a badly fitting part can jam the works of a machine. He and Elion set out to create such compounds.

In 1950 Elion invented 6-mercaptopurine(6-MP), which became one of the first drugs to fight cancer successfully by interfering with cancer cells' nucleic acid. It worked especially well against childhood leukemia, a blood cell cancer that formerly had killed its victims within a few months. When combined with other anticancer drugs, 6-MP now cures about 80 percent of children with some forms of leukemia.

Another compound Elion developed in her cancer research was called allopurinol. Because it can prevent the formation of uric acid, allopurinol has become the standard treatment for a painful disease called gout, in which crystals of uric acid are deposited in a person's joints.

Work on another breakthrough began in 1969, when Elion sent John Bauer, a researcher at the Burroughs Wellcome Laboratories in England, a new drug she had created that was related to a known virus-killing compound. At Elion's suggestion that he test her drug against a dangerous group of viruses called herpesviruses, he found that it stopped their growth. Elion and her coworkers then launched a search for variants of the drug that would kill herpesviruses even better than the original compound. In 1974, a Burroughs Wellcome researcher, Howard Schaeffer, synthesized a drug called acyclovir, which was 100 times more effective against herpesviruses than Elion's first drug.

As Hitchings and Elion developed drug after drug, they advanced together within Burroughs Wellcome. Finally in 1967, Elion was made head of her own laboratory, the newly created Department of Experimental Therapy. Although she had always enjoyed working with Hitchings, she was glad to have more independence. When Burroughs Wellcome moved to Research Triangle Park, North Carolina, in 1970, Elion moved with it. Her laboratory became a "mini-institute" with many sections.

Gertrude Elion officially retired in 1983, but her scientific legacy lived on. Workers from her team, using approaches she developed, discovered azidothymidine(AZT), the first drug approved for the treatment of acquired immunodeficiency syndrome (AIDS).

Elion received her greatest honors after her retirement. On October 17, 1988, she learned that she had won the Nobel Prize. In 1991, she was given a place in the Inventors' Hall of Fame, the first woman to be so honored. She also received the National Medal of Science that year. She is included in the National Women's Hall of Fame and the Engineering and Science Hall of Fame as well. In 1997 Elion received the Lemelson/MIT Lifetime Achievement Award. Elion died on February 21, 1999.

Recipient of the 1988 Nobel Prize in physiology or medicine, Gertrude B. Elion knew from the age of 15 that she wanted to do cancer research. *(Courtesy Burroughs Wellcome Company)*

Farnsworth, Philo
(1906–1971)
American
Inventor, Communications Industry

Philo Farnsworth was one of the pioneers of the television age, as he invented the mechanism for the electronic transmission of images. JOHN LOGIE BAIRD had invented mechanical television shortly before; Farnsworth's "image dissector" camera differed from Baird's "televisor," as the former had no moving parts—the scanner moved across the image electronically. Furthermore, Farnsworth's system cut the bandwidth necessary for transmission in half, as compared to mechanical television transmission, a consideration that takes on increasing significance as the airwaves become more crowded with the transmission of information.

Philo Taylor Farnsworth was born on August 19, 1906, in Indian Creek, Utah, near Beaver City, the community settled in 1856 by the grandfather after whom he was named. He was the eldest of five children born to Serena Bastian and Lewin Edwin Farnsworth, a farmer. When Farnsworth was 12 years old, the family moved to a ranch in Rigby, Idaho, equipped with its own power plant—which the boy maintained and improved, demonstrating his technical adeptness. The next year, he won a national invention contest.

In 1920, at the age of 14, Farnsworth began attending high school, where he impressed his chemistry teacher, Justin Tolman, with his scientific knowledge—Tolman later admitted that the youngster explained Einstein's theory of relativity more clearly than he himself understood it. Legend has it that Farnsworth conceived of his theory for the electronic transmission of images while plowing a potato field back and forth, row-by-row, when he realized that an electron beam could scan an image similarly, line-by-line. He explained his idea to an astounded Tolman, and even sketched him an image of what he later developed into the image dissector. By 1922, he had a design in mind.

Farnsworth took correspondence courses in physics from the University of Utah and subsequently entered Brigham Young University. However, the death of his father in 1924 required him to drop out in order to support the family—no matter, as he was largely an autodidact who understood more than most professors could teach him. After serving briefly in the navy, he took up a job as a canvasser for Salt Lake City's Community Chest, a philanthropic organization with fund-raising run by George Everson.

Farnsworth explained his television design to Everson and Leslie Gorrell, who combined to invest $6,000 to support his building of a television. A group of San Francisco venture capitalists, including bank president William W. Crocker and engineer Roy N. Bishop, further financed the furnishing of a research laboratory for Farnsworth, giving him a year to construct a prototype. On May 27, 1926, Farnsworth married Elma Pem Gardner, whom he had dated since college, and the next day the couple left for San Francisco.

A little more than a year later, on September 7, 1927(some three weeks before his deadline), Farnsworth gathered his investors at his 202 Green Street lab. There, his brother-in-law, Cliff Gardner, manned the image dissector, which scanned (slowly in one direction, quickly in the other) the image of Pem, Farnsworth's wife and assistant, then transmitted the image from the vacuum transmitter tubes to the receiver. "There you are," said Farnsworth, "electronic television." "The received line picture was evident this time," he wrote in his journal later that night—chronicling the first instance of electronic television broadcasting (John Logie Baird had broadcast a mechanical television signal two years earlier). Everson reacted more excitedly—"The damned thing works," he telegraphed to another investor, prompting them to support Farnsworth for another year.

Farnsworth demonstrated his electronic television system publicly in September 1928, and the next year he and his investors founded Television Laboratories, Inc., installing Farnsworth as vice president and director of research. This firm played David to the Goliath of RCA, the Radio Corporation of America, which tried to bully Farnsworth out of the television market. David Sarnoff, the infamous vice president of RCA, and his engineer, the Russian immigrant VLADIMIR ZWORYKIN, visited Farnsworth's laboratories to offer a meager $100,000 for the company, which the young inventor unhesitatingly refused. "There's nothing here we'll need," Sarnoff reportedly said, though it was only after this visit that Zworykin designed his iconoscope, which bore an uncanny resemblance to the image dissector.

Both the image dissector and the iconoscope used a cathode-ray tube for transmission—what distinguished the two was that the latter scanned the image with an electron beam, while the former used an "anode finger," a tube resembling its namesake with an aperture atop that received electrons containing the electronic image, emitted from magnetic coils. A patent war between these two devices ensued, but Farnsworth successfully defended his claim that he had designed his image dissector the year before Zworykin's 1923 patent application (in which he used a mechanical design, based on Nipkow spinning disks, not an electronic design). Farnsworth's former teacher, Tolman, testified to this effect, and Farnsworth won the 1939 suit, forcing Sarnoff (who had earlier claimed that "RCA doesn't pay royalties, we collect them") to pay cross-licensing royalties to Farnsworth Radio and Television, as the company was renamed in 1929. Papers reported the news as "Sarnoff's folly."

Farnsworth Radio and Television boomed during World War II, developing electronic surveillance equipment on government contracts. After the war, however, business fell off, forcing Farnsworth to sell his company to the International Telephone and Telegraph Company (ITT) in 1949. By this time, Farnsworth had moved to Maine and sunk into despondency over the besmirching of his claim as the inventor of television. He grew depressed and alcoholic and developed an addiction to painkillers. Even his later innovations—he made the first cold cathode-ray tube, developed the first simple electron microscope, and innovated a technological prototype to radar—could not resurrect him. He focused his later research on safe uses of atomic energy, but this too failed to buoy his spirits.

Farnsworth grew increasingly cynical of his invention as he aged—"he felt he had created kind of a monster, a way for people to waste a lot of their lives," his son Kent reported, adding that his father did not want television in his "intellectual diet." Farnsworth suffered a nervous breakdown and was subjected to shock therapy before he died of emphysema on March 11, 1971, in Holladay, a Salt Lake City suburb.

Although a statue in the U.S. Capitol Statuary Hall calls Farnsworth the "father of television," he died in relative obscurity. This was counteracted by the issuance of a commemorative postage stamp in 1983, his induction into the National Inventors Hall of Fame in 1984, and the establishment of the Philo T. Farnsworth Memorial Museum in Rigby, Idaho, in 1988. In 1999, *Time* magazine named him one of the 100 great scientists and thinkers of the 20th century.

⊗ **Fermi, Enrico**
(1901–1954)
Italian/American
Physicist

Enrico Fermi stands out from other physicists in that he excelled in both theoretical and experimental physics. In the former category Fermi developed a statistical system for quantum mechanics that Paul Adrien Maurice Dirac also arrived at independently, hence the name Fermi-Dirac statistics. Fermi also proposed a theory for beta decay. In the latter category Fermi discovered a method for inducing radioactivity with slow neutrons, though he did not recognize the process as nuclear fission. Later, working under the auspices of the Manhattan Project, he directed the first controlled nuclear chain reaction.

Fermi was born on September 29, 1901, in Rome, Italy, to Alberto Fermi, an administrator

Enrico Fermi served as a member of the Manhattan Project and directed the first controlled nuclear reaction. *(NARA, courtesy AIP Emilio Segrè Visual Archives)*

for the Italian railroads, and Ida de Gattis, a teacher. Fermi had an older sister, Maria; his older brother, Giulio, died when Fermi was 14 years old. In 1928, Fermi married Laura Capon, the daughter of an admiral in the Italian navy. The couple had two children—Nella in 1931 and Giulio in 1936.

In 1918, the Reale Scuola Normale Superiore awarded Fermi a fellowship on the strength of his application essay, which evinced his genius. In 1922, he graduated magna cum laude from the University of Pisa with a doctoral dissertation he wrote on research that he performed by using X-ray experimentation.

Fermi then received a fellowship from the Italian Ministry of Public Instruction to study at the University of Göttingen under Max Born and at the University of Leiden under Paul Ehrenfest. In 1925, Fermi returned to Italy as a lecturer at the University of Florence. During his tenure there, Fermi wrote the paper that applied Wolfgang Pauli's exclusion principle on electrons to atoms in a gas, a similar proposition to Dirac's a few months later, which resulted in Fermi-Dirac statistics. On the strength of this important advance Fermi won the new chair in theoretical physics established in 1927 at the University of Rome. The next year he published *Introduction to Nuclear Physics*, the first Italian textbook on the subject. In 1933, Fermi rounded out his work with theoretical physics by proposing his theory for beta decay, in which he proposed the existence of a new kind of force resulting from neutrons transitioning from higher to lower states of energy, later called weak force.

In 1934, Frédéric Joliot and IRÈNE JOLIOT-CURIE discovered artificial radioactivity by alpha bombardment. Fermi repeated the process, using neutrons instead of alpha particles to bombard stable isotopes as a means of converting them to unstable, radioactive elements. Fermi attempted this process with all of the elements and by accident discovered that bombarding through paraffin wax, which contained hydrocarbon molecules, slowed the neutrons enough to elicit more assured reactions. Uranium proved to cause the most interesting reaction, as it created a new element, with an atomic number one higher than that of uranium. In fact, Fermi had discovered nuclear fission, a fact that became apparent when German scientists replicated the process.

Upon acceptance of his 1938 Nobel Prize in physics for this discovery, Fermi and his family defected from Italy to the United States, where he became a professor of physics at Columbia University. In 1939, he drafted a letter carried to President Roosevelt by ALBERT EINSTEIN warning of the potential destructive power of nuclear weapons, especially in the hands of the Nazis. After Roosevelt instituted the Manhattan Project to build the first atomic bomb, Fermi played an instrumental role. He was responsible for the activation of an atomic pile on December 2, 1942, at 3:21 P.M., initiating a 28-minute self-sustaining nuclear chain reaction, the first step into the nuclear age.

The Fermis became U.S. citizens on July 11, 1944, and in 1946, the University of Chicago named him the Charles H. Swift Distinguished-Service Professor for Nuclear Studies. That year he received the Congressional Medal of Merit. After battling stomach cancer, he died on November 28, 1954, in Chicago. The U.S. Department of Energy awarded him the first Enrico Fermi Prize, named in his honor, posthumously. Element 100 was named fermium for him as well.

Ferranti, Sebastian Ziani de
(1864–1930)
English
Electrical Engineer, Energy and Power Industries

Sebastian de Ferranti designed the first large-scale system of electrical generation and distribution by a central power station. Ferranti's system used alternating current (AC) as a superior alternative to THOMAS EDISON's direct current (DC). The use of AC allowed for the generation of large amounts of voltage (up to 10,000 volts), which could be scaled down to appropriate levels for consumer use by a transformer. Ferranti's electrical system established the standards still used for power stations, as well as the grid system of power distribution.

Sebastian Ziani de Ferranti was born on April 9, 1864, in Liverpool, England. His father, César Ferranti, was a photographer whose studios were located in the family's home. At the age of 13, Ferranti installed electrical lighting in their house, a precursor of his career. The family then moved south to Ramsgate in Kent, where Ferranti attended St. Augustine's College. The faculty recognized his scientific promise by establishing a laboratory expressly for him, and he returned in kind by designing and installing electrical bells in the school.

At 17, Ferranti went to work at the Siemens brothers' Charlton works. Within a year there, he devised an electrical system for rotating and mixing molten steel, and he was consequently promoted to supervisor of the company's electrical lighting systems. Concurrently, he attended evening classes at University College, London, where he designed an improved dynamo that he patented in 1882, selling his first one for five pounds ten shillings, a high price for the day. His installation of this dynamo in the Cannon Street railway station brought him recognition as an innovative electrical engineer. He also collaborated with the distinguished physicist William Thomson (later dubbed Baron KELVIN) in designing the Thomson-Ferranti alternator; the pair established a company, in partnership with a solicitor named Ince, to market their alternator.

The Grosvenor Gallery Company of London bought one of Ferranti's dynamos and subsequently hired him as its engineer in 1886. In this position, Ferranti realized the efficiencies of scale that could be created by setting up one large power station to replace a series of smaller, 200-to 400-volt stations. His main obstacle was to overcome the assumption of 2,000 volts as the outermost limits of safety; Ferranti's power station would generate 10,000 volts of electrical energy.

Ferranti built this mega-station at Deptford, where fuel and water were more accessible than in the center of London. What distinguished Ferranti's system was his use of alternating current (as opposed to direct current), which he would feed through mains to the north of London. Spread out throughout London would be a series of step-down transformers, which would reduce the voltage to safe levels for household consumption. The implementation of this system represented the birth of the grid system of electrical distribution.

To implement this project, Grosvenor spun off the London Electric Supply Corporation Ltd., established in August 1887 with Ferranti as chief engineer. However, Parliament's Electricity Lighting Act of 1888 limited the scope of the Deptford plant's geographic coverage, forcing Ferranti to reduce electricity output to one-quarter of its intended capacity. That year, Ferranti married Gertrude Ince, the daughter of one of his business partners. Their son, Vincent Z. de Ferranti, followed in his father's footsteps to become an electrical engineer, as did his son, Sebastian J. Z. de Ferranti.

In 1892, Ferranti quit the London Electric Supply Company and consulted in the industry while planning his own business. Four years later, he established his own factory at Hollinwood, near Oldham in Lancashire, where he designed high-tension cables, circuit breakers, transformers, turbines, and spinning machines for manufacture. His company also generated electricity to supply to towns on contract, implementing on a smaller scale the electricity delivery scheme devised for the Deptford plant.

Ferranti served as president of the Institution of Electrical Engineers from 1910 through 1911. Addressing the organization, he argued for the transformation of coal energy (which operated at a mere 10 percent efficiency) into electrical energy, which proved to be a much more effective means of distribution. In 1924, Ferranti received the Faraday Medal, and in 1927, he was inducted into the fellowship of the Royal Society. Ferranti died while on vacation in Zurich, Switzerland, on January 13, 1930.

Fleming, Sir Alexander
(1881–1955)
Scottish
Bacteriologist

Sir Alexander Fleming discovered the antibacterial action of penicillin, though it was not until ERNST BORIS CHAIN and HOWARD WALTER FLOREY isolated and purified its active ingredient a decade later that penicillin became established as a wonder drug. Life before penicillin is now hardly conceivable, as bacterial infections are no longer life-threatening. The lives saved by penicillin number in the millions (if not more). Fleming cautioned against the indiscriminate use of penicillin, warning that bacteria could develop immunity to too-small doses, a phenomenon that has since happened.

Fleming was born on August 6, 1881, in the town of Lochfield, in Ayrshire, Scotland. He was the third of four children born to Grace Morton, the second wife of sheep-farmer Hugh Fleming, who brought to the union four children from his first marriage. Fleming's father died when the boy was seven years old. Fleming attended Louden Moor School, Darvel School, and Kilmarnock Academy before moving to London at the age of 13 to join his four brothers and one sister. There he studied business at the Regent Street Polytechnic Institute for three years, then worked in a shipping office for four years.

In 1900, Fleming enlisted as a private in the London Scottish regiment of the Territorial Army to fight in the Boer War, but he never saw action. When the death of his uncle the next year bequeathed him with 250 pounds, his brother Tom, who had a successful ophthalmology practice, encouraged "Alec" to enroll in medical school. He entered St. Mary's Hospital Medical School of London University on scholarship. He conducted vaccine therapy research as a junior assistant under Sir Almoth Edward Wright. He earned bachelor's and master's degrees with honors in 1908, conferred with the London University Gold Medal.

Although Fleming passed the Fellowship of the Royal College of Surgeons exam, qualifying him to practice surgery, he deferred to laboratory research in the Inoculation Service in order to remain at St. Mary's (where he ended up spending his entire career). He did, however, establish a profitable practice administering salvarsan (arsphenamine) to syphilitic patients—including prominent London artists, earning him the nickname "Private 606" (after the number of compounds Paul Ehrlich tested before discovering salvarsan, so-named to mean "that which saves by arsenic").

In 1914, St. Mary's appointed him as a lecturer, and later that year, he entered World War I as a captain in the Army Medical Corps, serving alongside his mentor, Wright, in Boulogne, France. In 1915, he married Sarah Marion McElroy, an Irish nurse who bore his only child, a son named Robert (later a physician like his father), in 1924. After the war, Fleming had returned to St. Mary's, which named him assistant director of the Inoculation Service in 1921 (the Royal College of Surgeons had named him Hunterian professor in 1919). There he conducted research on antibacterial agents, in response to his wartime treatment of wounded soldiers who died from simple bacterial infections.

In 1922, using mucous gathered from his own cold-ridden nose, Fleming discovered lysozyme, an enzyme capable of dissolving (or "lysing," hence the name) bacteria. Unfortunately, its efficacy as an antibacterial proved limited, though it did open the door to the potential existence of an agent that kills bacteria but does not harm the body. Fleming continued his search for an antibacterial agent. In 1928, the Royal College of Surgeons named him Arris and Gale lecturer, and London University appointed him as a professor of bacteriology.

Later that year, in September 1928, a series of fortuitous coincidences that can only be called serendipity converged. First, Fleming neglected to wash a stack of petri dishes harboring *Staphylococcus aureus* cultures before leaving on a two-week vacation. Second, a spore of the rare mold *Penicillium notatum* floated from a mycology laboratory downstairs, where it was being investigated as a potential asthma trigger, onto one of Fleming's uncovered culture dishes. Third, during Fleming's absence, London experienced an unusual cold spell, perfect conditions for inducing mold growth, then a heat wave, perfect conditions for fostering growth of the staph bacteria. Fourth, upon his return, Fleming noticed a clear halo surrounding the yellow-green mold growth—"That's interesting," he commented to professor Hare, a colleague visiting his lab. Fleming rightly deduced this to be an antibacterial zone where the mold inhibited the staph's growth. Although he did not realize the full significance of his discovery, he did realize that he had identified an antibacterial agent.

Fleming reported his finding in a 15-line article entitled "On the Antibacterial Action of Cultures of a Penicillium, with Special Reference to Their Use in the Isolation of B. influenzae," which appeared in a 1929 edition of the *British Journal of Experimental Pathology*. The scientific world took scant notice of his announcement. Unable to perform the challenging technical feats necessary for isolation of *Penicillium notatum*, Fleming abandoned penicillin research in 1932. Some seven years later, a sample of Fleming's penicillin strain made its way into the hands of Ernst Chain and Howard Florey, who isolated and purified the mold's active ingredient and established the powerful antibacterial action of the substance.

Recognition for the discovery of penicillin fell mostly on Fleming, due in part to Florey's reticence to lay claim to the "re-discovery." King George VI knighted both in 1944, and Fleming shared the 1945 Nobel Prize in physiology or medicine with Florey and Chain. Fleming retired to emeritus status in 1948, though he continued to work as a principal of the Wright-Fleming Institute of Microbiology, named that year after him and his mentor. The next year, his wife died, and four years after that, he married Dr. Amalia Koutsouri-Voureka, a fellow bacteriologist of Greek descent.

Perhaps the most interesting distinction Fleming received besides his Nobel Prize was his naming as Honorary Chief Doy-gei-tau by the Kiowa tribe. Fleming died of a heart attack on March 11, 1955, and is buried in the crypt of St. Paul's Cathedral.

Florey, Howard Walter
(1898–1968)
Australian/English
Pathologist

Howard Florey and ERNST BORIS CHAIN established the antibacterial efficacy of penicillin, the antibiotic discovered a dozen years earlier by ALEXANDER FLEMING. This trio shared the 1945 Nobel Prize in physiology or medicine for their contributions to the discovery and elucidation of penicillin as perhaps the most significant medical discovery of the 20th century. Before the advent of penicillin, people routinely died from simple bacterial infections; now, in the penicillin age, people recover from bacterial infections without even realizing the potential severity their condition posed in previous times.

Howard Walter Florey was born on September 24, 1898, in Adelaide, Australia. He was the only son (of three children) born to Bertha Mary Wadham, the second wife of Joseph Florey, a shoemaker and boot manufacturer who brought to his new family two daughters from his first marriage to Charlotte Ames. Florey's father amassed some wealth by establishing boot factories throughout Australia, though Florey still received a scholarship to St. Peter's Collegiate School. His chemistry teacher, "Sneaker" Thompson, inspired his interest in science (just as "Sneaker" had inspired another Nobel laureate, Laurence Bragg). Florey also secured a scholarship to the medical program at Adelaide University, graduating with a bachelor of science degree in 1921.

In 1922, Florey won a South Australia Rhodes Scholarship to Magdalen College of Oxford University, where he studied physiology under Sir Charles Scott Sherrington. He placed first in his class in the Honours School to earn his second bachelor of science degree in 1923. Sherrington, under whom Florey taught while pursuing his 1924 master's degree, nominated him for a John Lucas Walker studentship to Cambridge University, which he received. In 1925, Florey used a Rockefeller Traveling Scholarship to study physiology under Alfred Newton Richards at the University of Pennsylvania. In 1926, he returned to a fellowship at Gonville and Caius College and a Freedom Research Fellowship in pathology at London Hospital; that same year, he married Mary Ethel Hayter Reed, an Adelaide University classmate. Together, the couple had two children, Paquita Mary Joanna and Charles du Vé. He received his doctorate from Cambridge in 1927.

That year, Cambridge appointed Florey Huddersfield Lecturer in Special Pathology. During this period, he became familiar with the work of Alexander Fleming, who had identified lysozyme as an antibacterial enzyme. In 1931, the University of Sheffield appointed Florey to a chair of pathology. Four years later, in 1935, Florey returned to Oxford as the director of the new Sir William Dunn School of Pathology. He hired the German biochemist Ernest Chain, and together they embarked on research into antibacterial agents.

Chain's 1938 literature search led him to lysozyme, which in turn led him to Fleming's 1929 article in the *British Journal of Experimental Pathology*, "On the Antibacterial Action of Cultures of a Penicillium, with Special Reference to Their Use in the Isolation of B. influenzae." Incredibly, as he was reading the 15-line article, a woman walking the halls with a petri dish in her hand caught his eye; when he inquired what she was culturing, she replied *Penicillium notatum*, a strain donated to her lab by Fleming himself!

In 1939, Florey secured funding from the Medical Research Council in England and the Rockefeller Foundation in the United States. On Saturday, May 25, 1940, Florey's team (including Chain and Norman Heatley, among others) conducted one of the most important experiments in scientific history: they injected eight mice with a lethal dose of virulent streptococcus bacteria; one hour later, they injected four mice with penicillin. Sixteen-and-a-half hours after that, the four control mice were dead, while the four mice administered penicillin lived on. When the usually understated Florey phoned his assistant (and future wife) Margaret Jennings on Sunday to relay the news, he commented that "It looks like a miracle."

Florey set his lab into frantic action to prepare enough penicillin for a human trial; after using all the lab glass, they turned to using 16 enamel bedpans as well. By February 12, 1941, they had produced enough penicillin for an experimental injection into a human subject—police officer Albert Alexander, who got a bacterial infection after being scratched by a rose thorn. Within a day, he began to recover, while Florey and his team collected his urine to recycle the penicillin. Tragically, they couldn't produce enough penicillin to sustain his turnaround, and he eventually died.

Against Chain's wishes, Florey and Heatley risked a transatlantic flight (in a blacked-out plane) to seek companies interested in manufacturing penicillin in the United States, which had just entered World War II. Serendipitously, Florey's former mentor, Alfred Richards, helped them locate a Department of Agriculture laboratory in Peoria, Illinois, which was looking for applications for a by-product of starch production. This corn steep liquor turned out to increase penicillin production tenfold. What was more, a lab worker named Mary Hunt (nicknamed "Moldy Mary") discovered mold on a local rock melon (cantaloupe) that yielded 3,000 times more penicillin than Fleming's original strain!

Anticipating the vital significance of their penicillin research at the beginning of the war, Florey and his colleagues smeared *Penicillium* spores inside coat linings so that any one of them could continue research whatever the outcome of a German invasion. By late 1943, however, drug companies such as Merck, Squibb, and Pfizer were mass-producing penicillin, in time to treat soldiers wounded in the 1944 D-Day invasion. By the end of World War II, penicillin had already saved thousands of lives.

Recognition of the significance of Florey's "re-discovery" of penicillin came immediately: the prestigious Royal Society inducted him into its fellowship in 1942; King George VI knighted Florey in 1944; and in 1945, Florey shared the Nobel Prize in physiology or medicine with Chain and Fleming.

Later in his career, Florey discovered cephalosporin C, the basis for penicillin derivatives and antibiotic alternatives to penicillin. He also held a number of important administrative positions: from 1960 through 1965, he presided over the Royal Society (he was called the "Bushranger" president, as he was the first Australian to hold the post); Oxford University appointed him provost of Queen's College in 1962; and in 1965, he became the chancellor of Australian National University and president of the British Family Planning Association (concerned that his discovery led to overpopulation, he promoted contraception and abortion). Also in 1965, he received a peerage title as Baron Florey of Adelaide and Marston.

After years of unhappy marriage, Florey's first wife died in 1966. A year later, he married Margaret Jennings, his longtime assistant of 30 years. The last year of his life was perhaps his happiest, despite his declining health. Florey died of a heart attack on February 21, 1968.

Forrester, Jay
(1918–)
American
Computer Engineer

Working under contract to the U.S. Navy, Jay Forrester invented the Whirlwind computer, which became the backbone for the SAGE air defense system shielding the entire United States, a project that he directed. While building the Whirlwind, Forrester developed the memory system—random-access, coincident-current magnetic storage—that became the standard in most computers. His invention thus determined the direction the digital revolution would follow. However, Forrester did not rest on his laurels; before the SAGE system was even fully implemented, he shifted his focus by applying his expertise in computer systems to the field of management, and in the process he created the discipline of system dynamics, which uses computer-generated "what if" simulations to analyze how feedback loops affect the outcome of decisions.

Jay Wright Forrester was born on July 14, 1918, near Climax, Nebraska, where his parents, Ethel Pearl Wright and Marmaduke M. Forrester, owned a cattle ranch. As an adolescent, he used old car parts to build a wind-powered, 12-volt electrical system for the remote ranch. After graduating from high school in 1935, he received a scholarship to Nebraska's Agricultural College; three weeks before enrolling, he decided to switch his enrollment to the Engineering College of the University of Nebraska. In 1939, he graduated at the top of his class with a bachelor of science degree in electrical engineering.

The Massachusetts Institute of Technology (MIT) offered Forrester a $100-per-month research assistantship at its high voltage laboratory, sealing his decision to attend its graduate program in electrical engineering. The next year, he transferred to the newly established servomechanisms laboratory, where he worked under Gordon S. Brown developing servomechanisms to control radar antennae and gun mounts. When, in 1944, Forrester considered leaving the lab to start his own business, Brown persuaded him to stay by offering him *carte blanche* in choosing his next project.

Forrester chose to work on the U.S. Navy Bureau of Aeronautics's aircraft stability and control analyzer (ASCA), an analog computer program that would simulate flight based not on known airplane behavior but rather on data collected in wind tunnel tests of model planes. By the spring of 1945, Forrester had determined that an analog computer could not handle the complexity of the project. Perry Crawford of the Navy's Special Devices Center, who had written his master's thesis on digital computing solutions to fire control, suggested that Forrester abandon analog in favor of digital computing.

In 1946, Forrester earned his master of science degree and married Susan Swett, with whom he eventually had three children—Judith, Nathan, and Ned. Also in that year, MIT named Forrester director of its new digital computer laboratory, where he led the computer simulation project, which had changed its name to Whirlwind. Forrester demonstrated his genius by troubleshooting the Whirlwind's myriad problems: he extended the life span of the vacuum tubes a thousandfold by using a silicon-free cathode material in them, and he made the Whirlwind self-monitoring for potential breakdowns. His resolution of a problem with short-lived storage tubes not only solved a dilemma for the Whirlwind but also inaugurated a new paradigm in computer memory that still holds sway today.

After considering numerous memory alternatives (one-dimensional mercury delay lines; two-dimensional Williams Tubes; and MIT Storage Tubes, which he developed himself), Forrester jury-rigged his own system from a German material called Deltamax, which he used to make small rings that magnetized the reversible current flowing through them. After consulting with MIT graduate student William Papian, he subsequently replaced the Deltamax with magnetic ferrite cores coordinated with a wire grid to create a three-dimensional system known as random-access magnetic-core memory, the precursor to modern RAM memory. Forrester's invention of this computer memory system largely determined the direction of the digital revolution and established the standard by which most computers operate. In this sense, Forrester left his fingerprint on the lives of millions of computer users.

By the time the Whirlwind computer was constructed, all 3,300 electrostatic storage tubes and 8,900 crystal diodes housed in a 25,000-square-foot, two-story building, its original application for flight simulation had been abandoned. A new application saved it from similar abandonment: the Soviet detonation of an atomic bomb in August 1949 prompted the Navy to use the Whirlwind as the backbone of an automated real-time air defense system blanketing the entire country.

In 1951, the Navy and MIT installed Forrester as the director of Project SAGE (Semi-Automated Ground Environment), which networked Whirlwind computers in 23 direction centers, three combat centers, and one programming center strung out across North America. This system radar-tracked the movement of planes in U.S. airspace, allowing for the early detection of foreign bombers. Each four-story direction center housed a Whirlwind computer containing 80,000 vacuum tubes; remarkably, this system was operational 99.8 percent of the time over its 20-year life span, from its complete implementation in 1963 to its decommission in 1983.

At the suggestion of MIT president James Killian, Forrester left SAGE in 1956 to apply his knowledge of computer systems to more organic, human systems by joining MIT's Sloan School of Management, newly established with a $10-million grant from General Motors head Alfred Sloan. Officials from General Electric asked him to consider why its household appliance plants in Kentucky operated four shifts one year, then inexplicably had to lay off half its workers a few years later. In a notebook, he penciled in the various variables affecting the situation to visualize their dynamic interrelationship and thus gave birth to the field of system dynamics.

Computer programmer Richard Bennett invented a compiler, named SIMPLE (Simulation of Industrial Management Problems with Lots of Equations), that would input real-life variables and output computer simulations of real-life outcomes. Jack Pugh later extended this compiler into the DYNAMO series, programs that helped spread the influence of system dynamics. Forrester applied his theory of system dynamics to ever broader systems, disseminating his ideas in a series of influential books: *Industrial Dynamics* in 1961; *Principles of Systems* in 1968; *Urban Dynamics* in 1969; and *World Dynamics* in 1971. *Urban Dynamics* controversially asserted that low-cost housing, far from alleviating poverty, rather perpetuated it; *World Dynamics* analyzed global systems, concluding that industrialization and overpopulation threaten the earth's equilibrium.

Forrester directed Sloan's System Dynamics Program from its establishment in the early 1960s until 1989. He continues to promote the application of system dynamics to national economics, corporate culture, and precollege education as the Germeshausen Professor Emeritus of Management and senior lecturer at Sloan. Throughout his career, he received recognition for his pioneering ideas: he received the 1968 Inventor of the Year Award from George Washington University; the 1972 Systems, Man, and Cybernetics Society Award for Outstanding Accomplishment; and the 1974 Howard N. Potts Award from the Franklin Institute. In 1979, the National Inventors Hall of Fame inducted him into its membership.

Franklin, Benjamin
(1706–1790)
American
Physicist, Oceanographer

Although Benjamin Franklin is remembered mainly for his role in the American Revolution, he was also one of the preeminent scientists of his era. Self-taught, he investigated the then-mysterious phenomenon of electricity. Franklin proposed a "one-fluid" theory of electricity, which was underpinned by the concept that charges lost by one body must be gained simultaneously by another in equal amounts. Now known as the law of conservation of charge, it remains a fundamental scientific tenet. His famous kite experiment proved that lightning is an electric charge and spurred him to invent the lightning rod.

The 15th child of Josiah Franklin, Benjamin Franklin was born in Boston, Massachusetts, on January 17, 1706. Franklin's mother, Abiah

Benjamin Franklin's famous kite experiment proved that lightning is an electric charge. *(Smith Collection, Rare Book & Manuscript Library, University of Pennsylvania)*

Folger, was his father's second wife. In 1730 Franklin would marry Deborah Read.

After learning the printing trade in Boston and Philadelphia, Franklin worked as a printer in London from 1724 until 1726; there he made the acquaintance of several notable British scientists. In 1726 he returned to Philadelphia, where he published the *Pennsylvania Gazette* in 1729 and the wildly popular *Poor Richard's Almanac* in 1733. He entered public service in 1736, when he was appointed clerk of the state assembly.

Franklin was 40 when he embarked on his scientific career. Although not formally educated, he devoted tremendous effort to learning the theories of the day. He read the works of Robert Boyle and Sir ISAAC NEWTON and continued to correspond with the scientists he had met abroad. A leyden jar (a type of condenser—or capacitor—used for storing an electrical charge) donated to a Philadelphia library ignited Franklin's interest in electricity. From 1743 he made several important discoveries in the field. Franklin postulated that the movement or transfer of an electric "fluid" that comprised particles of electricity that could permeate other materials causes electrical effects. In addition, he theorized that whereas these particles repel each other, they are attracted to the particles of other matter. Franklin dubbed his concept the one fluid theory of electricity and introduced the terms *positive* and *negative* into scientific parlance. In Franklin's framework a charged body is one that has either lost or gained electrical fluid, thus becoming negative or positive. Implicit in Franklin's formulation was a concept that would later be named the law of conservation of charge, which holds that when one body loses a charge, another body must neutralize the effect by gaining an equal charge at the same time.

Franklin published his findings in his 1751 book, *Experiments and Observations on Electricity, Made at Philadelphia in America*. In 1752, Franklin undertook his famous (and dangerous) experiment to test whether lightning was an electric charge. By flying a kite in a thunderstorm, Franklin concluded that it was. The experiment also convinced him of the efficacy of lightning rods—which he invented and sold.

Franklin conducted other scientific endeavors as well. In the 1770s he pioneered the study of the Gulf Stream. In addition to measuring the ocean's temperature at different locations and depths, he collected data from sea captains and produced the first printed chart of that meteorological phenomenon. He also invented bifocal spectacles, the rocking chair, and the Franklin stove.

It is a testament to Franklin's many talents that his exceptional scientific career was only one facet of his life. After serving as the clerk of the state assembly until 1751, he became deputy postmaster for the colonies from 1753 until 1774. He was one of the signers of the Declaration of Independence in 1776 and was sent by the revolutionary government to France to seek military aid, which he single-handedly secured. Before he retired from public life in 1788, he was a delegate to the 1787 Constitutional Convention.

Franklin is credited with forging the branch of science dealing with electricity. After winning the Copley Medal in 1753, he was elected to the Royal Society in 1756. He was the chief founder of the American Philosophical Society, the colonies' first permanent scientific society, and played an important role in the creation of the University of Pennsylvania. Franklin died in 1790. His "one-fluid" theory influenced Count ALESSANDRO VOLTA, who would later apply many of Franklin's principles in producing the first battery.

Franklin, Rosalind Elsie
(1920–1958)
English
Chemist

Deoxyribonucleic acid, or DNA, carries the inherited information in the genes of most living things. In the early 1950s, scientists realized that the key to finding out how this information was stored and reproduced lay in the structure of DNA's complex molecules. Rosalind Franklin took X-ray photographs that gave two rival scientists, James Watson and FRANCIS CRICK, the clues they needed to work out the structure of DNA.

Rosalind Franklin was born on July 25, 1920, in London. Her father, Ellis, was a well-to-do banker, and her mother, Muriel, did volunteer social work while also raising five children. Rosalind decided at age 15 that she wanted to be a scientist. Her father objected, believing like many people of the time that higher education and careers made women unhappy, but she finally overcame his resistance. She studied chemistry at Newnham, a women's college at Cambridge University, and graduated in 1941.

As a way of helping her country during World War II, Franklin became assistant research officer at the Coal Utilization Research Association. She did research on the structure of carbon molecules, bringing, according to one professor, "order into a field which had previously been in chaos." She turned some of this work into the thesis for her Ph.D., which she earned from Cambridge in 1945.

Seeking new challenges, Franklin went to work for the French government's central chemical research laboratory in 1947. Friends later said that her three years there were the happiest of her life. She enjoyed an easy camaraderie with her coworkers, chatting at cafés and on picnics. She also learned a technique called X-ray crystallography, to which she would devote the rest of her career.

Many solid materials form crystals, in which molecules are arranged in regular patterns. In 1912, a German scientist named Max von Laue found that if a beam of X rays is shone through a crystal, some of the rays bounce off the crystal's atoms, while others pass straight through. When photographic film, sensitive to X rays, is placed on the far side of the crystal, the resulting photograph shows a pattern of black dots that can reveal important facts about the three-dimensional structure of the molecules in the crystal.

Chemists eventually also found ways to use X-ray crystallography on amorphous compounds, which did not form obvious crystals. Most of the complex chemicals in the bodies of living things are amorphous compounds. Molecular biologists were beginning to realize that the structure of these compounds revealed much about their function, and crystallography was a promising tool for revealing that structure. One of the molecules about whose structure scientists were most curious was DNA.

Scientists knew that the DNA molecule consisted of several smaller molecules. It had a long chain, or "backbone," made of alternating molecules of sugar and phosphate (a phosphorus-containing compound). Four different kinds of other molecules called bases were attached to the backbone. No one knew, however, whether the chain was straight or twisted, how the bases were arranged on it, or how many chains were in each molecule. Franklin and Maurice Wilkins, the researcher with whom she worked at King's College, hoped that Franklin's X-ray photographs would provide this information.

Franklin photographed two forms of DNA, a "dry," or crystalline, form and a "wet" form that contained extra water molecules. No one had photographed the wet form before. At the time, Franklin was not sure which type gave the more useful information. She took an excellent photograph of the wet form in May 1952, but she put it aside in a drawer and continued working with the dry form.

Two Cambridge scientists, a brash young American named James Watson and a somewhat older Britisher, Francis Crick, were also trying to work out the structure of DNA. Although Watson saw himself and Crick as competitors of the

King's College group, he and Wilkins became friends, and on January 30, 1953, he visited Wilkins at King's College. Without asking Franklin's permission, Wilkins showed Watson the photograph of "wet" DNA that she had made in May 1952. When he saw the photo, Watson wrote later, "my mouth fell open and my pulse began to race." He hurried back to Cambridge to describe the photo to Crick.

To Watson, the "GillSans Light"-shaped pattern of dots in Franklin's photo showed clearly that the DNA molecule had the shape of a helix. On the basis of this and other evidence, he and Crick concluded, as by this time Franklin also had, that the molecule consisted of two helices twined around each other. The backbones were on the outside and the bases stretched across the center. In other words, the molecule was shaped like a spiral staircase or a twisted ladder with the bases as steps or rungs.

Watson and Crick published a groundbreaking paper on the structure of DNA in Britain's chief science journal, *Nature*, on April 25, 1953. Neither then nor later did they fully credit Franklin for the important part her photograph had played in their discovery, and Franklin herself probably never realized its role. By the time the Cambridge scientists' paper appeared, she was no longer working on DNA. She had moved from King's College to Birkbeck, another college in the University of London, and was beginning an X-ray study of a common plant virus called tobacco mosaic virus. Almost nothing was known about the structure of viruses at that time. Franklin drew on her crystallography studies to make a model of the tobacco mosaic virus that was exhibited at the 1958 World's Fair in Brussels. The virus's inherited information was carried in RNA, a chemical similar to DNA. Franklin showed that the RNA molecule was also a helix.

In 1956, Rosalind Franklin discovered that she had ovarian cancer. The cancer proved untreatable, and she died of it on April 16, 1958.

Four years later, Watson, Crick, and Wilkins shared the 1962 Nobel Prize in physiology or medicine for their work on DNA. Supporters and critics still debate whether she would or should have been included if she had lived.

Freud, Sigmund
(1856–1939)
Austrian
Psychologist, Neuropathologist

Sigmund Freud founded the revolutionary theory of psychoanalysis, a method devoted to studying and analyzing psychological phenomena to diagnose and treat neuroses and mental conditions. Freud developed the psychoanalytic technique of free association—expressing, without censorship, the first thought that enters the mind. He used free association to study the unconscious mind and gain a better understanding of psychological disorders. Freud was also instrumental in promoting the analysis of dreams and formed pioneering and controversial theories regarding sexuality; he believed infantile psychosexual development played a critical role in adult psychological development.

Freud was born on May 6, 1856, in Freiberg, Moravia (now Příbor, Czech Republic). Freud was the eldest child of his wool merchant father's second family. Freud's mother, Amalie Nathanson, was 20 years younger than Jakob, his father, Freud's older half-brother, who was about the age of Freud's mother, had a child close to Freud's age. Making sense of this confusing family situation heightened Freud's intellect and curiosity. The family moved to Vienna in 1860 after the wool trade in Freiberg deteriorated, and Jakob was frequently unemployed.

Despite the family's lack of wealth, Freud's parents encouraged him in his studies and made financial sacrifices to further his education. Freud entered the University of Vienna in 1873 and graduated with a M.D. in 1881. He hoped for a career in biological research but chose instead to practice medicine at the Vienna General Hospital to better support his new wife, Martha Bernays. Freud worked in a number of departments at the hospital, remaining the longest in the nervous diseases department because of his interest in neuropathology.

After leaving the hospital in 1885, Freud went to Paris to study with neurologist Jean-Martin Charcot. It was during this four-month period, in which Freud worked with patients labeled as hysterics, that he began theorizing that neuroses may be caused not by organic disease but by psychological factors.

Freud returned to Vienna and started a private practice as a neuropathologist. Though he published several works on neuropathology, it was the publication in 1895 of *Studies in Hysteria* that marked the early stages of psychoanalytical theory. The publication, the result of a decade-long collaboration with physician Josef Breuer, introduced Freud's method of free association. An unpublished work, *Project for a Scientific Psychology*, written in 1895, presented a neurological approach to normal and abnormal psychology and also introduced Freud's definitions of the ego, the conscious, and the id, the unconscious.

Freud published what was considered to be his most important and original work in 1901—*The Interpretation of Dreams*. The publication theorized that the formation of dreams was influenced by unconscious experiences and desires. Another important work was *Three Essays on the Theory of Sexuality*, published in 1905. This book introduced his theories on the stages of psychosexual development and infantile sexuality. Freud believed that many disorders were a result of suppressed sexual wants. In 1923, he published a third work on psychoanalytic theory, *The Ego and the Id*. In this volume he further developed his concepts of id, ego, and superego, or conscience.

In 1902, Freud began holding weekly meetings at his home with colleagues to discuss psychoanalytic theory. The group evolved into the International Psychoanalytic Association in 1910, and participants included such notable members as Carl Gustav Jung and Alfred Adler. The association soon disbanded, however, due to disagreements and differences of opinion. By this time, Freud was already well known in Europe and discovering a highly receptive audience in the United States.

Freud and his wife had six children in the first 10 years of their marriage, creating considerable demand on Freud's time and energy, but he was undeterred in his pursuit to unveil the workings of the human mind. Though diagnosed with cancer in 1923, he continued to work and write. Freud was compelled to leave Vienna for London in 1938 when the Nazis gained control of Austria, but he proceeded to treat patients and work on his final book, *Moses and Monotheism*. Freud died in September 1939 but continues to be remembered as the father of psychoanalysis. He believed that psychoanalysis was more significant as theory than as treatment, but his work strongly influenced modern-day psychotherapy.

Frisch, Otto Robert
(1904–1979)
Austrian/English
Physicist

Otto Robert Frisch collaborated with his maternal aunt, the noted physicist LISE MEITNER, in deducing the process of nuclear fission, a term coined by Frisch to describe the splitting of the uranium nucleus. It was Frisch who confirmed the feasibility of using nuclear fission to induce a chain reaction capable of detonating a nuclear bomb, and thus he is primarily responsible for ushering in the atomic age whereby humans discovered the means to effect our own destruction.

Otto Robert Frisch collaborated with his maternal aunt, the noted physicist Lise Meitner, in deducing the process of nuclear fission, a term coined by Frisch to describe the splitting of the uranium nucleus. *(Niels Bohr Archive, courtesy AIP Emilio Segrè Visual Archives)*

Frisch was born on October 1, 1904, in Vienna, Austria. His father, Justinian Frisch, held a doctorate in law but worked as a publisher, and his mother, Auguste Meitner, was a retired pianist, composer, and conductor. Frisch graduated from high school in 1922 and matriculated at the University of Vienna, majoring in physics and minoring in mathematics. He wrote his doctoral dissertation on the discoloration of rock salts by cathode rays to earn his D.Phil. in 1926, at the age of 22.

Frisch first secured a job at an X-ray dosimeters manufacturing firm, but after a year, he moved to Berlin to accept a position in Carl Müller's optics laboratory at the Physikalisch Technische Reischsanstalt (PTR) developing a new measurement of brightness to replace candlepower. After three years there, he accepted a concurrent part-time position in the physics department at the University of Berlin helping Peter Pringsheim design a mercury-vapor detector. In 1930, he moved to the University of Hamburg to take up an assistantship under Otto Stern; together, the pair proved that beams of atoms act like waves (as do beams of electrons), and on his own, he conducted research on particle-like qualities of light.

After Hitler became chancellor of Germany in 1933, the university fired Frisch due to his Judaism, prompting him to immigrate to England. Stern recommended Frisch for an Academic Assistance Council grant supporting work on gamma ray emissions and cloud chamber construction under Patrick Blackett at Birkbeck College in London. He subsequently built an instrument on which he detected radioactive emissions from phosphorous and sodium.

In 1934, Frisch accepted Niels Bohr's invitation to work at Copenhagen's Institute of Theoretical Physics, where he worked with Czech physicist George Placzek measuring the absorption of slow-moving neutrons by cadmium, boron, and gold (using Nobel medals left with Bohr by German scientists fleeing Hitler).

In 1940, the Nazi invasion of Denmark forced Frisch to return to England, where the Australian physicist Marcus Oliphant offered him an auxiliary lectureship at the University of Birmingham. On his way there, he spent Christmas in Sweden with his aunt, Lise Meitner. The two physicists discussed the perplexing results of an experiment by Meitner's former colleague, Otto Hahn, and Fritz Strassmann—the inexplicable production of an isotope of barium resulting from the bombardment of the uranium nucleus with slow-moving neutrons. At first, Frisch considered Hahn's observations flawed, but after Meitner defended her former colleague's precision as an experimentalist, aunt and nephew struck upon the only explanation: the uranium nucleus must have been split in two. Frisch coined the term nuclear "fission" for this cleavage.

Accompanied by Meitner, Frisch hastily returned to Copenhagen to confirm their hypothesis experimentally before making his way back to England. In Birmingham, he collaborated with Rudolf Peierls pursuing Bohr's suggestion that uranium-235 would catalyze a chain reaction better than any other radioactive element or isotope. They not only confirmed Bohr's hunch, but they also discovered that it would require a mere few pounds of the radioactive material, not a few tons as was supposed, to start a chain reaction to detonate a nuclear bomb. Frisch reported this finding to Sir Henry Tizard, the British government's scientific adviser, which set in motion the forming of the Maud Committee to investigate the development of a nuclear bomb. Frisch could thus be identified as one of the first instigators of the process that resulted in the construction and detonation of the first nuclear weapon.

In August 1940, James Chadwick invited Frisch to Liverpool, where the Austrian oversaw the construction of a pulse height analyzer, or a "kicksorter," which analyzes uranium's alpha ray spectrum to determine its isotopic composition. In 1943, the British government granted him citizenship to facilitate his participation in the Manhattan Project, the secret atomic bomb building initiative taking place in Los Alamos, New Mexico. There, he led the Critical Assembly Group, in charge of the "dragon experiment," the trial run of a nuclear explosion, which he witnessed from a distance of merely 25 miles in order to conduct experiments.

After World War II, Frisch returned to England to serve as a deputy chief scientific officer and head of the Nuclear Physics Division of the Atomic Energy Research Establishment from 1945 through 1947. Cambridge University then named him Jacksonian Professor of Natural Philosophy and director of the nuclear physics department of the Cavendish Laboratory, a position he retained for the remainder of his career. In 1951, he married graphic designer Ulla Blau, a fellow Austrian, and together the couple had two children—Tony, who became a physicist, and Monica.

Frisch retired from Cambridge in 1972 but continued to work as chairman of Laserscan Ltd., the company founded to manufacture the "Sweepnik," the instrument he invented to track the path of charged particles passing through the liquid of a bubble chamber. He died on September 22, 1979, after sustaining injuries in a fall.

G

Gabor, Dennis
(1900–1979)
Hungarian/English
Physicist

Dennis Gabor is best known for inventing holography, a system of three-dimensional photography without lenses. Gabor devised this process in the course of trying to overcome the most significant limitation of early electron microscopes—that past a certain magnification the image would become distorted, thereby hindering thorough observation. Gabor hit upon the idea of recording on a photographic plate the phase patterns of the object under observation. When placed in a beam whose rays are all in the same phase, the plate produces a three-dimensional image of the object, which shifts as the observer alters perspective. Early versions of this technique were only modestly effective, as there was no reliable method by which to generate the requisite beam. With the invention of the laser in 1960, however, myriad applications for Gabor's discovery were found.

Gabor was born in Budapest, Hungary, on June 5, 1900, the eldest of his parents' three sons. His father, Berthold, was the grandson of Russian-Jewish immigrants and the director of the Hungarian General Coal Mines. His mother, Ady Jacobvits, had been an actress prior to giving birth.

After serving briefly in the Austro-Hungarian army at the end of World War I, Gabor enrolled in the Budapest Technical University, where he studied mechanical engineering. Rather than reenlist after being called up again during his third year of study, Gabor decided to leave the country, unwilling to serve what he viewed as an authoritarian government. He then moved to Berlin and entered the Technische Hochschule, from which he earned a diploma in 1924 and a doctorate in engineering in 1927.

Upon graduation Gabor took a position in the Siemens and Halske physics lab in Siemenstadt, Germany, where he invented the quartz mercury lamp. However, Gabor's contract was terminated shortly after Adolph Hitler assumed power in 1933, despite the fact that his family had converted to Lutheranism in 1918. Gabor went back to Hungary but moved to England in 1934 and began a long association with British Thompson-Houston Company (BTH). At BTH he first worked on developing a plasma lamp (a new sort of fluorescent lamp) that he had tinkered with in Hungary; he then shifted to electron optics. He also married a fellow BTH employee,

Dennis Gabor invented holography in 1947. *(AIP Emilio Segrè Visual Archives)*

Marjorie Louise Butler, in 1936. The couple had no children.

World War II limited Gabor's work, as he was restricted from access to much of BTH's military research by virtue of his foreign birth. After the war, however, he focused his attention on the electron microscope. In 1947, he had the insight that led to his developing the technique of holography. He published his theory the same year and coined the term hologram (from the Greek meaning "completely written") to describe it. The practical implementation of his insight was beset by technical problems, however, including a persistent double image produced by the photographic plate. Gabor left BTH in 1949 to become a reader (equivalent to an associate professor in the United States) at the Imperial College of Science and Technology at the University of London. In 1958, he was appointed professor of applied electron physics. With his students he built a flat television tube, an analog computer, and a Wilson cloud chamber.

The development of the laser in 1960 led to renewed interest in Gabor's work. The ability to concentrate a narrow light source with all waves in the same phase eliminated the problems that had nagged Gabor's earlier efforts. After retiring from Imperial College in 1967, Gabor continued his own research and demonstrated significant potential applications of holography in computer data processing and a number of other fields.

The importance of Gabor's work was widely recognized. He was awarded the Nobel Prize in physics in 1971 and became a member of several scientific societies. He also held more than 100 patents. Moreover, he became a popular speaker, in part as a result of two books he wrote on the importance of scientists using their knowledge to benefit society. Gabor died in 1979, but his legacy lives on. Holography remains an important tool in many aspects of modern life, including photography, mapmaking, computing, medicine, and even supermarket checkout scanners.

Galilei, Galileo
(1564–1642)
Italian
Astronomer, Physicist

Galileo Galilei made significant contributions in the fields of astronomy and physics, particularly in mechanics. His discoveries in astronomy, accomplished with the telescope, a relatively new invention at the time, changed the way the universe was viewed. Galileo was the pioneering force behind the application of mathematics to the analysis of mechanics, and he also demonstrated that falling bodies accelerate uniformly and independently of their weight. Galileo's support of Nicolaus Copernicus's theory that the planets revolve around the sun was contrary to the teachings of the Roman Catholic Church and caused years of conflict and unrest for Galileo.

Galileo was born February 15, 1564, in Pisa, Italy, the oldest of seven children. His mother was Giulia Ammannati of Pescia, and his father was Vincenzio Galilei, a musician and descendant of a Florentine patrician family. The family moved to Florence in 1575, and Galileo was educated at a monastery near Florence. When he entered the order as a novice in 1578, his father, who did not approve, removed him to Florence.

Galileo entered the University of Pisa in 1581 to study medicine but became fascinated with mathematics. He studied mathematics with a private tutor and left the university in 1585 without a degree. Galileo returned to Florence to teach and continued his studies privately. He secured a teaching position at the University of Pisa in 1589 and moved to Padua in 1591, where he gained a professorship in mathematics. Galileo returned to Pisa in late 1610, where he became the chief mathematician at the University of Pisa.

Galileo made one of his first discoveries in 1583 while still a university student. Observing that lamps swinging in the wind took the same amount of time to swing regardless of the degree of movement, Galileo proposed the use of pendulums for clocks. In 1586 Galileo created a

Galileo used this telescope and "discoverer" objective to observe in 1610 that Jupiter had four satellites. *(AIP Emilio Segrè Visual Archives, E. Scott Barr Collection)*

hydrostatic balance to calculate relative densities. Galileo also formulated the law of uniform acceleration for falling bodies through a series of experiments in which he rolled balls down sloped planes (Galileo did not drop weights from the leaning tower of Pisa as was widely reported). He later documented many of his observations on mechanics in *Dialogues Concerning Two New Sciences* (1638).

Galileo is perhaps best known for his contributions in astronomy. Though he did not invent the telescope, he improved upon the basic design to create an astronomical instrument and was the first to study the stars and planets. Galileo was amazed by the abundance of stars, invisible to the naked eye, and observed that the moon had mountains and craters, not a smooth surface as was traditionally thought. He also discovered sunspots. In 1610 Galileo viewed Jupiter and discovered that it had four satellites—a major finding. He quickly published his observations in *Starry Messenger* and enjoyed immediate acclaim.

Galileo's studies of the universe convinced him to adopt the Copernican principle that the earth revolved around the sun. This posed problems between Galileo and the Roman Catholic Church, and Galileo had no choice but to retract his support of the theory. Galileo tried to avoid controversial issues in subsequent years, but his thinly veiled support of the Copernican theory in *Dialogue on the Two Chief World Systems, Ptolemaic and Copernican* in 1632 was the last straw. In 1633 Galileo was condemned by the Inquisition and placed under house arrest.

Galileo is regarded as the father of mathematical physics for his extensive studies on mechanics, and his contributions to physics and astronomy render him one of the greatest scientists of all time. Galileo never married, but at the age of 35 he fathered two daughters and a son with a Venetian mistress, Marina Gamba. The older daughter, Virginia, became a cherished companion to Galileo in his later years. Galileo spent his final eight years under house arrest in Arcetri, near Florence, and for the last four he was totally blind. Galileo died on January 8, 1642, just a few weeks before his 78th birthday.

Gilbert, Walter
(1932–)
American
Molecular Biologist, Chemist

Walter Gilbert split half of the 1980 Nobel Prize in chemistry with FREDERICK SANGER for their independent but simultaneous discovery of DNA (deoxyribonucleic acid) sequencing, while the other half of the prize went to Paul Berg, a biochemist who pioneered "gene splicing." Leading up to his DNA sequencing, Gilbert had verified the existence of lac repressors, a trigger mechanism for DNA.

Gilbert was born on March 21, 1932, in Cambridge, Massachusetts. His mother, Emma Cohen, was a child psychologist, and his father, Richard V. Gilbert, was a professor of economics at Harvard University. The family moved to Washington, D.C., when Gilbert was seven, and there he became interested in science. He ground his own telescopic glass at the age of 12, and he studied nuclear physics independently at the Library of Congress while a senior at Sidwell Friends High School. After graduating, he matriculated at Harvard in 1949, majoring in chemistry and physics. In 1953, he graduated summa cum laude and married Celia Stone, with whom he later had two children—a son and a daughter. He remained at Harvard to earn his master's degree in physics the next year.

Gilbert traveled to England to pursue doctoral studies in theoretical physics under Abdus Salam at Cambridge University, where he also worked with the famous geneticists FRANCIS CRICK and James Watson. For his dissertation, he mathematically predicted the actions of elementary particles in "scattering" experiments. He earned his Ph.D. in mathematical physics in 1957 and returned to Harvard to work under Julian Schwinger as a National Science postdoctoral fellow. Within two years, Harvard had appointed him to an assistant professorship.

In 1960, Watson (who had since moved from Cambridge to Harvard) included Gilbert in his team of investigators using pulse labeling to locate messenger RNA, or ribonucleic acid, which transfers information from DNA to ribosomes. Proof of the existence of messenger RNA came from another laboratory, however, at the California Technological Institute. Gilbert then collaborated with the German postdoctoral fellow Benno Muller-Hill in attempting to isolate the lac repressor, which hypothetically acted as a kind of "on/off" switch for genetic transferal. Gilbert's team tracked the process in *Escherichia coli* bacteria by introducing "tagged," or radioactive, lactose, which elicited the anticipated result: the lac repressor bonded to the lactose, confirming the existence of the former.

In 1964, Harvard tenured Gilbert as an associate professor of biophysics, and four years later, it promoted him to a full professorship. In 1972, Harvard named him to an endowed chair as the American Cancer Society Professor of Molecular Biology. By then, he had located the site of lac repression, the lac operon, and he used this gene in his next line of investigation—the sequencing of DNA. Gilbert, in collaboration with his graduate student Allan Maxam, tagged the DNA radioactively (a similar tactic as in his lac repressor research), then passed broken fragments of the DNA strands through gel by zapping it with an electrical current—a process called gel electrophoresis. X-ray photographs of the segments revealed their chemical composure, allowing Gilbert and Maxam to identify the chemical sequence of DNA. Sanger innovated a similar procedure at about the same time, hence the dual awarding of the 1980 Nobel Prize in chemistry to Gilbert and Sanger, as well as to Berg.

Walter Gilbert shared half of the 1980 Nobel Prize in chemistry with Frederick Sanger for their independent but simultaneous discoveries of DNA sequencing. *(Courtesy The Biological Laboratories, Harvard University)*

Starting in 1978, Gilbert began to investigate the possibility of approaching the new field of biotechnology from a corporate perspective. By 1981, he had resigned from his Harvard professorship to help found the biotech firm Biogen N.V., which he served as chair of the scientific directors and then as chief executive officer. Within three years, the inherent conflict of interest between pure science and capitalistic science frustrated him, and in 1985, he returned to Harvard, which had named him the H. H. Timken Professor of Science in the cellular and developmental biology department. Two years later, he assumed the chair of the department as the Carl M. Loeb University Professor.

Gilbert has spent the later part of his career working on the human genome project, an attempt to map the entire DNA scheme of humans. Although the project received criticism for treading into God's territory, Gilbert defended its importance for the potential to discover new cures for disease. He convinced Congress to budget $2 million annually to support the project in his laboratories at Harvard through a National Institutes of Health grant. He even started a company to copyright the genetic code.

Besides the Nobel Prize, Gilbert received ample recognition for his career. He shared the 1979 Albert Lasker Basic Medical Research Award with Sanger and won Columbia University's 1979 Louisa Horwitz Gross Prize as well as the 1980 Herbert A. Sober Memorial Award of the American Society of Biological Chemists.

⊠ Goddard, Robert Hutchings
(1882–1945)
American
Physicist

Physicist Robert Hutchings Goddard was a pioneer in rocketry and spaceflight theory. Interested in space travel from an early age, Goddard thought of rockets as a means to such flight. He developed and launched the first liquid-fueled rocket in 1926 on his aunt's farm in Massachusetts. Many of Goddard's innovations in rocket technology were incorporated in the development of rockets and missiles for both weaponry and spaceflight programs. Goddard's contribution to modern rocketry was great, but it might have been greater still had the publicity-shy Goddard published more and effectively communicated his findings to other scientists and engineers.

Goddard was born on October 5, 1882, in Worcester, Massachusetts. His father, Nahum Danford Goddard, and his mother, Fannie Louise Hoyt Goddard, were of modest means. His father worked as a machine shop owner, bookkeeper, and salesman, but his interest in inventions influenced him to steer his son toward experimentation and invention. Goddard was a sickly boy and was frequently absent from school. Science fiction proved to be a solace, and Goddard often dreamed of inventions and spaceflight devices.

NASA's Goddard Space Flight Center is named for Robert Hutchings Goddard. *(NASA, courtesy AIP Emilio Segrè Visual Archives)*

Because of his ill health, Goddard did not graduate from high school until 1904 at the age of 21. He immediately enrolled at Worcester Polytechnic Institute, majoring in general science with a concentration in physics. He then studied physics at Clark University in Worcester, earning his master's degree in 1910 and his doctorate in 1911. Goddard was conducting research in physics at Princeton University when he fell dangerously ill with tuberculosis in 1913. After he recovered, he returned to Clark University in 1914. He became a full professor by 1919 and remained at Clark for the duration of his career.

Though Goddard had long been fascinated by rocketry, it was not until 1909, when he was a student at Clark, that he began to seriously research the subject. In 1914, Goddard secured his first two patents: one for a liquid-fuel gun rocket, and one for a two-stage powder rocket. Goddard made a number of additional discoveries in his laboratory at Clark. He was the first to successfully prove that rockets could be fired in a vacuum and did not have to react against air, which was the common belief at the time. Goddard's experimentation also allowed him to accomplish higher degrees of energy efficiency and exhaust velocities with his rockets than had previously been achieved. The Smithsonian Institution recognized his successes and provided Goddard with funding beginning in 1917, helping him to advance his research.

During World War I, Goddard helped create tube-launched rockets, which were later used in the design of bazookas, prevalent in World War II. Goddard wrote about his findings on solid-propellant rockets in a paper called "A Method of Reaching Extreme Altitudes." Published in 1919, the seminal work detailed many of the fundamental ideas of modern rocketry.

Goddard began to explore liquid propellants in 1921 after becoming frustrated with problems presented by solid propellants. He launched the first liquid-fueled rocket in 1926, and though it climbed only 41 feet, it was a significant event in the history of rocket flight and led to a generous grant from Daniel Guggenheim. Goddard then set up a research facility in Roswell, New Mexico, in the early 1930s to continue developing rockets. Among his creations were gyroscopic steering, fuel-injection systems, and the automatic launch sequential system. In 1935, he successfully shot a liquid-propelled rocket faster than the speed of sound.

During the later years of his life, Goddard turned to defense-related research and worked on jet-assisted takeoff devices and variable-thrust rockets, used for aircraft. Goddard died on August 10, 1945, in Baltimore, Maryland, and received significant posthumous recognition for his achievements. In 1960, his wife, Mrs. Esther C. Goddard, whom he married in 1924, was awarded $1 million by the federal government for rights to use Goddard's patents, which numbered more than 200. Also in 1960, Goddard received the Smithsonian Institution's Langley Gold Medal for excellence in aviation. The following year, Goddard was presented with a Congressional Gold Medal, and a building, NASA's Goddard Space Flight Center, was named in his honor.

Goodyear, Charles
(1800–1860)
American
Inventor, Rubber Industry

Charles Goodyear "accidentally" discovered the vulcanization process of transforming natural rubber into a stable, usable form when a sample flew out of his hand and onto a woodstove, where it sizzled and hardened. His invention became ubiquitous, most importantly used in automobile tires, and served as a precursor to the plastics revolution that characterized the 20th century.

Goodyear was born on December 29, 1800, in New Haven, Connecticut. His father was an inventor and hardware entrepreneur. Goodyear joined his father in the hardware business, but their Philadelphia storefront failed in 1830. His brother, Nelson Goodyear, patented hard rubber doll heads on May 5, 1831, and the raw ingredient, India rubber (or "caoutchouc"), fascinated Charles Goodyear for the rest of his life.

After seeing a store display of a rubber life preserver, Goodyear invented an improved rubber valve for life preservers. However, when he presented his new product to officials of the first American rubber manufacturer, Roxbury India Rubber Company, at the New York City retail store in the summer of 1834, they responded discouragingly, as the heat had melted their inventory (they buried $20,000 worth of it near the Roxbury, Massachusetts, factory). If he could come up with a curing process to prevent rubber from melting in heat and stiffening in cold, though, they would be all ears.

Inspired, Goodyear returned to Philadelphia, only to be jailed for debt, the first in a series of such imprisonments in his life. Upon his release, he commenced experiments with additives, hitting upon magnesia as a promising solution. However, the several hundred magnesia-dried rubber overshoes he produced with the help of his family ended up melting when exposed to heat. Hounded by neighbors sick of the noxious smells emanating from his home laboratory, he moved his family to a fourth-floor tenement in New York City.

There, Goodyear added quicklime to his rubber-magnesia combination and then boiled the concoction; the resulting compound won a medal at a New York trade show, bolstering his confidence. In 1836, while recycling old samples he had decorated, he applied nitric acid to remove bronze paint, but when the rubber blackened, he threw it away as ruined. The next day, however, he retrieved the sample from the trash, realizing that the resulting black substance was dry to the touch. In 1837, the United States Postal Service contracted him to produce 150 mailbags with the new rubber combination, but these, too, ended up melting when exposed to heat.

At this point, Goodyear collaborated with Nathaniel Hayward, a former employee of the Roxbury India Rubber Company. In February 1839, while visiting a Woburn, Massachusetts, general store to demonstrate a sample of rubber with sulfur and white lead as additives, Goodyear gesticulated wildly. Some of his rubber sample flew from his grasp and landed on the potbellied woodstove heating the room. After the sample sizzled to a hard finish, an apologetic Goodyear scraped it off the stove. Upon closer inspection later, he realized that the chemical composition in addition to the heat had created the effect he'd desired all along—dry and hard, but malleable.

While in Boston, he was jailed for ducking out on a $5 hotel bill, and then returned home to New York City to find his infant son dead (of his 12 children, six died in infancy). Bereft and sick with dyspepsia and gout, but encouraged by his recent discovery, he proceeded to conduct experiment after experiment in search of the optimal combination of circumstances to create the ideal form of rubber. After countless trials, he realized that steaming the combination under pressure for four to six hours at 270° Fahrenheit yielded the best results.

Chemically, Goodyear had discovered a kind of polymerization, as the sulfur and polyisoprene molecules cross-linked to form a three-dimensional lattice that solidified the consistency, while also allowing for a degree of elasticity. His brother-in-law, once skeptical of Goodyear's scientific skills and parental responsibility, backed a venture involving the new discovery, marketing ruffled shirtfronts (the fashion rage of the age), as this incarnation of rubber was perfectly suited to create the puffy effect desired by stylish men.

In 1844, Goodyear filed and received patent number 3,633. However, instead of securing his process, this development only represented the beginning of his struggle to profit from his invention. For example, when he went to patent his invention in England, he discovered that THOMAS HANCOCK, who had finally discovered the same process in 1843, after 20 years of experimentation, when he examined a sample of Goodyear's rubber, had patented his "vulcanization" process two weeks earlier. (A friend of Hancock's had dubbed it the vulcanization process after Vulcan, the Roman god of fire.) Hancock offered one-half of his patent to Goodyear if he would drop the suit he'd filed, but Goodyear refused to back down (and eventually lost). Goodyear was eventually involved in 32 infringement cases, diverting funds from his invention into the hands of lawyers.

Goodyear exhibited his rubber with much fanfare at the London and Paris World's Fairs of the early 1850s; after a technicality prevented him from securing French patent rights (which he was counting on to pay for his elaborate display in the French fair), he spent 16 days in the "hotel," or jail. There, Napoleon III visited to present him with the Cross of the Legion of Honor. In 1853–1855, he published *Gum Elastic and Its Varieties*.

When Goodyear died in New York City on July 1, 1860, he was $200,000 in debt. It was not until after his death that royalties began to flow in to his wife and children for his invention. He was inducted into the National Inventors Hall of Fame in 1976.

⊠ Greenfield, Susan
(1952–)
English
Neurobiologist

Susan Greenfield is one of the most visible scientists discussing the controversial theories of consciousness brought about by increasingly sophisticated research on the human brain. Greenfield distinguishes herself by her ability to explain complex theoretical notions in easy-to-understand language, thus disseminating scientific understanding to the masses through her frequent appearances in the popular press as well as on television. Her understanding and articulation serve as promotional tools for the role of women in science.

Greenfield's mother was a chorus girl, and her father was an electrician. Although her family kept few books in the house, Greenfield was a voracious reader as a child and became the first member of her family to attend college. She earned a first-class degree from St. Hilda's College of Oxford and then continued on in Oxford's Department of Pharmacology to earn her D.Phil. degree.

"Dissecting the human brain" inspired Greenfield to enter the field of neurobiology, and her first experiment involved sampling "the fluid that bathes the brain and spinal cord and [analyzing] it for chemicals released in the brain." She "found a particular chemical was released from the brain cells that are lost in Parkinson's disease." Greenfield subsequently focused much of her research on Parkinson's, as well as on Alzheimer's disease.

Greenfield held postdoctoral fellowships in Oxford's Department of Physiology, as well as at the Collège de France in Paris under Professor J. Glowinski and at the New York University Medical Center in New York City under Professor Rodolfo Llinas. In 1985, Lincoln College of Oxford University appointed her as a university lecturer in synaptic pharmacology, as well as a fellow and tutor in medicine. She also spent time that year as a visiting research fellow at the Institute of Neuroscience in La Jolla, California. In 1987, she published her first of a string of books about the brain, *Mindwaves: Thoughts on Intelligence, Identity and Consciousness*.

In 1994, Greenfield became the first woman to deliver the famous Royal Institution Christmas lecture, an appearance that demonstrated her ability to communicate complex scientific notions in commonplace language understood by mainstream audiences. She subsequently spent much of her time promoting science in the public eye through lectures and television appearances. In her public appearances, she rarely misses an opportunity to promote the entrance of women in the sciences and to push the improvement of science education for school-aged girls.

Also in 1994, Greenfield published *Journey to the Centres of the Brain*, followed up the next year by *Journey to the Centres of the Mind: Towards a Science of Consciousness* and *The Human Mind Explained* in 1996. That year, Queens University in Belfast, Ireland, hosted Greenfield as a visiting distinguished scholar, and later that year, Oxford appointed her as a professor of pharmacology. In 1997, Greenfield published *The Human Brain: A Guided Tour*, which addressed a general readership and became a best-seller.

In 1998, the Royal Institution of Great Britain appointed her as its first woman director in its 200-year history. While holding this prestigious administrative position, she continued to conduct primary research as head of a group of 18 scientists studying Parkinson's and Alzheimer's diseases, trying to develop strategies to halt the death of neurons in these diseases. She also continued to write, publishing four books in as many years: *Ego: The Neuroscience of Emotion* in 1999, *Brainpower: Working out the Human Mind* and *The Private Life of the Brain* in 2000, and *Ego: The Neuroscience of the Self* in 2001.

Greenfield has won much recognition for the significance of her career. She received the 1998 Faraday Medal of the Royal Society for making the most significant contribution to the public understanding of science. *Harpers & Queen* magazine ranked her 14th in its list of the "Fifty Most Inspirational Women in the World." She has also received honorary doctorates from Oxford Brookes, St. Andrew's, Exeter, Sheffield Hallam, and London Universities.

Greenfield has said that she is most proud of her work promoting science in the public realm. For example, in March 1999, she conducted a consultative seminar for the British prime minister, Tony Blair, and his wife. And in July 2000, she hosted *Brain Story*, a six-part series on the mind that was broadcast on BBC2. The series created some controversy by seeming to invalidate religion as a mere feeling, but Greenfield (though an atheist herself) took pains to remain objective in her approach to the religion/science divide.

Gurdon, Sir John Bertrand
(1933–)
English
Molecular Biologist

Sir John Gurdon was the first scientist to successfully clone an animal. Although many scientists were skeptical of his 1962 announcement that he had used the intestinal cells of one South African frog to create another frog, the scientific community soon accepted the reality of cloning. Gurdon thus opened the door to the possibility of profoundly beneficial medical applications of cloning, as well as profoundly thorny ethical questions as to the morality of cloning human beings.

John Bertrand Gurdon was born on October 2, 1933, at Dippenhall, in Hampshire, England. He studied classics at Eton College, having been advised that he was not fit to study science. He continued to study classics upon entering Christ Church College of Oxford University, but he finally transferred to the department of zoology, then headed by Sir Alister Hardy. After graduating with first-class honors in 1956, he remained in the zoology department to conduct doctoral research on nuclear transplantation in *Xenopus* under Michael Fischberg. He received his Ph.D. in 1960, then he traveled to the United States to conduct two years of postdoctoral research on bacteriophage genetics at the California Institute of Technology, where he served as the Gosney Research Fellow. In 1962, he returned to Oxford.

That year, Gurdon announced that he had "cloned" a female South African frog from intestinal-cell nuclei. Actually, it was not until the next year that the term *clone* was coined by the biologist J. B. S. Haldane in his speech, "Biological Possibilities for the Human Species of the Next Ten-Thousand Years." Nevertheless, the scientific community understood the implications of Gurdon's announcement immediately, and in fact many doubted his claim. Autogenesis, or the creation of an organism from nothing other than its own cells, seemed impossible. In order to verify his findings, Gurdon cloned a recessive albino frog from a non-albino mother.

Back in Oxford, Gurdon held a research fellowship at Christ Church College while simultaneously serving as a demonstrator in Oxford's zoology department. In 1965, Oxford promoted him to a lectureship, a position he retained for the next seven years. In 1971, Gurdon was inducted into the fellowship of the Royal Society at the age of 38, becoming one of the society's youngest members. Also in that year, he moved from Oxford to Cambridge, joining the Medical Research Council's Laboratory of Molecular Biology. By 1979, he had ascended to the head of the cell biology division. In 1983, Gurdon inherited the title of John Humphrey Plummer Professor of Cell Biology from Sir Alan Hodgkin. Two years later, he received the 1985 Royal Medal from the Royal Society, and that same year, the Royal Institution appointed him as Fullerian Professor of Physiology and Comparative Anatomy.

In 1991, the Duke of Edinburgh established the Wellcome Cancer Research Campaign Institute, installing it in a newly constructed building on the Cambridge campus. Gurdon was appointed chairman of the institute, where he continued to conduct research on nuclear transplantation and differentiation. After his announcement of the first successful cloning, Gurdon concentrated his research on nuclear transplantation, or the removal of cellular nuclei from one organism to implant them in another of the same organism, chiefly frogs. He continued to conduct nuclear transplantation research at Wellcome, focusing on the phenomenon of "signaling," or the process whereby genetic differentiation takes place at the cellular level.

On January 1, 1995, Cambridge University named Gurdon Master of Magdalene College, succeeding Sir David Calcutt, and in June of that year, Gurdon was knighted. The next year, he delivered the prestigious Rutherford Memorial lecture. After a decade as the chairman of the Wellcome Research Center, Gurdon decided to step down as of October 1, 2001, handing over both this position and his title as the John Humphrey Plummer Professor of Cell Biology to his successor, Jim Smith. During his tenure, the institute grew to encompass 17 independent groups, totaling about 200 scientists.

Gurdon has continued to receive honors for his work. In the year 2000, he received the Jean Brachet Memorial Prize from the International Society for Differentiation, and the next year, he received the Conklin Medal from the Society for Developmental Biology. He has also received honorary doctorates from numerous institutions: the University of Chicago in 1978, Université René Déscartes in Paris in 1982, the University of Oxford in 1988, the University of Hull in 1998, and the University of Glasgow in 2000.

H

Haber, Fritz
(1868–1934)
German
Chemist

Fritz Haber's achievements ranged across disciplines: physical chemistry, organic chemistry, physics, and engineering. His best-known accomplishment may have been his development of an economical process for synthesizing large amounts of ammonia. For this work Haber was awarded the Nobel Prize in chemistry in 1918. Haber considered the scientific community to be global and made significant advances in expanding communication among countries; after World War I, the Kaiser Wilhelm Institute for Chemistry in Berlin, Germany, which Haber managed, became a leading research center where scientists from around the world gathered.

Haber's father, Siegfried, was a wealthy merchant and manufacturer of chemical dyes and pigments, and Haber was encouraged to study chemistry to enable him to take over the business. An only child, Haber was born on December 9, 1868, in Breslau (now Wroclaw, Poland). His mother, Paula, died in childbirth, and his father remarried in 1877 and had three daughters. Though Haber did not have a close relationship with his father, his stepmother, Hedwig Hamburger, treated him kindly.

Haber enrolled at the University of Berlin in 1886, intending to study chemistry. After one semester he transferred to the University of Heidelberg, where he studied physical chemistry, physics, and mathematics under the direction of ROBERT WILHELM BUNSEN. After earning his doctorate in 1891, Haber entered the Federal Institute of Technology in Zurich, Switzerland, to study the latest advances in chemical engineering.

After a failed attempt to run the family business, Haber took a teaching position at the Technical Institute, Karlsruhe, in 1894. There he researched electrochemistry and thermodynamics and published several works, including the original *Thermodynamics of Technical Gas Reactions* in 1905. In this book Haber discussed the behavior of gases in relation to thermodynamic theory in order to set industrial requirements for creating reactions. The work influenced subsequent teaching and research in thermodynamics. His successes led to a position as professor of physical chemistry in 1898.

In 1905, Haber also began his work on the synthesis of ammonia. The production of large quantities of ammonia was becoming more critical as the growing population in Europe created

Fritz Haber's team of scientists invented chemical weapons and poison gas to aid the German war effort during World War I. *(E. F. Smith Collection, Rare Book & Manuscript Library, University of Pennsylvania)*

greater demands on agricultural production and, in turn, industrial fertilizer. Nitrates, used in fertilizer, required ammonia for their production. Haber developed an inexpensive and practical method for the large-scale synthesis of ammonia from nitrogen and hydrogen. The engineer Carl Bosch then developed Haber's process for industrial production, creating the Haber-Bosch ammonia process. Industrial output of the ammonia began in 1910.

Haber began his reign as director of the Kaiser Wilhelm Institute for Chemistry, located just outside Berlin, in 1912. When World War I broke out in 1914, Haber volunteered the institute's services to help the German war effort, and he and his team invented chemical weapons and poison gas. In 1916, Haber headed the Chemical Warfare Service, and his ammonia process was used to create explosives. After the war Haber continued as director of the Kaiser Wilhelm Institute; in 1919, he started a continuing research seminar, the Haber Colloquium. The seminars and the institute gathered together notable chemists, physicists, and other scientists. Haber traveled extensively to promote communication and cooperation among scientific communities, and in 1930 he helped found the Japan Institute, designed to foster mutual understanding and cultural interests.

Haber married Clara Immerwahr, a fellow chemist, in 1901, and together they had one son. Clara disapproved of Haber's work in poison gases, and after a fierce argument in 1915 she committed suicide. Haber married Charlotte Nathan in 1917 and had two children, but the couple divorced in 1927.

The rise of the Nazis altered Haber's life irreversibly. Haber, a Jew, resigned from his post in 1933 and traveled to England, where he worked for several months at the Cavendish Laboratory in Cambridge. Recovering from a heart attack, the ailing Haber was en route to a position in what is now Israel when he died on January 29, 1934, in Basel, Switzerland.

Though some scientists had protested Haber's Nobel Prize because of his involvement in chemical weaponry, his contributions to chemistry were widely recognized, well respected, and influential in the discipline for many years.

Hancock, Thomas
(1786–1865)
English
Inventor, Rubber Industry

Thomas Hancock invented the masticator, a machine that recycled rubber scraps into a solid mass of rubber, which he used in collaboration with Charles Macintosh in the production of rubberized raincoats, or "mackintoshes." Later in his career, Hancock took out an English patent on what he called the "vulcanization" process of stabilizing rubber for practical applications. CHARLES GOODYEAR, the inventor of this process, rightly (but unsuccessfully) contested Hancock's patent. Hancock became the world's largest manufacturer of rubber goods, an industry that transformed modern life with its practical applications.

Hancock was born on May 8, 1786, in Marlborough, in Wiltshire, England. He was the son of a timber merchant and cabinetmaker, and one of his younger brothers helped introduce steam road carriages. His four brothers later joined in his rubber-manufacturing venture, producing such goods as pneumatic cushions, mattresses, pillows and bellows, hose, tubing, solid tires, shoes, packing, and springs.

In April 1820, Hancock patented elasticized India rubber fastenings for clothing—including gloves, suspenders, shoes, and stockings. He produced these products by cutting strips of Para rubber from "bottles," a process that wasted much of this raw material, so he innovated a method for recycling the loose ends of material. "It occurred to me that if minced up very small the amount of fresh-cut surface would be greatly increased and by heat and pressure might possibly unite sufficiently for some purposes," he wrote in his journal.

The masticator earned its name from the function it performed: this wooden, hollow-cylindered machine "chewed" on the rubber scraps with teeth studding two hand-cranked rotors filling the cylinder's core. When he opened the masticator for the first time, Hancock was surprised to find a solid mass of rubber, perfectly suited for other applications, such as pressing into iron molds. However, Hancock did not call his machine by its proper name in public, instead referred to it as a "pickle" so as to mask its function. And instead of patenting this device, which would have revealed the design to a world eager to steal ideas in flagrant violation of laws, Hancock kept his masticator hidden from public view.

In 1821, Hancock collaborated with Charles Macintosh in a joint venture producing raincoats. Macintosh had devised a formula for waterproofing cloth by "painting" rubber that had been liquefied with naphtha (a by-product of his dye-making business) onto fabric. Macintosh sandwiched the rubber between layers of cloth, but he ran into trouble when he sewed these layers together, as the holes made by the stitches leaked. Hancock solved this problem by melting the rubber and using that heat as the binding force in place of stitching, creating a perfectly waterproof rubberized fabric. The team named their raincoats after the Scotsman—"mackintoshes," or "macks" for short.

By the time the masticator was installed in the mackintosh production factory in Manchester, Hancock had fortified its construction, with metal as its material, and powered the machine with steam. In 1828, he traveled to Paris to personally oversee the installation of his proprietary machine in a factory there. He finally patented the device in 1837, as Macintosh's patenting of his own waterproofing process opened up the possibility of outside discovery of the true design of the "pickle."

In the early 1840s, Hancock received a sample of Charles Goodyear's new type of rubber. Hancock immediately recognized the presence of sulfur in combination with the natural rubber, so he commenced experimentation to discover the procedure for creating this new rubber. Heating the mixture proved effective, and a friend of Hancock's dubbed the process "vulcanization" after the Roman god of fire, Vulcan. Hancock filed for an English patent covering the vulcanization process in November 1843, merely a few weeks before Goodyear himself applied for an English patent. Goodyear brought a suit against Hancock, who offered to share half of the proceeds, but Goodyear refused to settle on these terms; Hancock prevailed in the end.

From 1820 through 1847, Hancock was issued 17 patents for various rubber applications, such as the cutting of rubber into sheets and square thread, and the blending of rubber with pitch and tar. In 1857, he published his *Personal Narrative of the Origin and Progress of the Caoutchouc or India Rubber Manufacture in England*. Hancock died on March 26, 1865, in Stoke Newington, London.

⊠ Hargreaves, James
(1720?–1778)
English
Inventor, Textile Industry

James Hargreaves is generally regarded as the inventor of the spinning jenny, the first successful multi-spindle spinning machine, which appeared in England in 1764. The machine revolutionized the textile industry, as it required only one operator to spin eight thread, thus multiplying efficiency eightfold (subsequent design improvements increased the number of spindles to 80, multiplying efficiency eightyfold, and then to 120). The spinning jenny is attributed as one of several innovations that fueled the Industrial Revolution, as it exemplified the period's mechanization of the manufacturing process, thereby drastically reducing the amount of labor necessary to produce goods that traditionally had been handcrafted.

Hargreaves was born in Stanhill, a village within Ostwaldtwistle, in Lancashire, England, and baptized on January 8, 1720. His father was George Hargreaves, and he had one sister, Elizabeth, three years his junior. He never learned to read or write, having received no formal education, though he had a natural inclination toward engineering. He married Elizabeth Grimshaw at Church Kirk on September 10, 1740, and together the couple eventually had 11 children (one, Henry, died as an infant).

Hargreaves supported his family as a handloom weaver in the Blackburn region, which was famous for its Blackburn Checks and Greys. Before the mechanization of his trade, a weaver was dependent on several spinners to supply yarn (spinners outnumbered weavers four-to-one in Blackburn at the time—2,400 spinners to 600 weavers in a population of about 4,000). Hargreaves most likely relied on his children to spin the yarn that he wove. This system undoubtedly inspired him to consider alternative means of supplying himself with yarn. His inventiveness made itself manifest early on, as he purportedly invented a system to double the output in the carding process by hanging a series of two or three interconnected cards from the ceiling.

In 1761, the Society for the Encouragement of Arts, Commerce and Manufacturers of Britain offered a 50-pound prize for the improvement of the spinning machine, an enticement that may have figured into Hargreaves's designs. In any case, his reputation for inventiveness is evidenced by the fact that Robert Peel employed him at his Brookside Mill in 1762 to assist in the building of a carding engine with cylinders, a project that never reached fruition.

Hargreaves's invention of the spinning jenny in 1764 is surrounded in myth. Legend has it that, at the village inn (or at his home, depending on the account), he witnessed the accidental overturning of a spinning wheel, which continued to rotate horizontally, with its spindle vertically inclined. This sight inspired his realization that such an arrangement could accommodate a series of spindles in place of just one. His original design (supposedly first drawn on the hearthstone with a burnt stick) included eight spindles. Accounts differ as to the origins of the name—some say he named it after his wife or one of his daughters (none of whom was named Jenny)—but regardless, the name undoubtedly (and perhaps humorously) drew upon another connotation of the word "jenny," or "engine." A later account, from

1807, claimed that he used only a pocketknife to construct his prototype.

By 1767, Hargreaves had perfected the design of his spinning jenny to such a degree that children could operate it. Up until then, he had tried to keep his invention a secret, going so far as to move from Stanhill to Ramsclough. By then, however, he had sold several, arousing the wrath of local spinners who feared the machine would displace them of their craft. In 1768, a mob gathered at Blackburn Market Cross and ransacked Brookside Mill (Peel apparently owned some spinning jennies) and Hargreaves's home in Ramsclough, forcing a hammer into his hand to smash his own jennies.

Hargreaves moved to Nottingham, where the high demand for spun thread protected him from such retribution. In partnership with Thomas James, he established a cotton mill. Finally, after his design had already been copied many times over, he applied for a patent on July 12, 1770 (his spinning jenny had 16 spindles by then.) Unfortunately, such a measure was too little, too late, and his claim to the invention remained disputed throughout his life. By the time Hargreaves died in Nottingham in April 1778, the spinning jenny had grown to encompass 120 spindles (all run by one operator), and more than 20,000 of them had proliferated throughout Britain. It remained the primary machine for spinning from 1770 until 1810.

Harrison, John
(1693–1776)
English
Horologist

John Harrison was a working-class carpenter who rose to fame as a horologist by designing and building a succession of clocks, the last of which was capable of keeping its accuracy well enough to navigate ships through cross-oceanic voyages. Harrison spent the majority of his years at work on the problem and then spent numerous years convincing the British government to grant him the prize it promised for the proper solution. Harrison revolutionized sea-travel, allowing sailors to navigate more accurately than ever (while also keeping good time)!

Harrison was born in 1693, at Foulby, in Yorkshire, England, and baptized on March 31 of that year. His father, Henry, was an estate carpenter and surveyor who moved the family to Barrow-upon-Humber. When Harrison contracted smallpox at the age of six, his parents set a watch on his pillow for its ticking to keep him company, and it also peaked his horological interest.

Harrison received little formal education, but he followed his father's footsteps into carpentry, while also indulging his interest in timepieces by constructing his first at the age of 20 in 1713, a longcase clock made almost entirely of wood that now resides in the collection of the Worshipful Company of Clockmakers in Guildhall.

In 1714, the British Parliament announced what became known as the "Queen Anne Act," which offered a 20,000-pound prize (the equivalent of one million pounds now) for the solution to the "longitude problem," establishing the Board of Longitude to administer the submission of answers. The problem: sailors at sea could not calculate their longitude accurately without knowing the local time, which in practice required them to know a reference time— namely, Greenwich Standard Time. But no existing clock could keep accurate enough time while also withstanding the vicissitudes of temperature and humidity as well as the rolling waves at sea. In order to win the prize, the solution would have to keep the ship within half-a-degree (30 minutes in longitude, or two minutes in time).

In 1718, Harrison married his first wife, Elizabeth, who gave birth to their son, John, a few months later. Elizabeth died in May 1726, and by November of that same year, Harrison had married again—to another Elizabeth. Together, they had two children—William, born in 1728, and Elizabeth, born in 1732. By now, Harrison and his youngest brother, James, were gaining a reputation for their quality clockmaking. They designed a bimetallic pendulum clock, with the different coefficients of expansion of brass and steel rods maintaining a constant length (a swing-rate) for the pendulum. The team also constructed a turret clock at the Brocklesby Park stables that required no lubrication, a radical innovation.

In 1728, determined to "find the longitude" (a disparaging term applied to those foolish enough to try for the prize), Harrison traveled to London, where his design so impressed George Graham that this renowned horologist loaned Harrison the funds to construct his proposed clock. From 1730 through 1735, Harrison constructed H1, as it is now known, a clock completely counterbalanced so as to run independent of the direction of gravity. In 1736, on a trial trip to Lisbon on the *Centurion* and back on the *Orford*, the ships would have sailed off-course if not for H1's accuracy. However, the clock's accuracy did not satisfy the terms set by the Board of Longitude, so Harrison requested (and received) financial support to continue.

Harrison commenced work on his second design (H2) in 1737, but by 1740 he realized a design defect, and so he immediately started work on a third clock. After 15 years of work on it, he realized that it, too, would probably not meet the Board's requirements, so he requested support once more to build a new design (while continuing for four more years to work vainly on H3). For his fourth design, Harrison abandoned his clock designs completely in favor of a watch design. In the meanwhile, the Royal Society granted Harrison its 1749 Copley Medal.

H4 measured a mere 13 centimeters in diameter, but it outperformed his previous three efforts. His son, William, tested its accuracy on a voyage to the West Indies, departing on the *Depford* on November 18, 1761, and arriving in Jamaica on January 19, 1762, with H4 only 5.1 seconds behind Greenwich Standard Time. William then embarked on a second trial, to confirm the first, on March 28, 1764, aboard the *Tartar*. Forty-seven days later, the ship landed at Madeira in Barbados, with H4 only 39.2 seconds behind—three times as accurate as required by the Board of Longitude guidelines.

The Board, however, waffled on awarding the prize. Ostensibly, they believed the watch's accuracy to be a fluke, but in reality, the Board was trying to save the prize for Nevil Maskelyne, the incoming Astronomer Royal, who was also vying for the money. The Board made outlandish demands, asking to see Harrison's secret designs, and insisting on retaining all four timepieces (H1 through H4) before granting the first 10,000 pounds prize money; to earn the second half of the prize, Harrison had to provide two more watches capable of equaling H4's accuracy.

The two duplicates of H4 were constructed by Harrison's son, William, and by Larcum Kendall, a prominent horologist who probably contributed to the construction of H4. H5 (William's copy of H4) and K1 (Kendall's copy) were both completed by 1769 and inspected the next year, but the Board continued to resist, insisting that all copies must be made by the Harrisons. On January 31, 1772, Harrison sent a letter to King George III via Stephen Demainbray, the king's private astronomer. The king personally attended a trial that attested to the incredible accuracy of the H4 design, but the Board refused even the king's appeal.

Finally, an act of Parliament in June 1773 won Harrison another 8,750 pounds (he never received the full amount) and recognition that he had indeed solved the longitude problem. Captain Cook confirmed the clock's accuracy on a three-year voyage (during which K1 lost less than eight seconds per day), returning in July 1775. A year later, Harrison died on March 24, 1776, at his home in Red Lion Square in London.

Hero of Alexandria
(fl. A.D. 62?)
Greek
Mathematician

A number of written works on mathematics, geometry in particular, and physics have been attributed to Hero of Alexandria. Hero developed the formula for calculating the area of a triangle from its three sides, now known as Hero's formula, and invented the aeolipile, the first steam-powered engine. *Metrica*, Hero's most significant work on geometry, was lost until 1896.

Besides his works, which were written in Greek, nothing is known about the life of Hero. The earliest written mention of Hero appeared in a work by PAPPUS of Alexandria around A.D. 300, which quoted from Hero's book, *Mechanics*. The tentative date of A.D. 62 for Hero's life was derived from the description of an eclipse in one of Hero's works, found to correspond undeniably to the eclipse of A.D. 62. Because of the nature of the books written by Hero, it was believed that he taught at the University of Alexandria in such disciplines as mathematics, physics, pneumatics, and mechanics.

The treatise *Metrica* dealt with geometry and comprised three books. *Metrica* introduced his formula for finding the area of a triangle. The first book encompassed plane figures and surfaces of common solids. Hero explained how to calculate the area of quadrilaterals, circles, regular polygons, and ellipses. He also described how to find the surface area of cylinders, cones, spheres, and segments of spheres. Also in Book I was a method for estimating the square root of a number, a method later used in computers. Book II provided methods for determining the volume of such figures as the cone, pyramid, cylinder, prism, and sphere. The third book discussed procedures for dividing certain areas and volumes into parts of given ratios.

Hero's longest work was *Pneumatics*, considered to be a collection of notes for a textbook. The treatise tackled a number of mechanical problems and included discussion of the aeolipile, the steam-powered engine designed and built by Hero. The engine was made of a sphere with two nozzles positioned so that steam jets produced from the inside would cause the device to turn. *Pneumatics* also discussed the pressure of air and water, as well as the occurrence of a vacuum in nature. Treatment of siphons, coin-operated machines, a fire pump, and a water organ also appeared in the work. Hero also described a number of toys and playthings, such as puppet shows and trick jars that released wine or water separately.

Another important work by Hero was *Mechanics*, which was a textbook for engineers, builders, and architects that dealt with the mechanical problems of everyday life. The theory of the wheel and the theory of motion were covered in the work, and Hero also explained how to construct plane and solid figures in a given proportion to a given figure. He presented the theory of the center of gravity and equilibrium and the theory of the balance, using the ideas of ARCHIMEDES. *Mechanics* also discussed the five simple machines—the lever, the pulley, the winch, the wedge, and the screw. The work described cranes, devices for transport, winepresses, and screw-cutters, believed to have been invented by Hero.

Dioptra was about land surveying and contained a description of a diopter, a surveying instrument most likely developed by Hero. The work also described an odometer, another Hero invention, and dealt with astronomy, offering a method for calculating the distance between Alexandria and Rome through the simultaneous observation of a lunar eclipse in both cities. Yet another work, *Catoptrica*, offered the theory of reflection and provided instructions on how to make mirrors.

Hero's works have been widely studied, and evidence exists to suggest his texts enjoyed a large audience. *Pneumatics* was read by many during the Middle Ages and the Renaissance, and more than 100 copies have surfaced. His writings indicate that Hero was a man of great scientific knowledge and creativity.

⊗ Hertz, Heinrich Rudolf
(1857–1894)
German
Physicist

Though he died at the young age of 36, Heinrich Rudolf Hertz made significant advances in electrodynamics. Hertz was the first to produce radio waves artificially and succeeded in proving that radio and light waves were electromagnetic waves. A persistent and highly self-motivated researcher, Hertz studied various aspects of electricity, including cathode rays and dielectrics.

Born into a wealthy family, Hertz was encouraged by his parents, particularly his mother, to excel in his studies. He was born on February 22, 1857, in Hamburg, Germany, the son of Gustav F. Hertz, a barrister, and Anna Elisabeth Pfefferkorn. The eldest of five children, he was educated at a strict private school. At an early age he showed skill at woodworking, as well as a gift for languages. As a teenager he studied Greek and Arabic. In 1875, Hertz went to Frankfurt to gain hands-on experience in engineering but soon decided to pursue an academic and scientific career.

In 1877, Hertz entered the University of Munich to study mathematics and transferred a year later to the University of Berlin, where he studied under physicist Hermann von Helmholtz. Helmholtz proved to be a profound influence on Hertz, and Hertz flourished in Berlin's rigorous research environment. Hertz received his doctorate in 1880 and spent the next three years as Helmholtz's assistant. During that period Hertz succeeded in completing the research for 15 publications, the majority of which dealt with electricity. Hertz discussed electromagnetic induction, the inertia of electricity, dielectrics, and cathode rays. It was also during this time that Hertz, with Helmholtz's prompting, became interested in the electromagnetic theories of JAMES CLERK MAXWELL.

From 1885 to 1889, Hertz worked as a professor of physics at Karlsruhe Technical College. There he began the experimental trials that would drastically impact the field of electrodynamics. By late 1888, Hertz had successfully produced electromagnetic waves using an electric circuit. A metal rod in the circuit possessed a small gap in the middle, and when sparks traversed this gap, high-frequency oscillations resulted. To prove that these waves were transmitted through the air, Hertz set up a similar cir-

The first person to broadcast and receive radio waves, Heinrich Hertz went on to prove that radio and light waves were electromagnetic. *(Deutsches Museum, courtesy AIP Emilio Segrè Visual Archives,* Physics Today *Collection)*

cuit a short distance from the first and used the circuit to detect the waves. He was also able to demonstrate that these waves were reflected and refracted like light waves, and that, though they traveled at the same speed as light waves, these waves had a longer wavelength. These waves were called Hertzian waves but later came to be known as radio waves. Hertz published his findings in nine papers, which drew significant attention.

Hertz's discoveries secured him a position at the University of Bonn in 1889, where he continued his theoretical study of Maxwell's theory. He also returned to the topic of cathode rays and in 1891 turned his attention to a theoretical study of the principles of mechanics as discussed by Helmholtz in his work on the principle of least action.

Among the numerous awards presented to Hertz for his findings on electric waves were the Matteucci Medal of the Italian Scientific Study (1888), the Baumgartner Prize of the Vienna Academy of Sciences (1889), the La Caze Prize of the Paris Academy of Sciences (1889), the Rumford Medal of the Royal Society (1890), and the Bressa Prize of the Turin Royal Academy (1891). A unit of frequency, the hertz, was named in his honor. Hertz married Elisabeth Doll in 1886 and had two children. After several years of poor health, Hertz died in 1894 of blood poisoning. Had he lived another decade, Hertz would have been able to see GUGLIELMO MARCONI'S discovery that the transmission of radio waves could be used as a means of communication.

⊠ **Hodgkin, Dorothy Crowfoot**
(1910–1994)
English
Chemist, X-ray Crystallographer

Dorothy Crowfoot Hodgkin is considered the founder of protein crystallography for her indefatigable work mapping the molecular structure

Dorothy C. Hodgkin pioneered the use of X-ray crystallography, which she used to understand the structures of penicillin, vitamin B_{12}, and insulin. *(AIP Emilio Segrè Visual Archives,* Physics Today *Collection)*

of important compounds such as penicillin. She won the 1964 Nobel Prize in chemistry for her structural analysis of vitamin B_{12}, but the crowning achievement of her long career was her deduction of the structure of insulin, which she commenced in the 1930s and completed in the late 1960s.

Hodgkin was born on May 12, 1910, in Cairo, Egypt, where her father, John Crowfoot, worked as an archaeologist for the British Ministry of Education. Her mother, Grace, was an artist and an expert in Coptic textiles. Hodgkin grew up separated from her parents by World War I, as she remained in the safe haven of England while her mother accompanied her father to his new post in Sudan as the director of the ministries of education and antiquities. For her 16th birthday, her mother gave her a book on X-ray crystallography by its pioneer, William Henry Bragg, a gift that helped determine her future.

In 1928, Hodgkin entered Somerville, the women's college of Oxford University, and graduated with a bachelor's degree in chemistry in 1932. She proceeded to Cambridge University, where she conducted postgraduate research under J. D. Bernal, who was working with X-ray crystallography. Hodgkin and Bernal applied X-ray crystallography first on pepsin, which they fully analyzed by 1934. That year, Hodgkin began to suffer from rheumatoid arthritis, a condition that afflicted her for the rest of her career.

Also in 1934, Hodgkin returned to Somerville as a researcher and teacher with her own lab space, though it was submerged in the basement of the Oxford Museum. She pursued her doctorate, writing her dissertation on her X-ray crystallography study of cholesterol, which she completed in 1937 to earn her Ph.D. On December 16 of that year, she married African studies specialist Thomas L. Hodgkin. Together the couple had three children, raised by family and nannies while their father taught in the north of England and their mother worked at Oxford. He finally joined her on the Oxford faculty in 1945.

Hodgkin spent the years during World War II studying penicillin, one of the first antibiotics, which ERNST BORIS CHAIN had brought to her for analysis. From 1942 through 1946, Hodgkin and her graduate student, Barbara Rogers-Low, deciphered penicillin's molecular structure. In honor of this significant achievement, the Royal Society inducted Hodgkin as a fellow, only the third woman so honored by 1947. The next year, she commenced analysis of vitamin B_{12}, an even more complex molecule necessary to prevent pernicious anemia in humans. Hodgkin and her team worked for six years collecting data; as they were preparing to interpret this data, Hodgkin met Kenneth Trueblood of the University of California at Los Angeles, where he had programmed a computer to calculate crystallographic readings. Trueblood and Hodgkin collaborated cross-continentally to arrive at the molecular structure of vitamin B_{12} in 1956. A year later, Oxford finally appointed Hodgkin as a reader (the equivalent of a full professorship in the United States), and in 1958, they finally equipped her with adequate laboratory facilities.

In 1964, Hodgkin became the first British woman to win a Nobel Prize in the sciences. The next year, she received Britain's Order of Merit, the second woman so honored, after Florence Nightingale. Hodgkin spent the remainder of the 1960s deciphering the 777 atoms of insulin by analyzing 70,000 X-ray spots, and in 1969, she announced her results (which were refined in 1988 with the help of advanced computers). Hodgkin retired in 1977. She suffered a stroke and died on July 30, 1994.

Hollerith, Herman
(1860–1929)
American
Engineer

Regarded as the father of information processing, Herman Hollerith's main achievement was the creation of a system for recording and retrieving information on punched cards. The system was used for data processing and gained widespread usage after successfully handling the computation of the 1890 U.S. census. Hollerith went on to found the company that eventually became International Business Machines Corporation (IBM).

Hollerith was born on February 29, 1860, in Buffalo, New York. His parents were German immigrants, and his father was a teacher of classics. Equal emphasis was placed on course work and practical work at the Columbia School of

Mines in New York, where Hollerith was an engineering student. He studied physics and chemistry, as well as surveying, geometry, drawing, and assaying. The school's requirement that engineering students visit places of industry to observe practical methods likely drew Hollerith to machine and metallurgy shops, which would later play a significant role in his career.

After graduating in 1879, Hollerith worked with his instructor, W. P. Trowbridge, on the 1880 U.S. census. By then, the census had become a labor-intensive, time-consuming project. Though the census itself took only a few months to carry out, tabulating and analyzing the 1880 data was projected to take nearly a decade. A mechanical solution was needed. Dr. John Shaw Billings, the head of the division of vital statistics, recognized potential in Hollerith and discussed with him the possibility of a tabulating system that would solve the problem the

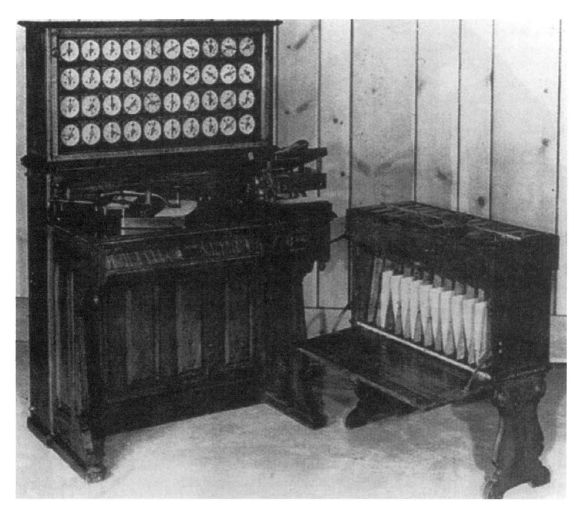

Herman Hollerith, who founded the company that became IBM, invented this punch card tabulator for the 1890 census, a precursor to the modern computer. *(Courtesy Department of Electrical and Computer Engineering, University of Maryland)*

census bureau faced. Thus, after Hollerith left the census bureau to teach at the Massachusetts Institute of Technology (1882–1884) and work at the U.S. Patent Office in Washington, D.C., he continued to investigate the census problem.

By 1884, Hollerith had developed a first design and applied for a patent. The system used punched tape that ran over a metal drum and under brushes. Whenever the brushes encountered a hole, a circuit was completed and the data recorded. The use of electricity increased the speed and efficiency of the system. Hollerith continued to make design modifications to the machine, including the use of punched cards rather than paper. The machine-readable cards, which came to be known as the Hollerith cards, were more easily replaced and corrected than tape. By 1887, he had developed a machine designed for the census that could handle up to 80 cards per minute. The code used to record the alphanumeric information on the punch cards was called the Hollerith code.

When the census bureau began preparing for the 1890 census, it held a competition to find the best system. Competitors were to input and tabulate data from one city, and each method was timed. The systems of competitors William Hunt and Charles Pidgin took more than 100 hours to input the data, while Hollerith's system took just more than 72 hours. For the computation, Pidgin's system took more than 44 hours and Hunt's more than 55. Hollerith's machine tabulated the data in five hours and 28 minutes and won the competition. Using Hollerith's system, the 1890 census took three years to process, compared to the seven years it took for the 1880 census to be completed.

Hollerith adapted his machine for commercial use after the 1890 census, and in 1896 he established his own company, the Tabulating Machine Company, which manufactured and marketed his systems. His business was successful, and his machines, because of their versatility and ability to work with data of almost any kind, were used in a variety of trades. In 1911, Hollerith sold his share of the company, and the company became IBM in 1924. Hollerith's census methods persisted into the 1960s and helped move the field of data processing into the computer age.

Holmes, Arthur
(1890–1965)
English
Geologist

Arthur Holmes transformed two important geologic concepts with theories that have withstood the test of time. First, he proposed an estimate of the earth's age using the rate of radioactive decay to establish the first estimate based on chronological (as opposed to stratigraphic) dating. Throughout his career, he revised his estimate, arriving at a more accurate figure near the end of his career. Second, he proposed a mechanism to explain the contentious theory of continental drift, explaining that radioactive convection could shift the plates in the earth's mantle. This theory was vindicated in the early 1960s, when geologic evidence supported Holmes's hypothesis.

Holmes was born on January 14, 1890, in Hebburn-on-Tyne, in Newcastle, England. His mother, Emily Dickinson, was a teacher, and his father, David Holmes, was a cabinetmaker; both parents came from farming stock. At Gateshead High School, Holmes's physics teacher encouraged him to read WILLIAM THOMSON KELVIN's *Addresses*, introducing him to the notion of the geologic age of the earth. In 1907, Holmes matriculated at Imperial College in London to study physics on a scholarship. He earned his bachelor of science degree in mathematics and physics in 1909.

Holmes, who had commenced geologic study under Robert Strutt (later the fourth Lord of Raleigh) at Imperial College, continued this line of study at the Royal College of Science, graduating as an associate in 1910. He then conducted postgraduate research with Strutt, applying the recently discovered phenomenon of radioactivity to geology as a means of dating the age of rocks. Physicist Ernest Rutherford had suggested that radioactive decay, which occurs at a set rate, can accurately record the passing of time; Strutt used this application to determine the age of rocks by comparing the amount of uranium in rock samples to the amount of the by-product of its radioactive decay—namely, helium. In his very first paper, published by the Royal Society in 1911, Holmes devised a helium/uranium ratio for calculating geologic time, and used this ratio to estimate the earth's age—1,600 million years, 40 times more than Kelvin's estimate of 40 million years.

Also in 1911, Holmes embarked on a geologic expedition to Mozambique, where he contracted a case of malaria that prevented him from fighting in World War I. The next year, he returned to Imperial College to teach geology. He published his first book, *The Age of the Earth*, in 1913, and the next year, he married Margaret Howe; together, the couple had one son. In 1920, he published his second book, *The Nomenclature of Petrology*, and the next year, he published *Petrographic Methods and Calculations*; these two books had a profound influence on his field of specialization, petrology, or the description and classification of rocks.

In 1920, Holmes left academia for an industry job leading an oil exploration expedition in Burma as chief geologist for the Yomah Oil Company. Four years later, after failing to locate oil deposits, he returned to England to head Durham University's geology department, newly established by Irvine Masson.

In 1915, the German geophysicist Alfred L. Wegener had proposed the theory of continental drift, expounding that continents actually migrate. Geologists resisted this hypothesis for lack of an explanation of the mechanism whereby such huge landmasses might move. In a 1929 presentation to the Geological Society of Glasgow, Holmes supplied an explanation: radioactive convection. He theorized that radioactive heat in the earth's mantle created convection currents capable of shifting the continents. Humbly, he stressed that his ideas were "purely speculative" and would "have no scientific value until they acquire support from independent evidence." Harry Hess provided such evidence in 1962 (following up on his 1960 proposition attributing continental drift to seafloor spreading), when he provided the geologic mechanism responsible for such radical continental shifting—the eruption of molten magma, which could push the continents apart and then cool to create new oceanic flooring. This theory validated and elaborated Holmes's theory.

In 1938, Holmes's first wife died, and the next year, he married Dr. Doris Livesey Reynolds, a fellow geologist on the Durham faculty. In 1942, Edinburgh University named him to the Regius Chair of Geology, and the Royal Society simultaneously inducted him into its fellowship. In 1944, he published the book for which he is best remembered, *Principles of Physical Geology*, which crystallized in simplicity and clarity the field that had previously been plagued by confusing jargon.

Throughout his career, Holmes continued to refine his estimate of the earth's age, addressing the issue in more than 50 papers; in 1959, the *Transactions of the Edinburgh Geological Society* published his final estimate—four billion years. In testament to the accuracy of Holmes's calculations, the current estimate for the commencement of the Cambrian period, 590 million years ago, jives with Holmes's estimate of 600 million years ago.

The year that Holmes retired (1956), he received both the Wollaston Medal from the Geological Society of London and the Penrose Medal from the Geological Society of America.

In retirement, he remained at Edinburgh as a professor emeritus until 1962. In 1964, he received the Vetlesen Prize. In 1965, he published a completely revised edition of his classic text, *Principles of Physical Geology*. Holmes died in London on September 9 (or 20), 1965.

Hoobler, Icie Gertrude Macy
(1892–1984)
American
Biochemist

Trained as a physiologist and chemist, Icie Macy Hoobler combined these two disciplines to study the biochemistry of nutrition in humans, especially women, infants, and children. Her studies helped establish the minimum requirements for many vitamins and minerals in the human diet at a time when this information had been inadequate. During years of detailed work, she also examined metabolism during a woman's reproductive cycle and illuminated the relationship between diet, nutrition, and healthy pregnancies in women.

Born on July 23, 1892, in Daviess County, Missouri, Icie Gertrude Macy was the daughter of Perry Macy and Ollevia Elvaree Critten Macy. Macy was a ninth-generation descendant of Thomas Macy, a Baptist minister who was one of the first settlers of Nantucket, Massachusetts. She grew up on the family farm. From 1926 to the mid-1930s, Macy raised her two nieces after the death of their mother, Macy's sister Ina. In 1938, when she was 44, she married B. Raymond Hoobler, a physician.

Even though neither of Macy's parents was well educated, they encouraged their children to pursue an education. She received her primary education in a one-room Missouri schoolhouse and her secondary education at a boarding school. Her interest in science became clear during her secondary studies, but to satisfy her father, who had decided that she was going to be a musician, she studied music in high school and at the teacher training college she attended in Missouri. After receiving an A.B. from Central College for Women in 1914, Macy enrolled in the University of Chicago in 1916 where, with the encouragement of chemist Mary Sherrill, she began to study chemistry. Macy earned a B.S. in chemistry from the University of Chicago in 1916.

For her graduate education, Macy attended the University of Colorado, where she studied biochemistry under the direction of Robert C. Lewis. She received her M.S. in chemistry from the University of Colorado in 1918 and began her Ph.D. studies in that field at Yale the same year. At Yale, she studied with Lafayette B. Mendel of Yale's Sheffield Scientific School. Mendel encouraged Macy in her studies of the nutritional value of cottonseed flour, which was being substituted for wheat flour during World War I. Government authorities were concerned about using cottonseed flour as a food staple because of so-called cottonseed meal injury, the physical deterioration of animals who consumed this product. Macy determined that cottonseed was not deficient in vitamins or nutritional value, but it contained the poisonous compound gossypol, which caused harmful side effects. For her work on this and other studies, she was awarded her Ph.D. in 1920.

After finishing her Ph.D., Macy worked for a while at a hospital in Pittsburgh and at the Department of Household Science at the University of California at Berkeley. In 1923, she took a job at the Merrill-Palmer School for Motherhood and Child Development in Detroit, Michigan. She also worked as a researcher at the Children's Hospital of Michigan in Detroit. She remained at Merrill-Palmer until her retirement in 1959.

At Merrill-Palmer, Hoobler worked on a 25-year study of nutrition and human physiochemistry that was underwritten by the Children's Fund. As the leader of this research team,

Hoobler focused research on six topics: the metabolism of women during their reproductive cycle, the composition of human milk, infant growth and development, childhood growth and development, the nutrition of meals given to children in Michigan's child-care institutions, and blood studies related to health and disease.

Perhaps Hoobler's greatest contribution to the health of American mothers and their children was the discovery that vitamin D and vitamin B needed to be added to milk to prevent rickets, a bone disease that was then common among children. For her work, Hoobler was awarded the University of Colorado's Norlin Achievement Award in 1938, the American Chemical Society's Garvan Medal in 1946, and the Modern Medicine Award in 1954. She returned to her native Missouri in 1982 and died in Gallatin, Missouri, the town nearest her childhood farm, on January 6, 1984. She was 92 years old.

Hopkins, Donald
(1941–)
American
Physician

Donald Hopkins has devoted his career to the eradication of infectious diseases with uncanny success. He commenced his career heading up the newly established Smallpox Eradication Program (SEP) in Sierra Leone, contributing significantly to the purging of smallpox from the human population within a decade. He next focused his efforts on eradicating the guinea worm disease from the earth, achieving similarly successful results in a similarly short span of time. He keeps one guinea worm larva, named "Henrietta," in a jar of formaldehyde on his desk, calling it the last sample of this disease-inducing species.

Hopkins was born on September 25, 1941, in Miami, Florida. He attended Morehouse College and spent time abroad in 1961 on exchange at the Institute of European Studies of the University of Vienna; while there, Hopkins visited Egypt, where he first witnessed the havoc wreaked by an infectious disease, in this case, the eye infection trachoma. Then and there, he decided to dedicate his career to public health. He graduated from Morehouse with a bachelor of science degree in 1962 and then pursued his medical degree at the University of Chicago School of Medicine. He earned his M.D. in 1966, gaining board certification in pediatrics, and the next year he married Ernestine Mathis. He later continued his education at the Harvard School of Public Health, where he earned a master's degree in public health in 1970, gaining board certification in public health.

In 1966, the World Health Organization (WHO) passed a resolution to eradicate smallpox from the face of the earth, a goal dating back to EDWARD JENNER's 1796 discovery of the smallpox vaccine, which he derived from cowpox inoculation. From 1967 through 1969, Hopkins directed the Smallpox Eradication/Measles Control Program in Sierra Leone, part of the global Smallpox Eradication Program established in response to the WHO resolution. The SEP proved incredibly successfully, meeting its goal of smallpox eradication in a mere decade: the last natural case of smallpox was reported in Merka Town, Somalia, in October 1977 (another case escaped from a research lab in 1978). A global commission certified smallpox eradication in 1979, and the World Health Assembly ratified it in 1980—smallpox was thus the first disease eliminated by medical science. In 1983, Hopkins published *Princes and Peasants: Smallpox in History,* a book that earned him a Pulitzer Prize nomination.

In 1977, Hopkins moved from an assistant professorship in tropical public health at the Harvard School of Public Health back to the Centers for Disease Control and Prevention, where he had served as a medical officer from 1972 through 1974. When he returned to the

CDC, he served as assistant director of operations and later as deputy director, as well as acting director for the year of 1985. In 1987, he became a senior consultant at the Carter Center, a research institute started by former President Jimmy Carter and housed at Emory University in Atlanta, Georgia.

As head of the Global 2000 Guinea Worm Eradication Program, Hopkins implemented a program that met with similar success as his smallpox eradication initiative. In 1985, two years before he joined the Carter Center, three-and-a-half million people were afflicted with *dracunculiasis*, or guinea worm disease, which is passed along through drinking water infested by guinea worm larvae. By distributing water filtration systems and education, Hopkins's program was able to cut this number to two million by 1989. By 1996, the eradication program had decreased the prevalence of the disease by 95 percent, leaving a mere 150,000 cases and effectively eradicating guinea worm disease from all but Sudan.

In a final push for complete eradication, Hopkins and Carter organized a donor conference in Nigeria that raised $10 million. In addition, they secured water-filtering materials donated by the DuPont Co. and Precision Fabrics Group Inc. and convinced American Home Products (formerly American Cyanamid) to donate Abate, a nontoxic chemical for killing guinea worm larvae. As the associate executive director for the Control and Eradication of Disease Programs at the Carter Center, Hopkins also heads the effort to control river blindness (*onchocerciasis*) worldwide.

Hopkins has received much recognition for his work in reducing human suffering by eradicating diseases from existence. In 1991, the National Research Council (in conjunction with the National Academy of Sciences, the National Academy of Engineering, and the Institute of Medicine) named him an Outstanding African American. In June 1995, he received a John D. and Catherine T. MacArthur Foundation Fellowship, granting him $320,000 over a five-year period. In 1997, the American Academy of Arts and Sciences inducted him as a fellow, as did the National Academy.

Hopper, Grace Murray
(1906–1992)
American
Computer Scientist

Grace Hopper was a pioneer in the field of computer science, helping to invent the first compiler (A–O) as well as the first computer languages (Flow-Matic and COBOL). She got her start with computers in the U.S. Navy during World War II, and she remained in the service until her retirement, after which the navy promoted her to rear admiral, the first woman to hold this rank.

Grace Brewster Murray was born on December 9, 1906, in New York City to Mary, a homemaker, and Walter Murray, an insurance broker. She followed in the footsteps of her great-grandfather, a navy admiral, and her grandfather, a civil engineer who showed her how to survey on field trips together. Murray commenced her long relationship with Vassar College as an undergraduate studying mathematics and physics for her bachelor's degree in 1928. She proceeded to Yale University for her master's degree in 1930 and then returned to teach at Vassar while continuing on doctoral work at Yale. During that time, she married Vincent Foster Hopper, an English literature teacher. She received her Ph.D. in 1934 when Vassar promoted her to an instructorship. By 1939, the college had raised her to the rank of assistant professor.

After World War II broke out, Hopper enlisted in the U.S. Navy's Women Accepted for Voluntary Emergency Services, which assigned her to the Board of Ordnance computing project at Harvard University working under HOWARD HATHAWAY AIKEN. When she reported to duty on

Grace Murray Hopper is considered by many to be the mother of the computer and computer languages (Flow-Matic and COBOL). *(Special Collections, Vassar College Libraries)*

July 2, 1944, he assigned her to the Mark I computer, the prototype of the modern personal computer, although it measured 51 feet long and eight feet high. Hopper and her colleagues were responsible for fixing the machine when it broke, including when a moth flew into its gears; from then on, they called the process of fixing breakdowns debugging. In 1945, Hopper and her husband divorced, and in 1946, Hopper resigned from Vassar to continue on at Harvard even after the war effort ceased, working with the Mark I and its successors, the Marks II and III.

Realizing the vast potential of the computer, she also realized the need for people to transform this potential into understandable programs, so she joined the Univac division of the Remington Rand Corporation in 1949 as a systems engineer. One of her first innovations was a compiler, which allowed the computer to generate its own programs. In 1957, she essentially translated computer languages into plainer English (both for the machine and for the human operating it), resulting in first the Flow-Matic language and later in COBOL (common business-oriented language) in 1959. She climbed the ranks in the company from director of automatic programming to chief engineer to staff scientist by the time she retired in 1971. She continued to climb the ranks simultaneously in the navy, rising to the post of commander by the time she retired in 1979.

That year, the Department of Defense awarded Hopper its highest mark of distinction, the Distinguished Service Medal. Six years later, the navy further recognized Hopper by promoting her one final time, to rear admiral, the highest rank ever attained by a woman. She received a host of other honors for her achievements, including the 1946 Naval Ordnance Development Award, the 1973 Legion of Merit, induction into the Engineering and Science Hall of Fame in 1984, and the National Medal of Technology in 1991. She died at the age of 85 on January 1, 1992.

Hounsfield, Godfrey Newbold
(1919–)
English
Physicist

Sir Godfrey Hounsfield invented computerized axial tomography scanning technology, commonly known as the CAT scan, a medical diagnostic tool that uses X-ray crystal detectors to photograph cross sections of the body that computer programs then compile and reconstruct into an image of internal organs and tissue. Doctors use this technology to diagnose previously undetectable diseases and conditions. Hounsfield defended the machine against early criticism of its seemingly exorbitant price, arguing that it ultimately saved money by preventing unnecessary surgical procedures. For his innovation, he received the 1979 Nobel Prize in physiology or medicine.

Godfrey Newbold Hounsfield was born on August 28, 1919, in the village of Newark in Nottinghamshire, England. He had two older brothers and two older sisters. His father was a steel-industry engineer who had become a farmer by the time Hounsfield was born. Hounsfield attended Newark's Magnus Grammar School, but much of his formative education was self-initiated: he built his own electrical recording devices, experimented with flying in homemade gliders off haystacks, and investigated jet propulsion by lighting acetylene under tar barrels filled with water.

When World War II broke out, Hounsfield volunteered as a reservist in the Royal Air Force, which assigned him as a radar mechanic instructor at the Royal College of Science and then at Cranwell Radar School. At the same time, he advanced his own education at London's City and Guilds College, where he passed the examination in radio communications and also received a certificate of merit for his construction of a large-screen oscilloscope. After the war, Air Vice-Marshal Cassidy recommended Hounsfield for a grant to London's Faraday House Electrical Engineering College, where he earned a degree in electrical and mechanical engineering in 1951.

That year, EMI (Electrical and Musical Industries—now Thorn EMI) hired Hounsfield to work on radar systems and guided weapons and then appointed him to run a small design laboratory in Middlesex. In 1959, he and his design team built the EMI-DEC 1100, the first all-transistor computer constructed in Britain, made possible by Hounsfield's innovation of a magnetic-core transistor to achieve a degree of speed equal to valve-driven computers.

EMI then transferred Hounsfield to its Central Research Laboratories in Hayes, where he designed and patented a high-capacity, immediate-access thin film computer-storage system that proved commercially unviable in the 1967 marketplace. Instead of immediately assigning him to a new project, EMI granted him latitude to formulate his own next undertaking. On a country stroll, inspiration struck him: he conceived the notion of a medical instrument that would take a series of X-ray cross sections, or "slices," that a computer program could compile to reconstruct a visual representation of internal physiology in much greater detail and clarity than achievable through X-ray photography alone. He called this technology tomography, or computerized transverse axial scanning, popularly known as "CAT scanning" (for "computerized axial tomography").

Hounsfield recruited the radiologists James Ambrose and Louis Kreel to work on the X-ray component of his proposed machine, while he worked to advance computer technologies capable of handling the demands of his complex conception. The British Department of Health and Social Services helped finance the project, which went into a trial phase in 1971 with the installation of a prototype at Atkinson Morely's Hospital in Wimbledon. Doctors initially performed CAT scans on the brain exclusively; as head of the

Medical Systems section at EMI, Hounsfield continued to develop CAT technologies, leading to the introduction of the first "whole body scanner" in 1975. That year, the Royal Society inducted him into its prestigious fellowship.

In 1979, Hounsfield shared the Nobel Prize in physiology or medicine with Allan M. Cormack, an American nuclear physicist who in 1963 and 1964 had calculated and published mathematical formulae for computer reconstruction of images, thus establishing a theoretical basis for CAT scanning independent of Hounsfield. Interestingly, neither held a doctorate, and their degrees were not in the fields honored by the award. In addition to the Nobel, Hounsfield also received Britain's highest engineering honor, the MacRobert Award, and in 1981, the Queen knighted him.

In 1977, EMI had promoted Hounsfield to the rank of senior staff scientist, and he continued to improve CAT scanning technology while also researching an alternative technology—nuclear magnetic resonance imaging. He retained this post until 1986, when he scaled back to consultant status with the company. A lifelong bachelor, Hounsfield committed himself wholeheartedly to his work and neglected even to establish a permanent residence until late in life, when he settled in Twickenham.

⊠ Huygens, Christiaan
(1629–1695)
Dutch
Physicist, Astronomer

Christiaan Huygens fueled the Scientific Revolution with his wide-ranging scientific contributions, from astronomical observations to theories of mechanical physics. By the age of 25, he had made a major astronomical discovery (rings and satellites of Saturn). He also invented the pendulum clock and published a renowned book on the oscillatory motion exemplified by the pendulum.

His contribution that exerted the greatest influence on society was his proposal of the wave (or pulse) theory of light, which overturned the particle theory of light.

Huygens was born on April 14, 1629, in The Hague, Holland. He was the son of Constantin Huygens, a Renaissance man who raised Huygens in an intellectually stimulating atmosphere, entertaining the likes of René Descartes in his salon. Huygens commenced his study of mechanics at the age of 13, and in 1645, he enrolled in the University of Leiden to study mathematics and law. He transferred to the College of Breda two years later, graduating with a degree in jurisprudence.

In 1651, at the age of 21, Huygens published his first paper, on the quadrature of certain curves. He and his older brother discovered a new method for grinding and polishing telescopic lenses, and in 1655, he made his first major scientific discovery—the existence of rings encircling the planet Saturn. As well, he discovered one of Saturn's moons, Titan. The next year, he identified components of the Orion nebula. In 1657, he invented the pendulum clock while studying planetary motion. In 1659, he published *Systema Saturnium*, gathering together his astronomical observations, including the above discoveries as well as a demonstration of the period of Mars.

In 1661, Huygens visited London for the first time; in a lecture before the Royal Society, he demonstrated his laws of collision. Duly impressed, the Royal Society inducted him as a foreign member in 1663. The *Académie des Sciences* elected him a foundation member in 1666, upon his moving to Paris to accept a pension from King Louis XIV. He remained there until 1681, installing himself at the Bibliothèque Royale. In 1672, he instructed GOTTFRIED WILHELM LEIBNIZ in mathematics and arranged for the presentation of his first paper, on differential calculus, to the *Académie des Sciences* in 1674.

Christiaan Huygens discovered the rings and satellites of Saturn. *(Original engraving by Edelnick, Copyright: Rijksmuseum voor de Geschiedenis der Natuuringtenschappen, courtesy AIP Emilio Segrè Visual Archives,* Physics Today *Collection)*

In 1673, Huygens published *Horologium Oscillatorium*, which Florian Cajori considered second only to Sir ISAAC NEWTON's *Principia* in scientific significance. In it, Huygens presented his theories of harmonically oscillating systems, including his formula of the period of oscillation of a simple pendulum. In the appendix, he communicated his laws of centrifugal force of uniform motion in a circle, which he considered similar to gravitational force (proofs of this work was first published posthumously).

In a 1678 communication to the *Académie des Sciences*, he introduced his undulatory theory of light, also known as the wave or pulse theory of light. Arguing against the particle theory of light, Huygens theorized that light vibrates through the ether until it reaches the eye. Huygens's principle posits the existence of two "wave fronts," one primary and one secondary, which accounted for both reflection and refraction of light. Interestingly, he did not publish these ideas for a dozen years, finally doing so in 1690 in *Traité de la Lumière*.

Huygens opposed his contemporary (and perhaps his only intellectual peer), Sir Isaac Newton, on numerous scientific issues. The wave theory of light differed from Newton's explanation, in that Huygens conceived the velocity of light as directly proportional to the sine of the angle of refraction, while Newton considered the proportion inverse. It was not until 1851 that this disagreement was settled experimentally by J. B. L Foucault and A. H. L. Fizeau. Huygens settled his own score with Newton when he met him personally on his second trip to London, in 1687—in a lecture before the Royal Society, he publicly argued against Newton's theory of gravity, pointing out several points that Newton conspicuously avoided.

Also on this trip, Huygens presented optical lenses with unusually large focal lengths to the Royal Society (where they remain on display). Upon his return to his homeland, where he remained for the rest of his life, he invented a telescopic eyepiece for reducing chromatic aberrations; this achromatic eyepiece still bears his name. Huygens suffered from a protracted and painful illness and finally died on June 8, 1695, in The Hague.

Hyatt, Gilbert
(1938–)
American
Electrical Engineer, Computer Scientist

Gilbert Hyatt received recognition as the inventor of the integrated circuit microprocessor more than two decades after he first conceived of it, in 1968. Almost 20 years expired between Hyatt's first patent application, in December 1970, and the United States Patent and Trademark Office's awarding of his patent in July 1990. In the meanwhile, engineers at Intel and Texas Instruments had developed the same technology, which fueled the personal computer revolution of the 1980s by reducing the size of the necessary hardware to run a computer to the size of a single silicon chip. Subsequent appeals obscured whether Hyatt could legitimately lay claim to inventing the "computer-on-a-chip" technology (he lost on a technicality), but he retained his claims to the invention of the integrated circuit.

Gilbert P. Hyatt was born in 1938. He received his master's degree in electrical engineering in the early 1960s, and then worked his way up to the position of research engineer with Teledyne. In 1968, he left Teledyne and formed Micro Computer, Inc., to develop his idea of shrinking computer technology onto a single chip, capable of performing all the functions of a computer except interfacing and storing memory. He called his chip the "microcomputer," the same as his company's name.

Hyatt contracted Intel and Texas Instruments to build his chips, while simultaneously raising venture capital from a group that included Dr. Noyce of Intel. Micro Computer also developed applications for its technology, including the operation of a machine tool control system and an integrated circuit drafting system. On December 28, 1970, Hyatt filed his first patent application covering his microcomputer with the United States Patent and Trademark Office.

In early 1971, the investors attempted a coup but realized they could not take the technology with them, as Hyatt had built in protections. However, the revolt crippled Micro Computer, which closed its doors in September 1971, before it had a chance to manufacture its computer chip. Interestingly, Intel introduced its Intel 4004 computer chip, designed by Stanford University graduate and Intel engineer Marcian E. "Ted" Hoff, at this same time. Also at this time, Gary Boone and his colleagues at Texas Instruments were developing single-chip technology. On July 4, 1971, they built the first computer-on-a-chip, and they filed a patent two weeks later, on July 19, 1971. Texas Instruments introduced the TMS100 later in 1971.

After the demise of his company, Hyatt started consulting for the aeronautics industry to finance his continuing computer research, while also educating himself how to prosecute patent applications. This latter interest arose as his patent applications languished in the U.S. Patent Office. In the meanwhile, Gary Boone of Texas Instruments filed his patent for single-chip microprocessor architecture on September 4, 1973. Hyatt filed multiple follow-up applications, including one dated December 14, 1977.

It wasn't until July 1990, some two decades after Hyatt's initial patent application, that he was issued patent number 4,942,516—"Single chip integrated circuit computer architecture." News of this granting was not made public until late August, when it hit the computer industry like a time bomb. Its implications were far-reaching, as it gave Hyatt rights to the technology used in most computer applications, especially those that fueled the personal computer revolution of the 1980s. Industry watchers commented that if Hyatt charged a modest licensing fee, he could collect a handsome sum without bankrupting computer-chip producers. By 1992, Hyatt had received some $70 million from American, Japanese, and European companies. He also made a licensing arrangement with the Phillips Corporation, a multinational concern.

Of course, companies such as Texas Instruments, which was witnessing the rewriting of history that had credited it as originating the microprocessor, appealed Hyatt's patent claim. A patent, such as Hyatt's, that takes years to receive confirmation while the industry proceeds to develop and market the very product under question, is commonly called a "submarine" patent. The Patent Office's Board of Appeals and Interferences heard the case and overturned Hyatt's claim to the single-chip technology, deciding that his December 28, 1970, application did not describe this technology in accordance with regulations (his December 14, 1977, application appended the necessary description), while Boone's September 4, 1973, application did.

This judgment was only a partial defeat for Hyatt. As he explained, his patent, "Single chip integrated circuit computer architecture" (abbreviated "516" after the last three digits) included about 50 claims, only half of which concerned "single-chip" technology, while the other half advanced integrated circuit claims. Boone's "interference" addressed only single-chip claims, and he won on a narrowly defined interpretation of what constituted a legitimate description of the technology according to regulations. As of 1996, the patent office attached a notice to patent 516 canceling its claim to single-chip microcontroller rights. However, all of Hyatt's other claims, namely, those involving integrated circuit technology, remained standing.

Hyatt, John Wesley
(1837–1920)
American
Inventor, Plastics Industry

John Wesley Hyatt was a prodigious inventor best remembered for his development of celluloid, the first synthetic plastic. Along with his brothers, he was also an enterprising entrepreneur, establishing companies for the manufacture and marketing of all his major inventions. This combination of scientific inventiveness and business acumen helped launch the plastics industry, which radically transformed society in the 20th century by replacing natural materials with synthetic materials, thus conserving natural resources and saving money. Celluloid is best known as the material used for the thin sheets of film used in photography and cinematography, though celluloid's flammability led to its replacement by other newly developed plastics in these and other applications.

Hyatt was born on November 28, 1837, in the small hamlet of Starkey, New York. At the age of 16, he moved to Illinois to become a printer. However, he did not remain in this profession, as his active mind turned to invention. He filed his first patent—for a knife sharpener—in 1861. In 1869, he patented a new method for making dominoes and draughts.

In the 1860s, the Phalen and Collender Company of New York announced a contest for identifying an inexpensive substitute for ivory suitable for making billiard balls. Although this challenge undoubtedly interested Hyatt, what enticed him more was the award: $10,000 in prize money. In collaboration with his brother, Isaac Smith Hyatt, he experimented with different combinations of materials in attempts to synthesize a substance that mimicked the traits of ivory.

In 1868, Hyatt happened upon the mixture of pyroxylin, a partially nitrated cellulose, with camphor, which he then heated under pressure. The result: a transparent, colorless, hard substance that seemed perfectly suited to the purpose of making billiard balls (among a host of other applications).

Ironically, Hyatt did not win the prize, even though his invention, celluloid, replaced ivory as the raw material in the billiard-ball manufacturing industry. Intent on reaping financial rewards from celluloid, he joined forces with his brothers (Charles was the third Hyatt brother) to found the Albany Billiard Ball Company in Albany, New York (where the brothers also ran the Albany Dental Plate Company and the Embossing Company). They also established the Celluloid Manufacturing Company to benefit from making the raw material of his discovery. In 1872, the brothers moved this company to Newark, New Jersey.

Celluloid lent itself to numerous commercial and industrial applications. For example, combs were manufactured at that time out of tortoise shells and animal bones—expensive materials. Bernard W. Doyle of Leominster, Massachusetts, founded the Viscoloid Corporation in 1900 to manufacture celluloid to supply the numerous comb-manufacturers in Leominster, the center of comb-making in the United States (in 1925, the DuPont Company absorbed Viscoloid).

Perhaps the most famous application of celluloid was not developed until about 20 years after its discovery. Then, the burgeoning fields of photography and cinematography required thin film upon which to develop negatives; celluloid proved to be a perfect material, as it could readily be manufactured in sheet form. However, celluloid was also relatively flammable, a trait that led to its eventual replacement by other, newer synthetic materials. Ping-Pong balls are one of the few applications that still use celluloid.

Hyatt made many other inventions and developed them entrepreneurially. In the early 1880s, he innovated the Hyatt filter, which purified moving water chemically, and in 1881, he and his brothers established the Hyatt Pure Water Company. A decade later, he developed the Hyatt roller bearing, and in 1891 or 1892, the brothers founded the Hyatt Roller Bearing Company in Harrison, New York. In 1900, he invented a multiple-stitch sewing machine, capable of sewing 50 lock stitches simultaneously. Other inventions included a mill for converting by-products from cane sugar refining into fuel and a machine for straightening steel rods.

Throughout his career, Hyatt took out more than 200 patents, placing him beside THOMAS EDISON in the prodigious output of original ideas and applications. In 1914, the Society of Chemical Industry awarded Hyatt its Perkin Medal. Hyatt died on May 10, 1920, in Short Hills, New Jersey. Seven years later, his Celluloid Manufacturing Company was bought out by the Celanese Corporation, later known as Hoechst-Celanese, one of the largest U.S. manufacturers in the plastics industry that Hyatt helped to found.

J

Jacquard, Joseph-Marie
(1752–1834)
French
Inventor, Textile Industry

Joseph-Marie Jacquard was an inventor whose innovation had much wider application than he had originally intended. He invented the Jacquard loom as a means of expediting the arduous task of hand-looming intricate patterns; instead, Jacquard programmed the pattern into a series of punch cards that symbolized the pattern through holes punched in the card. This hole functioned as a type of on/off switch, telling the loom whether or not to weave. This system of communicating with machines grew into the binary system of 0s and 1s, indicating to the machine whether to switch on or off, and lent itself perfectly to the use of transistors, which function as on/off switches. The binary system evolved into the most basic language of computers.

Jacquard was born in 1752 in Lyon, France. He apprenticed in several different trades—bookbinding, cutlery making, and typefounding—before committing himself to the trade of weaving. Jacquard based this decision mostly on the fact that he had just inherited a small weaving business from his family. He entered the profession wholeheartedly, devoting himself to producing the most intricate and aesthetically pleasing designs. However, weaving these complex designs proved to be time-consuming, making the business a losing proposition. Jacquard's weaving business failed, and he moved back to cutlery making, but his mind remained fixed on weaving.

By 1801, Jacquard had devised a means of increasing the efficiency of the weaving process exponentially: by replacing the weaver with an automated system. Jacquard's system relied on punch cards whose holes corresponded to weaving instructions. However, this system reduced the role of weavers in the weaving process from hand manufacturers to mere designers of a product manufactured by machine. Weavers of the day revolted against their potential obsolescence by destroying Jacquard looms, but once the technology became available, it was impossible to stay its growth into the position previously occupied by human workers. Not only did this invention help incite the Industrial Revolution, but it also presaged the technological revolution of more than a century later.

Between 1801 and 1804, Jacquard improved upon the design and construction of his looms as well as perfecting his punch-card system at the Paris Conservatoire des Arts et Métiers. His demonstration of the Jacquard loom and its attendant punch-card system in Paris in 1804 so impressed Napoleon that the dictator awarded Jacquard a medal, a patent, and a pension. By 1812, there were 11,000 Jacquard looms in France alone, and they were spreading into other countries.

Jacquard died in 1834, but the significance of his work lives on. His invention of the punch-card system acted as a harbinger of the advent of computers under the binary system. In fact, CHARLES BABBAGE, who theorized and built a calculating machine utilizing punch cards, had already realized this application in the 1830s. Though Jacquard's invention was extremely progressive theoretically, its immediate practical implications were less positive, as the machine put many weavers out of work and foretold the future mechanization of the labor process.

Jenner, Edward
(1749–1823)
English
Physician

Edward Jenner discovered the process of vaccination by observing that exposure to the rare disease of cowpox, even as a mere inoculation, made people immune to smallpox. He thereby established the practice of vaccination and in the process helped found the fields of virology and immunology. The number of lives saved by vaccination is incalculable.

Jenner was born on May 17, 1749, at Berkeley, in Gloucestershire, England. The son of the vicar of Berkeley, he was educated there until 1761 when, at the age of 13, he went to Sodbury to apprentice under a surgeon. He moved to London in 1770 to study anatomy and surgery as John Hunter's first boarding student at St. George's Hospital. He returned to Berkeley in 1773 to set up a medical practice and remained there the rest of his life.

Jenner first distinguished himself as a naturalist and bird-watcher, as the Royal Society published his observations of the cuckoo in 1788 and inducted him into its fellowship that year on the strength of this work. It was commonly known that cuckoos laid their eggs in the nests of hen hedge sparrows, who, it was believed, removed their own eggs to make room for the developing cuckoo. Jenner witnessed the young cuckoo heaving the hen hedge sparrow eggs out of the nest, thus revising the conventional wisdom.

Also in 1788, a smallpox epidemic plagued Gloucestershire, and local physicians responded by employing the variolation technique of inoculating the public with live smallpox taken from people suffering a mild case of the disease. This practice of scoring the arm veins of the healthy and exposing them to smallpox matter gathered from pustules on the affected was imported from Turkey in 1721 by Lady Mary Wortley Montagu, the wife of the British ambassador to the Ottoman Empire. The Dutch physiologist Jan Ingenhousz achieved much success with this method, but most physicians could not replicate his results, instead experiencing high mortality rates as patients developed not immunity but smallpox itself.

While performing such inoculations, Jenner noted that those who had survived cowpox did not fall prey to smallpox—in fact, they did not even suffer a mild attack of smallpox, as did others, in response to variolation. Jenner developed an hypothesis over his two decades of working with pox that cowpox inoculation (which did not render patients immune from cowpox) might inoculate patients against smallpox, and by 1796, he was ready to test his theory experimentally.

Jenner administered the cowpox inoculation to James Phipps, an eight-year-old patient: he made two small cuts in the boy's arm and introduced a speck of cowpox. The usual reaction resulted, as Phipps experienced a slight fever a week later, from which he recovered promptly. Two months later, Jenner administered variolation, and Phipps remained completely healthy. Jenner concluded that cowpox inoculated against smallpox, and he named his procedure "vaccination" (from the Latin for cowpox, *vaccinia*).

Jenner reported his findings to the Royal Society, which advised him against publishing something "so much at variance with established knowledge." He disregarded this advice, choosing instead to privately publish a memoir in 1798 entitled *An Inquiry into the Causes and Effects of the Variolae Vaccinae, a disease discovered in some of the Western Counties of England, particularly Gloucestershire, and known by the name of Cow Pox*. The practice of vaccination spread throughout Europe—within two years, more than 100,000 people had been vaccinated. In 1802, Parliament awarded Jenner 10,000 pounds in recognition of the contribution his vaccination made to public health, and it granted him yet another 20,000 pounds five years later.

For his final contribution, Jenner returned to his interest in naturalism to write *On the migration of birds*, which he published the year of his death. Jenner died of a stroke at his birthplace on January 24 (or 26), 1823. After his death, vaccination continued to gather momentum, as the practice of variolation was outlawed in 1850, and in 1853, vaccination was made compulsory. However, this law was not enforced until 1872, when the smallpox death rate was 90 per million, down from 3,000 to 4,000 per million the previous century. Smallpox was eradicated in the 20th century and now exists only in vaccination form.

Jobs, Steven
(1955–)
American
Computer Engineer, Entrepreneur

Steve Jobs helped revolutionize contemporary culture by introducing the personal computer—first the Apple I and II, and then the Macintosh. He thus aided in the paradigm shift from the industrial age to the information age, as society based itself less on material goods and more on intellectual property. Later in his career, Jobs acted as CEO of Pixar, a studio that transformed the movie industry through computerized animation.

Steven Paul Jobs was born on February 24, 1955. He was adopted by Clara and Paul Jobs, a machinist at Spectra-Physics, and raised in Mountain View, California. When Jobs reached adolescence, the family moved to Los Altos, California, where he attended Homestead High School. During a freshman year field trip to a Hewlett-Packard computer factory in Palo Alto, he saw his first desktop computer. He subsequently telephoned William Hewlett personally for assistance with a school project, a bold move that resulted in a job with HP that summer.

There, he met a fellow worker named STEPHEN WOZNIAK, who had dropped out of the University of California at Berkeley. Woz, as he was called, had developed a "blue box" that erased long-distance telephone charges, which Jobs helped to market. He graduated from high school in 1972 and matriculated at Reed College in Oregon to study physics and literature. He dropped out after only one semester, preferring to hover in the counterculture on the periphery of campus.

In 1974, he landed a job as a video game designer for Atari. After a trip to India in search of spiritual enlightenment, Jobs returned to California and started attending meetings of Wozniak's Homebrew Computer Club. Still working for HP, Wozniak was also designing a personal computer, which he constructed in Jobs's garage. Jobs marketed the machine to a local electronics equipment seller, who ordered 25. The pair sold their prized possessions—Jobs a Volkswagen microbus and Woz an HP scientific calculator—to raise $1,300 in start-up funds to found a computer company, which they named Apple after

Jobs's favorite fruit, which he picked as an orchard worker in Oregon.

Jobs continued to market the Apple I computer, which sold for $666 in 1976. Over the next year, the company sold $774,000 worth of Apple I computers. The next year, Woz designed the Apple II computer, and Jobs encouraged independent programmers to create software for it, resulting in the development of 16,000 programs. Within three years, the Apple II earned $139 million, representing growth of 700 percent. In 1980, the underwriting firm Hambrecht & Quist, in collaboration with Morgan Stanley, took Apple public at $22 a share. The price shot up to $29 the first day of trading (valuing Jobs's holdings at $165 million), and by December 1982, the share price was about $30. That year, Apple sales reached $583 million, up 74 percent over the previous year. Apple's compounded growth from 1978 through 1983 was 150 percent per year.

International Business Machines introduced its first personal computer in 1981 and grabbed almost one-third of the market within two years by dominating the corporate segment. Apple responded by developing the Macintosh, a project that Jobs oversaw personally. "We have thought about this very hard and it would be easy for us to come out with an IBM look-alike product, and put the Apple logo on it, and sell a lot of Apples. Our earning per share would go up and our stock holders would be happy, but we think that would be the wrong thing to do," said Jobs. Instead, he conceived of the introduction of the Mac as a revolutionary act, emphasizing its user-friendly interface as a kind of democratizing influence on the world of computers.

Apple launched the Mac in 1984 with a commercial broadcast on Super Bowl Sunday, evoking drab Orwellian imagery that exploded symbolically with the arrival of the Macintosh. A piece of marketing genius, the spot lodged itself in the cultural consciousness and fueled sales. And the Macintosh itself achieved its revolutionary goal of becoming a "mass market appliance," establishing the personal computer as a practical tool accessible to mainstream users.

However, internal strife plagued Apple, as Jobs's hired gun, president John Sculley (formerly president of the Pepsi-Cola Company), railroaded Jobs to the sidelines in an attempt to discipline the company known in the industry as "Camp Runamok." Jobs responded by resigning in late 1985 to start up an educational hardware and software firm, the NeXT Company. Sculley complicated the transition by suing Jobs for stealing Apple research and employees, which Jobs settled out of court.

Although the hardware division of NeXT flopped, its software division inadvertently flourished. Out of pragmatism, not by design, Jobs chose Mach software from Carnegie Mellon University, which ran object-oriented programming (OOP), as the operating system for NeXT hardware. OOP turned out to be the perfect choice, as it allowed programmers to lift preexisting chunks of code in creating their own software, thus saving the effort of generating new code from scratch. Just as Jobs had revolutionized personal computing, so too did he revolutionize corporate and academic computing, by accident.

In 1986, Jobs bought a majority stake in Pixar, which LucasFilms (owned by George Lucas, creator of the *Star Wars* series and the Indiana Jones movies) spun off. Three years later, Pixar won an Academy Award for its animated short, *Tin Toy*. In 1989, he won the Software Publishers Association Lifetime Achievement Award. In 1991, he married Laurene Powell, with whom he has two children.

In the late 1990s, Pixar hit stride by creating three of the top six grossing domestic animated films ever—*Toy Story* (1995), *A Bug's Life* (1998), and *Toy Story II* (1999). Jobs contributed to these achievements by overseeing Pixar's partnership with Walt Disney Pictures, one key to to success. In 1996, Jobs returned to

Apple as its CEO and began to split his time between the two enterprises.

⊗ Joliot-Curie, Irène
(1897–1956)
French
Nuclear Physicist

Irène Joliot-Curie collaborated with her husband, Jean Frédéric Joliot-Curie, in the discovery of artificial synthesis for radioactive material and the discovery of the neutron. For these advances, the wife and husband couple was awarded the 1935 Nobel Prize in chemistry.

Curie was born on September 12, 1897, in Paris, the daughter of the Nobel laureate physicists Pierre and Marie Curie. She followed in their footsteps by studying under the Faculty of Science at the University of Paris. She served in World War I as a nurse radiographer. After the war, she joined her mother as an assistant at the Institute of Radium. While working in that laboratory, she conducted her research on alpha rays of polonium that served as the locus of her doctoral dissertation. She was hooded as a doctor of science in 1925.

Irène met Frédéric Joliot when he was also an assistant at the Institute of Radium, and the couple married in 1926. They had two children together, one daughter named Helene and one son named Pierre. They fused not only their personal lives but also their professional lives, conducting much of their research in tandem. Together, they discovered that stable elements could be destabilized to create radioactive material artificially. In 1933, they bombarded alpha particles at the stable element boron, which created a radioactive compound of nitrogen. They jointly published their results in the 1934 paper, "*Production artificielle d'éléments radioactifs: Preuve chimique de la transmutation des éléments.*" The Nobel Committee recognized the profound implications of their discovery by awarding both Joliot-Curies the Nobel Prize in chemistry for 1935.

The Faculty of Science in Paris had appointed her as a lecturer in 1932, and in 1937 it conferred on her the title of professor. The year before, the French government had named her undersecretary of state for scientific research. In 1938, Joliot-Curie's research on the behavior of heavy element neutrons opened up one door on the way to the discovery of nuclear fission. The next year, the Legion of Honor inducted her as an officer. During World War II, Frédéric led the underground resistance movement as the president of the Front National.

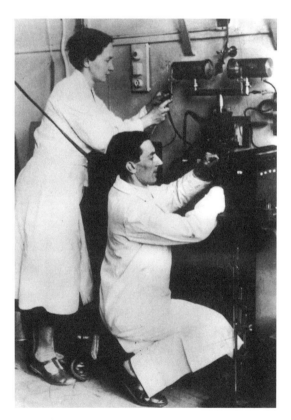

Irène Joliot-Curie and her husband, Frédéric, were jointly awarded the 1935 Nobel Prize in chemistry for their discovery of artificial radioactivity. *(Société Française de Physique, Paris, courtesy AIP Emilio Segrè Visual Archives)*

After the war, Irène succeeded her mother as the director of the Institute of Radium. Also in 1946, she was named a commissioner for atomic energy, a position she retained for six years, during which France amassed its first atomic stockpile. After the war, she worked to promote peace as a member of the World Peace Council. She also contributed her energy to women's rights as a member of the Comité National de l'Union des Femmes Françaises.

As with her mother, Marie Curie, who died from sustained overexposure to radiation, Irène Joliot-Curie also died from overexposure to radiation throughout her entire life. She contracted leukemia and died in Paris on March 17, 1956. Her husband succeeded her as director of the Institute of Radium. He also continued to oversee the construction of the 160 MeV synchrocyclotron at the new center for nuclear physics in Orsay, a project she had commenced before dying.

⊠ Joule, James Prescott
(1818–1889)
English
Physicist

One of the leading experimentalists of his time, James Prescott Joule's main contributions to science were the discovery of the first law of thermodynamics, the law of the conservation of energy, and his findings concerning the mechanical equivalent of heat. Joule also collaborated with Baron WILLIAM THOMSON KELVIN to develop the Joule-Thomson effect, which stated that the temperature of an expanding gas cooled if the gas did not perform external work.

The second of five children, Joule grew up in a wealthy brewing family. Born on December 24, 1818, in Salford, near Manchester, England, to Benjamin and Alice Prescott Joule, Joule was a shy and frail child. He and a brother were tutored at home, and from 1834 to 1837 the

James Prescott Joule discovered the first law of thermodynamics. *(AIP Emilio Segrè Visual Archives)*

brothers learned mathematics and science from chemist JOHN DALTON, known for his work on atomic theory. Joule developed an interest in physics at an early age and set up a laboratory near the brewery to conduct experiments.

Though Joule did not receive a formal education or a college degree, he made significant discoveries, most before reaching the age of 30. At the age of 19, Joule began independent research under the guidance of William Sturgeon, an amateur scientist. Joule was influenced by Sturgeon's interest in electromagnetic theories. At about the same time, Joule began to investigate the problems of heat, particularly the heat developed by an electric current. He found that the heat produced in a wire by an electric current was connected to the current and resistance of the wire. Joule announced his findings in 1840 in a paper entitled *On the Production of Heat by Voltaic Electricity*.

From 1837 to 1847, Joule studied the mechanical equivalent of heat and other forms of energy and established the principle of conservation of energy. He systematically studied the thermal effects caused by the production and passage of current in an electric current, and in 1843 Joule determined the amount of mechanical work needed to produce a given amount of heat; his discovery was guided by precise experiments in which he measured the degree of heat produced by rotating paddle wheels, powered by an electromagnetic engine, in water. Joule presented his observations in 1849 to the Royal Society in *On the Mechanical Equivalent of Heat*. A year earlier Joule had also written a paper on the kinetic theory of gases. The paper included the first estimation of the speed of gas molecules.

Joule worked with Thomson from 1852 to 1859 on experiments in thermodynamics. Their most significant discovery was that an expanding gas's temperature cooled under certain conditions. This became known as the Joule-Thomson effect and provided the basis for the development of a large refrigeration industry in the 19th century.

In 1850 Joule was elected to the Royal Society and enjoyed broad recognition and a strong reputation. He continued to carry out experimental investigations, but his findings failed to match the accomplishments of his early years. Joule married Amelia Grimes of Liverpool in 1847; she died in 1854, leaving him to raise their two children. Though Joule never received an academic appointment, his work on thermodynamics and the mechanical equivalent of heat were widely accepted and helped advance the sciences. The joule, a unit of energy, was named in his honor. Joule died on October 11, 1889, in Sale, Cheshire, after a lengthy bout with a degenerative brain disease.

Kapitsa, Pyotr Leonidovich
(1894–1984)
Russian
Physicist

Pyotr Kapitsa established his reputation as a scientist in cryogenics, or the study of low-temperature physics, as he discovered the phenomenon of superfluidity in helium II. However, Kapitsa left perhaps a greater mark on society through his determined stance in defense of intellectual freedom and moral decision-making in the face of totalitarian suppression under the Stalinist regime in the Soviet Union. In the absence of Kapitsa's principled resistance of Soviet efforts to dictate the course of scientific research, Soviet scientists would have enjoyed much less latitude.

Pyotr Leonidovich Kapitsa was born on July 8 (or 9), 1894, in Kronstadt, an island near St. Petersburg, Russia. His mother was Olga Ieronimovna Stebnitskaya, a renowned folklorist who also worked in higher education, and his father was Leonid Petrovich Kapitsa, a lieutenant general in the engineering corps. He married Nadesha Tschernosvitova in 1916. He studied under A. F. Ioffe in the electromechanics department of the Petrograd Polytechnic Institute, where he collaborated with Nikolai N. Semenov to propose a method for measuring atomic magnetism, which was later employed in Otto Stern's famous experiments in 1921. Kapitsa graduated in 1918 with an electrical engineering degree, and remained as a lecturer at the institute.

The violence and famine of the Bolshevik Revolution laid claim to the lives of his wife and young child. In 1921, Kapitsa traveled to England (at the suggestion of Ioffe and through the influence of Maxim Gorky) to work under Ernest Rutherford at Cambridge University's Cavendish Laboratory. In 1923, he placed a cloud chamber in a strong magnetic field to observe the bending paths of alpha- and beta-particles emitted by radioactive nuclei. That year he received his doctorate and won the JAMES CLERK MAXWELL fellowship, and the next year, Cambridge named him assistant director of magnetic research at Cavendish and a fellow of Trinity College. Also in 1924, he developed methods for creating very strong but short-lived magnetic fields. In 1927, he married Anna Alekseevna Krylova, and together the couple had two sons, Sergei and Andrei, both of whom became scientists

In 1929, the Royal Society inducted Kapitsa into its fellowship, and the next year the prestigious society appointed its first foreign member in over 200 years as its Messel Research Professor. By then, he had turned his attention

Pyotr Leonidovich Kapitsa established his reputation through research in cryogenics, but his greatest impact may have come from his lifelong defense of intellectual freedom under the Stalinist regime in the Soviet Union. *(AIP Emilio Segrè Visual Archives)*

to low-temperature physics, having invented a means of liquefying helium in much greater quantities than was previously possible. At the urging of Rutherford and under funding from the German chemist Ludwig Mond, the Royal Society built the Mond Laboratory expressly to host Kapitsa's magnetics and low-temperature cryogenics research, installing the Russian as the lab's director in 1933.

The next year, Kapitsa returned to the Soviet Union, his sixth visit to his homeland in 13 years, to attend the International Congress marking the hundredth anniversary of D. I. Mendeleyev's birth. On September 25, 1934, the Soviet government detained Kapitsa from returning to England, commencing his ongoing struggle for scientific (and personal) freedom. Kapitsa used the power derived from his international prominence and his scientific reputation to resist governmental control—for example, he refused to conduct any research, thereby forcing the state to purchase all of the equipment from the Mond Laboratory, transport it to Russia, and use it to outfit a lab for him at the S. I. Vavilov Institute for Physical Problems, which the government also established expressly for him. However, Kapitsa continued to play his political advantage by refusing the directorship of the institute for almost a year.

In his new lab, Kapitsa focused his cryogenics research on helium II, investigating the properties of this form of liquid helium that exists near absolute zero. When cooled below 2.17 degrees Kelvin, helium II reaches a state of viscosity that Kapitsa described as "superfluid," an even better medium for conducting heat than copper. He published classic papers on superfluidity in 1938 and 1942. In 1939, he invented a turbine for producing liquid oxygen.

After World War II, Kapitsa again asserted his power by refusing to contribute to the Soviet development of atomic weaponry, prompting the government to place him under house arrest. His insistence on maintaining intellectual freedom and moral reasoning in science is perhaps a more significant legacy than his actual scientific contributions. In 1955, two years after Stalin's death, Kapitsa finally returned to his directorship at the institute, where he focused his research on plasma physics, devising a continuous high-pressure method for heating plasma to temperatures of 2 million degrees Kelvin, causing the so-called fourth state of matter to lose its electrons to become electrified ions. He also investigated high-power microwave radiation, inventing the planotron and nigotron generators.

Throughout his career, Kapitsa received honors internationally in recognition of his important contributions: he received a dozen honorary degrees from almost as many countries, and he belonged to the primary scientific societies in practically every important country. He also received numerous awards: in 1965, Kapitsa was finally allowed to leave the Soviet Union (for the first time in more than three decades) to accept the Niels Bohr International Gold Medal from the Danish Society. In 1978, his pioneering work in cryogenics and his discovery of the superfluidity of helium II were finally recognized by the Royal Swedish Academy of Sciences, which granted him the Nobel Prize in physics, in conjunction with Arno Penzias and Robert W. Wilson, who discovered cosmic microwave background radiation. Kapitsa died in Moscow on April 8, 1984.

Kay, John
(1704–1764?)
English
Inventor, Textile Industry

John Kay invented the flying shuttle, an innovation that doubled production in loom-weaving while also raising the quality and consistency of the cloth. Known in some history books as "Kay of Bury" to distinguish him from the clockmaker who assisted Sir RICHARD ARKWRIGHT in investing the spinning frame, he was never appreciated for his invention: workers, whose livelihoods were threatened, ransacked his house; manufacturers similarly abused him, refusing to pay royalties for the use of his patented invention. Ironically, Kay died in relative obscurity in France, while his invention was one of the key elements that advanced the Industrial Revolution by mechanizing the production process.

Kay was born on July 16, 1704, at Park Walmersley, near Bury, in Lancashire, England. He was the 12th child born to a woolen manufacturer. Little is known of his youth; it is believed that he received his education on tour throughout Europe—some accounts specify France. When he returned from his travels, his father tried to install him as overseer of one of the family's woolen factories. While Kay may have filled this position, he reportedly preferred to work on his own, producing reeds (resembling combs) for looms, which maintained even spacing between yarn strands in the cloth manufactured.

Kay continued to focus his attention on the textile production process, searching for ways to improve the tools and procedures of the trade. For example, he filed English patent number 515 for his invention of an engine that manufactured and stored woolen thread. Another invention consisted of a machine that thrashed wool cloth to remove dust mites. He also developed other improvements in the dressing, batting, and carding machinery involved in textile manufacturing.

Growing up in Lancashire, the heart of textile country, and managing his father's woolen mill, Kay became intimately familiar with all the aspects of the manufacturing process. His incisive mind also identified inefficiencies in production techniques, of which there were many in the preindustrial era, when demand was increasing but the supply was still produced by traditional methods. The weft shuttle was one such glaring inefficiency.

Loom weaving necessitates the passing of a shuttle back and forth to create the weft, or the cross-weaving of threads that run perpendicular to the warp threads. The existing timeworn method required the weaver to pass the shuttle from one side of the loom to the other—from one hand to the other. The weaving of broader cloth required mill owners to employ two separate workers to do nothing other than throw the shuttle back and forth between each other. Clearly, Kay perceived, there had to be a better way.

Kay set about to devise an improved method for shuttling the weft. On May 26, 1733, he received English patent number 542 for a "New Engine or Machine for Opening and Dressing

Wool"—his flying shuttle. Kay carved a groove in a lathe beneath the weaving area of the loom, then mounted the shuttle on wheels that raced along this track back and forth between two shuttle boxes on each end of the loom. Kay called this wooden guide the "race-board," which allowed for the extension of the loom's width by a foot on either side. He attached cords to a "picking peg," which operated the shuttle boxes—the lone weaver held the cords in one hand, and with a flick of the wrist, he activated the "throwing" of the shuttle from one side to the other, and then back again with another flick.

Kay's ingenious invention significantly increased the efficiency of the looming process, doubling the amount of cloth produced while also improving the quality and consistency of the weave. However, this innovation created trouble for him on both sides of the production process. Producers joined forces to form the "Shuttle Club," a coalition dedicated to the use of the flying shuttle without paying royalties to Kay, as he was due by law. Kay bankrupted himself by mounting legal battles to protect his patent, as the factory owners essentially bribed the legal system in their favor.

On the other end of production, Kay's invention enraged textile workers, who feared for the security of their jobs. The prospect that their labor could be replaced by mechanization understandably scared them, and they responded by forming a mob in 1752 or 1753 to ransack Kay's house in Bury. He supposedly hid under wool and then fled to Manchester.

Such antipathy from the workers and exploitation by the manufacturers caused Kay to withhold other inventions. For example, he invented a machine-run loom with Joseph Stell of Keighley, the first "power loom," but he did not market it commercially. In fact, the discouraged Kay left the country for France, where he spent the rest of his days in poverty, desperately trying to sell his innovations. He died in 1764 (or 1780 according to some sources). His son, Robert Kay, continued his legacy by inventing the drop-box, essentially a refinement of the flying shuttle, which allowed the weaver to choose between three differently colored shuttles, to create design variety in the weft.

Kelly, William
(1811–1888)
American
Inventor, Steel and Iron Industries

William Kelly developed the pneumatic process of steelmaking that fueled the Industrial Revolution in the United States. Although Sir HENRY BESSEMER is generally attributed as the inventor of this technique—blasting air into molten pig iron ore to burn the carbon impurities—Kelly actually came up with the method several years prior to Bessemer. A protracted legal battle followed the introduction of the Bessemer process from Britain into the United States, as both inventors held a patent to the same essential process. Eventually, the two competing factions joined forces, and thereafter, steel production mushroomed in the United States, fueled by the westward expansion of the railways and the raising of skyscrapers, among many other steel applications.

Kelly was born on August 21, 1811, in Pittsburgh, Pennsylvania, the son of a wealthy landowner. He married the daughter of a prosperous tobacco merchant. He established a partnership in a dry goods and shipping company, McShane & Kelly, and in the early 1840s, while on a buying trip, he became interested in establishing an ironworks in Eddyville, Kentucky. He enlisted his brother as a partner to buy 14,000 acres of woodland with iron ore deposits, and a Cobb furnace, and thereby founded the Eddyville Iron Works.

Kelly's company (also reported as the Sweanee Ironworks and Union Forge) produced sugar-boiling kettles from wrought iron, which was converted from raw pig iron by a simple charcoal process. As this process depleted his land of both timber, used as fuel for smelting, and carbon-free iron, Kelly began to investigate ways to reduce fuel consumption and to use the lower-quality, carbonated iron ore remaining on his land. Observing his workers forging the ore, he noticed that when they reheated the pig iron, air drafts created a white-hot glow that burned off excess carbon, contributing to the creation of higher-quality wrought iron.

As early as 1847, Kelly conceived of a notion that dually benefited the forging process: blasting air into the molten pig iron would not only burn off the carbon impurities, converting them into oxides that could be sloughed off as slag, but also the burning carbon would act as fuel to generate heat internally, preventing the need to continue burning wood to heat the smelter and consequently saving on fuel consumption. When his father-in-law heard this scheme, he doubted Kelly's sanity and submitted him for examination by a doctor who, as it turned out, understood the logic of Kelly's plan and became one of his strongest supporters.

From 1851 through 1856, Kelly worked in secret developing a converter. However, in 1855, the British industrialist Henry Bessemer filed for a patent in the United States covering his Bessemer process of converting pig iron to steel by introducing air into the smelting. When Kelly discovered Bessemer's patent application in 1856, he gathered the necessary documentation to file a counterclaim, asserting priority since his design dated back to 1847. In the midst of the panic of 1857, Kelly filed the claim, while also filing for bankruptcy. In order to raise funds, he sold the patent to his father.

In 1859, Kelly renewed experiments on his Kelly process at Cambria Iron Works in Johnstown, Pennsylvania. In 1862, he established a steel plant at Wyandotte, Michigan, which produced the first batch of commercial steel by the Kelly process within two years. In 1863, he established the Kelly Pneumatic Process Company, and the next year, a group in Troy, New York, established a plant using the Bessemer process. Throughout this period, Kelly and Bessemer fought in the courts, each claiming primacy to the patent of the air-infusion process of generating steel from iron.

The contest between the Bessemer and Kelly processes was ultimately decided by the economy, which could not support the competition, forcing the two companies to pool their resources in 1866. Thereafter, U.S. steel production boomed, fueling the American Industrial Revolution. Steel made by the pneumatic process was used in the rails that laid the train tracks throughout the United States, in the ships that increased transatlantic commerce, and in the skyscrapers that housed the burgeoning business world.

In the cultural memory, Bessemer beat out Kelly, as the process ultimately bore the former's name in history books, with Kelly relegated to relative obscurity. In fairness, although Kelly conceived of the idea before Bessemer, Kelly could not succeed in perfecting the process, as did Bessemer, who added the steps of cutting off the air blast at just the right moment to maintain some carbon content, and of deoxidizing the steel before pouring it.

It was not until 1871, when Kelly managed to get his patent extended, that he finally realized financial compensation equal to the significance of his invention. He spent the end of his life in Louisville, Kentucky, where he established an ax-making business. Kelly died on February 11, 1888, in Louisville.

Kelvin, Baron (William Thomson)
(1824–1907)
Irish
Mathematician, Physicist

William Thomson was an ingenious scientist who contributed to the theoretical understanding of science, though he gained more renown for his practical applications of science. He is perhaps best remembered by the title bestowed upon him, Baron Kelvin, a name born by the absolute scale of temperature he devised earlier in his career. He is also renowned for his role in the laying of the first successful transatlantic telegraphic cable, a development of immense cultural significance, as it instantly connected the European and the American continents.

William Thomson was born on June 26, 1824, in Belfast, Ireland. He lived there until the age of six, when his Scottish mother died, and his father, James Thomson, moved the family back to Scotland, where he was appointed professor of mathematics at the University of Glasgow. Thomson's father educated him, and at the age of 10, he entered his father's university, beginning university-level work at 14. He studied astronomy and chemistry his first year, and natural philosophy (or physics) his second, winning a gold medal from the university for his "Essay on the Figure of the Earth."

Thomson entered Peterhouse College of Cambridge University in 1841 and published his first paper, "Fourier's expansions of functions in trigonometrical series," that year. Four years later, in 1845, he graduated as second wrangler (second of those graduating with a first-class degree) and first Smith's prizeman, and the next year Peterhouse College inducted him into its fellowship. He spent an intervening postgraduate year in Paris, working in Henri-Victor Regnault's laboratory.

Also in 1846, the University of Glasgow's chair of natural philosophy opened up, and by unanimous vote, the university elected him to the position, which he retained for the remaining 53 years of his career. Two years later, in 1848, he developed the absolute scale of temperature named after him (after the British government bestowed the title Baron Kelvin of Largs upon him in 1892). Thomson developed the theoretical underpinnings of the scale in three influential papers on heat published over the next three years. He gathered these together in the seminal 1851 text, "On the Dynamical Theory of Heat," in which he elucidated the Kelvin scale of temperature (it was not until the advent of the laws of the conservation of matter that theory could fully account for the function of the scale).

In 1853, Thomson published his investigations into oscillatory discharges of a Leyden jar. This work was not particularly distinguishing until he invented the mirror galvanometer as an instrumental application of the theory of electrical oscillation to telegraphy. Coincidentally, plans were afoot to lay telegraphic cable across the Atlantic Ocean from Ireland to Newfoundland; Thomson served as a technical adviser to the project and as a member of its board of directors.

William Thomson introduced the absolute scale of temperature, which uses the kelvin as the unit of measure. *(AIP Emilio Segrè Visual Archives)*

Cables laid in 1857 broke; the next year's attempts to lay cable succeeded, though signal transmission proved problematic. The project's electrical engineer, E. O. W. Whitehouse, insisted upon employing his own transmission schematic over Thomson's mirror galvanometer system. When the signal improved dramatically, it was discovered that Whitehouse had secretly replaced his technology with a mirror galvanometer. The third laying of cables in 1865 deployed the mirror-galvanometer system completely, resulting in the clearest possible signal.

The next year, in 1866, Thomson was knighted for his contributions to the transatlantic submarine telegraphic cable project. Up until this point, Thomson made nothing but significant contributions to science; in the second half of his career, he was not so infallible: he did not accept the atomic theory, Darwin's theory of evolution, and Rutherford's discovery of radioactivity, and what is more, he miscalculated the ages of the earth (400 million) and sun (50 million). By this point, however, Thomson had established his cultural significance with his concrete contributions of the Kelvin scale and the technology for cross-oceanic telegraphy, though other of his scientific work may have been more profound.

Throughout his career, Thomson served his field, and he is considered one of the most important voices advancing the understanding of physics in the 19th century. He presided over the Royal Society of Edinburgh three times (from 1873 through 1878, from 1886 to 1890, and from 1895 until he died). In the intervening five years between 1890 and 1895, he presided over the Royal Society of London, whose fellowship granted him two important prizes over the years: the 1856 Royal medal and the 1883 Copley medal. He published 661 papers, and he held 70 patents. Thomson died in his home at Netherhall, in Ayrshire, Scotland, on December 17, 1907, and was buried at Westminster Abbey.

King, Mary-Claire
(1946–)
American
Geneticist

Some geneticists merely analyze deoxyribonucleic acid (DNA) in test tubes, but Mary-Claire King saw genetics as closely tied to politics. "I've never believed our way of thinking about science is separate from thinking about life," she told an interviewer, David Noonan, in 1990. She has used genetics to study human origins, discover why some people are more susceptible to breast cancer or acquired immunodeficiency syndrome (AIDS) than others, and identify the lost children of people murdered by repressive governments. As the reporter Thomas Bass has noted, "Any one of her accomplishments could make another scientist's full-time career."

Mary-Claire King was born in Wilmette, Illinois, a suburb of Chicago, on February 27, 1946. Her father, Harvey, was head of personnel for Standard Oil of Indiana. Clarice, her mother, was a homemaker. A childhood love of solving puzzles drew Mary-Claire to mathematics, which she studied at Carleton College in Minnesota; she graduated cum laude in 1966.

King then went to the University of California (UC) at Berkeley to learn biostatistics, or statistics related to living things. Eventually King went to work for Allan C. Wilson, who was trying to trace human evolution through genetics and molecular biology, in his molecular biology laboratory. Wilson assigned King to compare the genes of humans and chimpanzees. At first she thought she must have been doing something wrong because "I couldn't seem to find any differences," but her results proved to be accurate. She proved that more than 99 percent of human genes are identical to those of chimpanzees. This startling research not only became the thesis that earned her a Ph.D. in genetics from Berkeley in 1973 but also was featured on the cover of *Science* magazine.

In 1975, King turned to a quite different aspect of genetics: the possibility that women in certain families inherit a susceptibility to breast cancer. Such women have the disease more often and at a much earlier age than average. Scientists at the time were discovering that all cancer grows out of damaging changes in genes, but usually those changes occur during an individual's lifetime and are caused either by chance or by factors in the environment, such as chemicals or radiation. The damaged genes involved in only a few rare cancers were known to be inherited. King eventually proved that about 5 percent of breast cancers are inherited.

In 1990, King localized the breast cancer gene, which she called BRCA1, at a position halfway down the lower arm of the 17th of the human cell's 23 chromosomes. By then she had become a professor of epidemiology at UC Berkeley's School of Public Health (in 1984) and a professor of genetics in the university's Department of Molecular and Cell Biology (in 1989).

King's most unusual genetic project, tied to her lifelong concern for human rights, was helping to reunite families torn apart during the "dirty war" waged in Argentina between 1976 and 1983. During that time the country's military government kidnapped, tortured, or murdered an estimated 12,000 to 20,000 citizens. Babies born in prison or captured with their mothers were sold or given away, thus becoming lost to their birth families.

In 1977, while the military government was still in control, a group of courageous older women began gathering every Thursday on the Plaza de Mayo in Buenos Aires, opposite the government's headquarters, to protest the loss of their sons and daughters and demand the return of their grandchildren. They called themselves Abuelas de Plaza de Mayo (Grandmothers of the Plaza of May). When a more liberal government took power in 1983, the group stepped up its campaign to locate the missing children. Even when the children were found, however, the families who had them usually refused to admit that they were adopted or give them up.

Knowing that genetic tests could show, for instance, whether a man was the father of a certain child, two representatives of the grandmothers' group traveled to the United States and asked for a geneticist to help them prove their relationship to the disputed children. They were sent to Luca Cavalli-Sforza, a renowned Stanford geneticist with whom King had worked; he in turn referred them to King. King began working with the group in 1984; she now calls Argentina her "second home."

At first King used marker genes to test for relationships in the Argentinian families, but later she adapted a better technique that Allan Wilson developed in 1985. Most human genes are carried on DNA in the nucleus of each cell, but small bodies called mitochondria, which help cells use energy, also contain DNA. Unlike the genes in the nucleus, which are from both parents, mitochondrial DNA is passed on only through the mother and therefore is especially useful in showing the relationship between a child and its female relatives. King says that mitochondrial DNA examination "has proved to be a highly specific, invaluable tool for reuniting the grandmothers with their grandchildren." As a result of King's work some 50 Argentinian children have been reunited with their birth families to date. The same technique has since been used to identify the remains of people killed in wars or murdered by criminals.

In 1995, King moved to Seattle (which she called "the Athens of genetics") to head a laboratory at the University of Washington. Her laboratory has pursued a number of projects, including continuation of the human rights work of identifying military murder victims and children separated from their families by war; this work, under the direction of Michele Harvey, now encompasses Bosnia, Rwanda, and Ethiopia as well as Argentina. The lab is also investigating the genetic characteristics of inherited deafness and genetic variations that may determine why some people are more readily infected with the human immunodeficiency virus (HIV) after exposure and develop full-blown AIDS more quickly after infection than others. The laboratory's chief focus, however, continues to be the study of BRCA1 and other genes involved in breast and ovarian cancer, including noninherited forms of the disease. "Our goal is to eliminate breast cancer as a cause of death," King says.

King's work has earned awards such as the Susan G. Komen Foundation Award for Distinguished Achievement in Breast Cancer Research (1992) and the Clowes Award for Basic Research from the American Association for Cancer Research (1994).

Mary-Claire King believes that women contribute a special gift to science. "Women tend to tackle questions in science that bridge gaps," she said. "We're more inclined to pull together threads from different areas, to be more integrative in our thinking."

Kipping, Frederic Stanley
(1863–1949)
English
Chemist

Frederic Stanley Kipping discovered silicon polymers, compounds valued for their slippery properties. Ironically, despite his three decades of research, Kipping did not establish the utility of silicon himself but instead considered his discovery useless for practical applications. Quite the opposite, silicon has proven invaluable as a lubricant and has served as a necessary material in many computer applications. In this sense, then, Kipping's discovery was a vital precursor to the computer revolution of the latter half of the 20th century, though the scientist himself could not have foreseen this development.

Kipping was born on August 16, 1863, at Higher Broughton, near Manchester, in Lancashire, England, the son of a banker. He attended Manchester Grammar School and the Lycée de Caen before enrolling at the University of London in 1879. However, he ended up matriculating at Owens College in Manchester (now Manchester University), where he earned a bachelor of science degree in chemistry in 1882 (awarded by the University of London.)

For the next four years, Kipping worked as a chemist at the Manchester Corporation Gas Works, until he realized the lack of promise in his future, so he pursued graduate study at the University of Munich, where he worked in Adolf von Baeyer's laboratory under William H. Perkin Jr. In 1887, he earned his Ph.D. with highest honors from the University of Munich, and that same year, he became the first person to earn a D.Sc. degree from the London University based solely on research.

Kipping followed Perkin to Heriot-Watt College in Edinburgh, working as a demonstrator his first year; the next year, he was promoted to a lectureship in agricultural chemistry and an assistant professorship in chemistry. The two collaborated in writing *Organic Chemistry*, which became a standard textbook in the field upon its publication in 1894 and remained a classic for more than a half-century.

In the meanwhile, Kipping had returned to London in 1890 to become chief demonstrator in chemistry at the City and Guilds Institute (now Imperial College of Science and Technology). In 1897, University College in Nottingham (now Nottingham University) appointed him as a professor of chemistry. That same year, the Royal Society inducted him into its fellowship, and the next year, the Institute of Chemistry did the same. Throughout this early part of his career, Kipping focused his research on nitrogen compounds and optically active camphor derivatives.

In 1900, Victor Grignard invented new magnesium-based organometallic compounds, called Grignard reagents. Kipping experimented with these reagents, and in the process he prepared the first organosilicon polymers. He coined the term *silicon* to describe the "glue-like" compound he discovered, but he failed to recognize the significance and utility of his discovery.

Kipping gained many honors and sat in positions of responsibility and authority throughout his career. Twice he served as vice president of the Chemical Society of London, first from 1908 through 1911, and then a decade later, from 1921 through 1923. The society granted him its 1909 Longstaff medal, and the Royal Society awarded him its 1918 Davy medal. In 1935, he retired to emeritus status, though he continued to conduct research and publish his results.

In 1937, Kipping published a paper entitled "Organic Derivatives of Silicon" in the *Proceedings of the Royal Society*, in which he stated that "the prospect of any immediate and important advance in this section of organic chemistry does not seem to be very hopeful." In 1944, he published the last of his 51 papers that appeared in the *Journal of the Chemical Society*, describing silicon chemistry. By this time, scientists had filed patents on the silicon chemistry processes for synthesizing lubricants, hydraulic fluids, varnishes, greases, synthetic rubbers, and waterlogging compounds, among other applications. What made silicon so useful was its chemical inertness and its ability to retain stability at high temperatures.

In his retirement, Kipping moved his family—consisting of just his wife and daughter by then, as his two sons had commenced their careers as chemist and chess strategist, respectively—to Criccieth, Wales, where he died on May 1, 1949. A decade later, in 1960, the Dow Corning Corporation endowed the American Chemical Society with the funds to establish the Frederic Stanley Kipping Award in Silicon Chemistry, a $5,000 award and lectureship granted biennially.

Kolff, Willem J.
(1911–)
Dutch/American
Physician

Willem Kolff is considered the "father of artificial organs," as he invented not only the first artificial kidney but also the first artificial heart implanted in a human being. He also contributed one of the first heart-lung machines, which inaugurated the era of open-heart surgery, as well as an intra-aortic balloon pump designed to treat cases of acute myocardial distress, and the first wearable artificial kidney.

Willem J. Kolff was born on February 14, 1911, in Leiden, Holland. His father, Jacob Kolff, was director of the Beekbergen Tuberculosis Sanitorium. As a child, he spent every Saturday afternoon for seven years learning carpentry, but he eventually followed in his father's footsteps by studying medicine at the University of Leiden starting in 1930. He served as an assistant in pathological anatomy from 1934 through 1936 and received his medical degree in 1938.

Kolff pursued postgraduate studies at the University of Groningen, where he first conceived the idea of an artificial kidney. He studied under the chief of the medical department, Professor Polak Daniels, who along with his wife committed suicide upon the Nazi occupation of the Netherlands. Kolff responded to the occupation by establishing continental Europe's first blood bank. The next year, he removed to the small town of Kampden to work in the municipal hospital.

Here, he scavenged materials—wooden drums, cellophane tubing, and laundry tubs—from a local factory to construct the first crude artificial kidney in 1943. The apparatus removed blood from the patient's body to perform the kidney's function of removing toxins from the bloodstream before pumping it back into the patient. His treatment added a few days to the lives of the first 15 patients. Then, in September 1945, a woman who had collaborated with the Nazis needed hemodialysis treatment. Despite the townspeople's urging to let the woman die, Kolff believed that the role of a physician is not to judge morality but to treat all patients. Hers was the first life saved by the artificial kidney.

After the war, Kolff earned his Ph.D. in internal medicine summa cum laude in 1946. Over the next four years, he continued to work in Kampden while lecturing at the University of Leiden. He also continued to improve his artificial kidney, sending free dialysis machines to researchers in England, Canada, and the United States. In 1950, the Cleveland Clinic Foundation invited Kolff to head its department of artificial organs as a professor of clinical investigations. There, he built one of the first heart-lung machines, which made open-heart operations possible. He also became a citizen of the United States in 1956.

In 1957, Kolff implanted the first artificial heart into a dog, which lived for an hour-and-a-half with the device in its chest. Despite this success, Kolff met resistance from the medical establishment in accepting the notion of an implantable artificial heart, as most journals refused to publish his papers on the topic. He therefore diverted his attention temporarily, designing the intra-aortic balloon pump to treat cases of acute myocardial distress. First introduced in 1961, the apparatus saw extensive use.

In 1967, the University of Utah hired Kolff as director of its Institute for Biomedical Engineering and as professor of surgery and research professor in the school of engineering. There, in 1975, be improved upon his artificial kidney by inventing a wearable version of it, featuring an eight-pound chest pack and an 18-pound auxiliary tank.

Kolff collaborated with a medical student, Robert Jarvik, and a veterinary surgeon, Don Olsen, to implant the Jarvik-5 mechanical heart in a calf that survived 268 days. By 1981, Kolff felt ready to attempt an artificial heart implantation in a human being, so he applied to the Food and Drug Administration for legal permission, which was granted. On December 2, 1982, the surgeon William DeVries implanted the Jarvik-7 artificial heart into a 61-year-old retired dentist named Barney Clark. Three more operations were necessary to replace a defective value and tweak the heart. Clark died after four months, though not of heart failure but of unrelated complications.

In 1986, Kolff retired at the age of 75. Throughout his career, he published some 600 scientific papers, as well as the book *Artificial Organs*. He also received numerous honors, including the 1947 Amory prize and the 1964 Cameron prize for practical therapeutics from the University of Edinburgh. That same year, he was voted one of the top 10 physicians in the United States, and in 1969 he won the Valentine medal and award for "outstanding contributions to the field of urology." Most significantly, the Dutch queen Juliana named him a commander in the Order of Oranje-Nassau. The American Academy of Achievement inducted him into its membership in 1971, and the Inventors' Hall of Fame inducted him in 1985. In 1990, *Life* magazine listed Kolff as one of the 100 Most Important Americans of the Twentieth Century.

Perhaps the most significant honor, however, was the fact that some 55,000 people are saved from almost certain death from end-stage renal disease now, thanks to the artificial kidney.

Krupp, Alfred
(1812–1887)
German
Inventor, Steel and Iron Industries

Alfred Krupp was named the "cannon king" for his introduction of the steel cannon, which is considered one of the first events in the founding of modern warfare. Krupp's contributions to social welfare counterbalanced his contributions to warfare, as he established many practices that honored and supported his workers, such as instituting sickness and death insurance and establishing worker housing, stores, schools, and hospitals.

Krupp was born on April 26, 1812, at Essen, in the grand duchy of Berg (now Germany). His father, Friedrick Krupp, had established a steel business with two partners on November 20, 1811. By 1816, Friedrich was the sole proprietor of the business when he developed a mass-production process for making high-quality crucible steel, from which he manufactured files and cast steel bars, and later he added tanner's tools, coining dies, and roll blanks to his company's product list. In 1826, both of Krupp's parents died, leaving him in charge of the family business at the age of 14. When he took over, the company was failing, employing a mere five workers.

Krupp revived the company like a Phoenix rising from the ashes, which he achieved in part by guaranteeing quality workmanship, and in part by treating his workers with utmost respect. He expanded operations into the manufacture of steel rolls, inventing a spoon roll for making spoons and forks. He also employed rolling mills for use in government mints.

Krupp made his name on the strength of his armaments manufacturing. In 1847, he introduced the first steel cannon (previously, cannons had been cast in bronze), which proved to be very successful. Initially, German law prohibited him from selling these cannons to Prussia (despite a deal Krupp signed with the Prussians in 1834). However, he got orders from elsewhere in the world, signing contracts with Egypt in 1856, Belgium in 1861, and Russia in 1863. The performance of his cannons in the Franco-Prussian War of 1870 through 1871 solidified his reputation as the "cannon king."

Krupp also secured his reputation by displaying a two-ton (4,300-pound) steel ingot, demonstrating the best steel casting to date, at the London Exhibition of 1851. He followed up on this by introducing the seamless forged and rolled railroad wheel in 1852. Krupp had discovered a process that reinforced the wheel to withstand the increasing speeds achieved by trains. Unlike previous steel wheels, his did not fracture under pressure.

In 1862, Krupp was the first to introduce HENRY BESSEMER's steelmaking process in continental Europe, and in 1869 he introduced the Siemens Martin open-hearth steelmaking process to his works. As his demand for raw materials increased, he recognized the efficiency of incorporating all aspects of the manufacturing process into his own company, so he acquired ore deposits and coal mines throughout Germany and also iron and steel works to expand his operations.

Krupp's social welfare policies modeled the policies enacted in social legislation by the German government in the 1880s. As early as 1836, he instituted a voluntary sickness and burial fund, which became compulsory in 1853 as sickness and death benefit insurance. In 1855, he introduced a pension fund for retired and incapacitated workers. The next year, he opened hostels for unmarried workers. In 1858, workers launched a bakery that they themselves ran. In 1861, Krupp built housing settlements for his workers (who numbered 16,000 by 1871), adding schools and retail store branches in 1863 and a hospital in 1872. His social agenda may not have been purely altruistic, for he generated a fanatic loyalty among his workers that kept them from organizing into unions and striking for more control of their lives.

By the time Krupp died on July 14, 1887, in his birthplace of Essen, he had armed 46 nations, and the manufacture of weapons represented 40 to 60 percent of the firm's overall output. The involvement of his great-grandson, Alfried Krupp von Bohlen und Halback, with the Nazi movement (employing some 70,000 workers in slave labor, a war crime for which he was imprisoned between 1948 and 1951) stained the Krupp reputation. The establishment of the Krupp von Bohlen und Halbach charitable foundation upon Alfried's death in 1967 did help to restore the family name.

Lancefield, Rebecca Craighill
(1895–1981)
American
Bacteriologist

Working at the Rockefeller Institute for Medical Research, Rebecca Craighill Lancefield made her scientific reputation in the study of streptococcus, a multifaceted bacterium whose workings and groupings were little understood when she began her career. During six decades, Lancefield identified more than 50 types of streptococcus and discovered how various strains reacted with the human body. Her research helped doctors to better understand diseases caused by this agent and to develop treatments to heal patients suffering from it.

Born on January 5, 1895, at Fort Wadsworth in Staten Island, New York, Rebecca Craighill was the daughter of William Edward Craighill, an officer in the U.S. Army, and Mary Byram Craighill, a homemaker. During her childhood and youth, Craighill traveled around the United States as her father was based at different army stations. She was encouraged to excel in her studies by both parents but especially by her mother, who had been influenced by Julia Tutwiler, an advocate for education for women. The third of six daughters, Rebecca Craighill was the first one to attend college. In 1918, she married Donald Lancefield, a biologist she met while they attended Columbia University. She had one child.

In 1912, Rebecca Lancefield traveled to Massachusetts to enroll as an undergraduate in Wellesley College, a college for women. She first intended to study French and English but found herself instead attracted to science when she took her first zoology course. She graduated from Wellesley with a B.A. in zoology in 1916. Lancefield briefly taught science at a girls' boarding school in Vermont, but wanting to pursue a career in science, she enrolled in Columbia University in New York City. Working intensively with Hans Zinsser, a noted bacteriologist, she earned a master's degree in bacteriology from Columbia in 1918.

Lancefield wanted to begin work on a Ph.D. in bacteriology immediately, but the entry of the United States into World War I delayed her plans. Her husband was drafted into the army and assigned to work at the Rockefeller Institute. Lancefield also applied for a job at the Rockefeller Institute and was hired to assist OSWALD THEODORE AVERY and Alphonse Dochez in their early studies of the streptococcus bacterium. After the war, she followed her husband for a one-year teaching stint in Oregon and

Rebecca Lancefield identified more than 50 strains of the streptococcus bacterium during her six decades of research. *(Photo by RI Illustration Service, Don Young, Courtesy of Wellesley College Archives)*

then returned with him to New York. In 1922, she began working again at the Rockefeller Institute, this time with Homer Swift, while also working on her Ph.D. dissertation in bacteriology at Columbia.

Lancefield won a Ph.D. from Columbia in 1925. Her dissertation, "The Immunological Relationships of Streptococcus viridans and Certain of Its Chemical Fractions," summarized the studies she had made at the Rockefeller Institute that showed there was no link between rheumatic fever and a species of streptococcus called *Streptococcus viridans*. Lancefield continued her work on streptococcus, and by 1933, she had devised a precipitin test by which she found five different strains, or subvarieties, of this bacterium. Lancefield later isolated the cause of rheumatic fever, which was by group A streptococcus, and determined that the virulence of a particular group A streptococcus was caused by a protein that protects that strain from being enveloped by human white blood cells, which normally kill invasive organisms.

For her contributions to the understanding of bacteriology, Lancefield won many awards, including the American Heart Association's Achievement Award (1960) and the Medal of the New York Academy of Sciences (1970). She served as president of the Society of American Bacteriologists in the late 1940s and, in 1970, was elected to the National Academy of Sciences. Rebecca Craighill Lancefield died in New York City of complications from a broken hip on March 3, 1981. She was 86 years old.

Lanchester, Frederick William
(1868–1946)
English
Inventor, Transportation Industry

Frederick Lanchester was an eclectic inventor and theorist, who is best remembered for introducing the first motorcar in Britain, and he started a line of fine automobiles (which ultimately went under, demonstrating Lanchester's lack of business sense, though he continued to contribute his technical knowledge as a consultant at Daimler). He may have had a deeper influence on military strategy, though, as he contributed theories on aerial combat as well as on ground deployment of forces. Although his theory offered a seemingly self-evident equation between the numerical force of a side and its ability to defeat the opponent, his ideas exerted a profound influence, not only on military situations, but also on corporate structure and management practices.

Frederick William Lanchester was born on October 23, 1868, in London, England. He was privately educated at the Hartley Institution in Southampton, after which he won a national scholarship to the Royal College of Science. He also studied at Imperial College in London.

He commenced his work career in Birmingham, working as an assistant manager and then as a manager at Messrs. T. B. Baker of Saltey, and, in 1889, he joined the Forward Gas Engine Company. After four years there, he established his own workshop, where he built his first experimental motorcar in 1894. Over the next half-decade, he continued to experiment and innovate new features for his gas engine and for his automobiles, such as rack-and-pinion steering, fuel injection, and power steering—he even patented disk brakes in 1901.

In 1896, Lanchester introduced Britain's first motorcar, which featured a single-cylinder five-horsepower engine and a chain drive. His second car completed a thousand-mile tour in 1900. The year before, he had founded Lanchester Engine Company, acting as manager for its first five years. During that time period, the company produced 350 cars (all of which he constructed with interchangeable parts) and then met with financial straits. He stepped down to a designer and technical adviser position for the next decade, but the company ultimately went bankrupt. He continued his involvement in the automobile industry as a consultant for Daimler from 1909 through 1920.

Lanchester next turned his interest to the theory and practice of crewed flight. He was one of the first theorists to understand that aircraft offered significant strategic advantages in military campaigns. In 1908, just prior to the outbreak of World War I, he published his ideas in the text *Aerial Flight*, and in 1909, Prime Minister Asquith invited him to join the Advisory Committee for Aeronautics (which he continued to serve through 1920). In 1911, he codesigned an experimental aircraft that crashed on its trial flight. However, despite this failure of his practical application on aeronautics, his air combat theories were proven correct when airplanes and aerial combat figured significantly in the outcome of the war.

Lanchester also developed a dynamical model of armed conflict, taking into consideration such factors as numerical strength, firepower, strategy, and attitude. Some military officials consider his theories brilliant, while other critics call his logic into question. For example, the battle of Iwo Jima in World War II demonstrates Lanchester's equations perfectly; however, an analysis of 601 battles from 1600 on did not support Lanchester's equations, though this may have more to do with a misapplication of his equations than an outright flaw in his methodology.

Interestingly, W. Edwards Deming and Japanese management theorists further developed Lanchester's combat theories in the 1960s, resulting in the so-called New Lanchester Strategy. This theory predicted that when a business captures 80 percent of market share, it attracts competition from big companies in parallel markets and from small start-ups, thus cutting into its monopoly of market share. Along similar lines, Lanchester is considered the patron saint of operations research.

Lanchester also invented numerous other sundry innovations. For example, he patented a loudspeaker and other audio equipment. In 1917, he patented the pneumatic roof, which came into use in World War II. In 1925, he founded Lanchester's Laboratories, to formalize his penchant for research and development consulting. He also continued to consult on the side, serving as a consulting engineer to the Breadmore diesel department from 1928 through 1930. Lanchester died on March 8, 1946.

Lavoisier, Antoine-Laurent
(1743–1794)
French
Chemist

Widely regarded as the founder of modern chemistry, Antoine-Laurent Lavoisier was a versatile and talented scientist who made numerous contributions not only to chemistry but also to geology, physiology, economics, and social reform. Through experimental observations Lavoisier discovered oxygen and its role in combustion. He also isolated the major components of air and introduced the method of classifying chemical compounds. Lavoisier was also instrumental in the development of thermochemistry.

The father of Lavoisier, Jean-Antoine, was a wealthy Paris lawyer who inherited his family's estate in 1741. He married Émilie Punctis, the daughter of an attorney, in 1742; on August 26, 1743, she gave birth to Lavoisier. She died when he was five years old. He was educated at the Collège des Quatre Nations, commonly known as the Collège Mazarin, beginning in 1754, when he was 11. The school had an outstanding reputation, and Lavoisier gained a thorough classical and literary education and the best scientific training available in Paris.

In 1761, Lavoisier began to study law, intending to follow the family tradition. He earned his degree in 1763 and received his license to practice law the following year. The pull of science was strong, however, and his friend Jean-Étienne Guettard, a geologist, encouraged him to study geology, mineralogy, and chemistry, which Guettard considered to be imperative for the analysis of rocks and minerals. From 1762 to 1763, Lavoisier took courses in chemistry given by the popular Guillaume François Rouelle. He also expanded his knowledge of geology under Guettard's guidance. Lavoisier completed the first geological map of France, which earned him an invitation to the Royal Academy of Sciences in 1768. Also that year, he invested in a tax-collecting company, which was prosperous enough to allow him to build his own laboratory.

Lavoisier is best known for his studies of combustion. During that time it was believed that combustible matter contained a substance known as phlogiston, which was emitted when combustion occurred. The primary flaw of this theory was that substances sometimes increased in weight as a result of combustion. Lavoisier sought to solve this problem and carried out a range of experiments. He burned phosphorus, lead, and a number of other elements in closed containers and observed that the weight of the solid increased, but the weight of the container and its contents did not. In 1772, he found that burning phosphorus and sulfur caused a gain in weight when combined with air.

Lavoisier then heard of Joseph Priestley's discovery that mercury oxide released a gas when heated and left behind mercury; Priestley referred to the gas as dephlogisticated air. Expanding on Priestley's findings, Lavoisier determined in 1778 that the gas that mixed with substances during combustion was the same gas that was emitted when mercury oxide was

Antoine-Laurent Lavoisier is regarded by many as the founder of modern chemistry. *(Smith Collection, Rare Book & Manuscript Library, University of Pennsylvania)*

heated. Lavoisier called the gas oxygen. He also recognized the existence of a second gas, which later was named nitrogen.

In 1776, Lavoisier worked at the Royal Arsenal, in charge of gunpowder production. It was there that Lavoisier collaborated with Marquis Pierre-Simon de Laplace to create an ice calorimeter and measure the heats of combustion and respiration. This work marked the beginning of thermochemistry. In 1787, Lavoisier worked with three other French chemists to propose a method of classifying chemical compounds. This system is still used today. Lavoisier published an important and influential work in 1789 called *Elementary Treatise on Chemistry*. It summarized his observations, discussed the law of conservation of mass, and provided a list of the known elements.

In addition to his combustion studies, Lavoisier explained the formation of water from hydrogen and sought to discover a method for improving the water supply to Paris. He also worked toward social reform, endorsed scientific agriculture, and was a member of the commission that promoted the metric system in France. In 1771, Lavoisier, at age 28, married Marie Anne Pierrette Paulze, who was 14 years old. The couple had no children, but Lavoisier's wife became his close collaborator. Lavoisier's life was cut short in 1794, when he was executed during the Reign of Terror.

Lawes, Sir John Bennet
(1814–1900)
English
Agriculturist

Sir John Bennet Lawes invented artificial fertilizer when he discovered superphosphates, his name for the combination of rock phosphate with sulfuric acid. The synthesis of fertilizer had profound implications on agricultural practices, as it freed farmers from absolute dependence on animals to produce manure to feed and nourish their crops. Lawes famously stated that his discovery of synthetic fertilizer established that agriculture can be an artificial process, or one that is not completely bound to the vagaries of nature. In the practical realm, his discovery led to the establishment of the fertilizer industry, which became an important segment; in the scientific realm, Lawes collaborated with Sir Joseph Henry Gilbert to found the Rothamsted Experimental Station (RES), where the pair performed their "classical experiments" on the effects of artificial fertilizers on soil conditions and crop yields.

John Bennet Lawes was born on December 28, 1814, at Harpenden, in Hertfordshire, England. His father owned the Rothamsted estate, which Lawes inherited in 1822. After several unsuccessful attempts at obtaining a university education (studying at University College, London under A. T. Thomson, at Eton College, and at Oxford's Brasenose College), Lawes returned to farm the family manor in 1834.

Over the next eight years, Lawes experimented with both organic and inorganic fertilizers. Traditionally, farmers fertilize with manure, thus making them dependent on animals to produce this natural fertilizer. Seeking to free farmers from this dependence, Lawes experimented with the use of ground-up animal bones, which proved to be an excellent fertilizer. He subsequently discovered that sulfuric acid, a cheap byproduct of many industrial applications, could perform the same function as grinding at much less expense. His next innovation was to substitute rock phosphate, derived from the petrified residues of bird excreta, for the animal bone, which had the same limiting effects as manure.

In 1842, Lawes patented his phosphate-sulfuric acid mixture as "superphosphate," the first artificial fertilizer. Within the next year, he had established a superphosphates manufacturing facility in Deptford, importing Chilean nitrates for the necessary nitrogen content, and he later founded the Lawes Chemical Company Ltd., which manufactured other agricultural chemicals in addition to superphosphates. Lawes thus founded the artificial fertilizer industry, a segment that would have profound effects upon the future of agriculture.

Also in 1843, Lawes commenced his collaboration with Joseph Henry Gilbert, who he appointed as the chemist at Rothamsted Laboratory, as he dubbed his manor, as the first agricultural experimental station in the world. Lawes and Gilbert continued their partnership over the next 57 years at what came to be known as the Rothamsted Experimental Station, which continues to this day the research that Lawes and Gilbert initiated.

Lawes and Gilbert commenced nine long-term experiments that became known as the Rothamsted Classical Experiments (one project was abandoned in 1878, leaving eight ongoing experiments). Lawes and Gilbert endeavored to track the effects on crop yields of the elements known to be contained in manure—namely, nitrogen, phosphorous, potassium, sodium, and magnesium. The experiments compared trial plots fed different combinations and concentrations of these minerals with plots fertilized with manure. Lawes conscientiously recorded the weight of the produce yielded by each crop for future comparison, and Gilbert sampled the soil for chemical analysis. Their efforts held practical implications for farmers, who fertilized their crops according to the results reported from the RES.

Lawes gained recognition for his pioneering efforts, as the Royal Society inducted him into its fellowship in 1854 and awarded him and Gilbert its royal medal in 1867. In 1878, he became a fellow of the Institute of Chemistry, and in 1882, the title of baronet was bestowed on him. The Royal Society of Arts awarded him its 1894 Albert medal, and Lawes received honorary degrees from Cambridge, Oxford, and Edinburgh Universities, representing quite a distinction for a man who never earned a university degree on his own.

Lawes died on August 31, 1900, at the Rothamsted manor. More than a decade before this, though, in 1889, he established the Lawes Agricultural Trust to fund the ongoing efforts of the Rothamsted Experimental Station and to support the continuation of the classical experiments, which continued after his death with very few modifications.

Lear, William
(1902–1978)
American
Inventor, Communications Industry and Transportation Industry

Although William Lear's name is associated closely with the jet he designed in the 1960s and marketed successfully to corporate executives, he also made two other significant contributions to society—the car radio and the eight-track tape. The former continues to affect people's lives, while the latter mushroomed in popular use in the late 1960s and early 1970s before fading out into oblivion, the technological equivalent of bell-bottoms. Lear also made a host of other inventions, including a universal radio amplifier, the first reliable aeronautical radio compass, and an automatic pilot system that locks into a radio signal.

William P. Lear was born on June 26, 1902, in Hannibal, Missouri. When he was still young, his family moved to Chicago, where he attended public school through the eighth grade. He was only 16 when the United States entered World War I, so he lied about his age to enlist in the U.S. Navy. In the service, he learned radio electronics, one of his fields of specialization thereafter. When the war ended, he learned to fly while working as an airplane mechanic at Grant Park Airport back in Chicago, and aviation became another of his fields of specialization.

In 1922, Lear founded Quincy Radio Laboratory, the first of his many companies. Within two years, he had invented the company's first major product (and the first of his many inventions)— a practical and affordable car radio, the first of its kind, However, sales were not his strong suit, so in 1930 he signed his rights over to Paul Galvin, president of the Galvin Manufacturing Corporation, which marketed the "model 5T71" radio at about $120 for installation into most car makes. Galvin later dubbed the radio the "motorola" (a combination of "motor" and "Victrola"), the name the company adopted in 1947.

In 1934, Lear invented the all-wave radio receiver, which he sold to the Radio Corporation of America. RCA installed this radio amplifier in its entire line of radio products, a testament to the universality of Lear's design. In 1939, Lear formed Lear Inc. to manufacture aircraft instrumentation, and he subsequently founded Lear Corp. and LearAvia Corporation to fill contracts amounting to more than $100 million with the U.S. military during World War II.

In 1940, Lear invented "Learmatic Navigator," an automatic pilot system that locked into a radio signal as a means of steering the airplane. In recognition of the importance of this invention, Lear received the Frank M. Hawks Memorial Award. After the war, Lear refined his miniature autopilots for fighter jets, and he invented the first fully automatic landing system. For this last achievement, he won the Federal Aviation Administration's Collier Trophy, which President Harry Truman granted him in 1950.

In 1962, he demonstrated the first ever completely blind landing of a passenger flight on a French Caravelle jetliner, a feat honored by the French government. Also in 1962, he sold his 23 percent interest in Lear Inc. for $14.5 million, which he used to fund the founding of Learjet Inc. This company fulfilled his vision of marketing small jets to corporate executives for comfortable and convenient air travel, freeing them from commercial airline schedules and hassles. On October 7, 1963, the first Learjet took off, leading immediately to yearly sales of $52 million. By 1975, 500 Learjets had been delivered, making it the model of choice for the corporate world.

In the early 1960s, Lear invented eight-track tape technology—a continuous magnetic tape loop with four stereo programs running simultaneously on eight parallel tracks. The Ford Motor Company installed eight-track players in its cars, thus fueling popularity in eight-track sound for a short time, until the fad wore off and other technologies replaced the eight-track in the popular imagination.

By the time Lear died, on May 14, 1978, he had received about 150 patents. Posthumously, he was inducted into the International Aerospace Hall of Fame (in 1981) and the National Inventors Hall of Fame (in 1993). Interestingly, Lear believed in the eventual development of "teleporting" technology, akin to "beaming" on *Star Trek*, whereby a person could travel across time and space. He also believed in UFOs and provided explanations rational enough to make the Pentagon nervous.

Leblanc, Nicolas
(1742–1806)
French
Chemist

Nicolas Leblanc discovered a commercially viable process for producing soda, or sodium carbonate, a necessary raw material in many industrial applications in the late 18th and 19th centuries. Leblanc never realized a profit from his invention, though, as the French Revolution brought about the seizure of his soda factory and the publication of his method, which other manufacturers used without paying royalties to Leblanc, whose patent was rendered moot by the revolution. Leblanc's method of soda production remained in use for more than a century, as subsequent scientists made improvements to his methodology.

Leblanc was born an orphan at Ivoy-le-Pré, in Indre, France, probably on December 6, 1742. His guardian was a physician who encouraged Leblanc to follow in his footsteps, so the boy apprenticed to an apothecary before studying at the École de Chirurgie in Paris to qualify to practice medicine. In 1780, Louis-Philippe-Joseph (later known as Philippe-Égalité, the duc d'Orléans) appointed Leblanc as the Orléans family surgeon, a position that afforded him the time and financial backing to conduct his own independent research.

In 1775, the French Académie des Sciences had offered a monetary prize for the development of a commercially viable process for making soda from salt (both of which derive from sodium). The existing processes relied on resources such as potash, barilla shrubs from Spain, or seaweed, which were becoming scarce, whereas salt was plentiful. Soda was used to produce glass, soap, paper, and porcelain, industries that were expanding at the time, thus requiring more raw materials. New processes existed at the time, but none that was efficient enough for industrial application.

As early as 1783, Leblanc discovered his technique for creating soda from salt, or sodium chloride, which he would dissolved in sulfuric acid to form sodium sulfate, releasing hydrogen chloride in the process. He then roasted the salt cake (as sodium sulfate was known) with powdered coal and calcium carbonate (crushed chalk or limestone) to create "black ash." He then dissolved this dark residue in water to extract the sodium carbonate, which he reconstituted into crystal form by heat, leaving a waste by-product of calcium sulfide, known as "galligu."

The Académie des Sciences neglected to award Leblanc the prize money, though he did file a 15-year patent in 1791. By then, he had already established a company in partnership with the duke, as well as J. J. Dizé and Henri Shée, and built a factory at St. Denis that produced some 350 tons of soda its first year. However, neither his patent nor his ownership of the company lasted long, interrupted by the eruption of the French Revolution.

In November 1793, Leblanc's patron, the duc d'Orléans, was guillotined. The soda factory, which he had financed, was nationalized, and the Committee of Public Safety published Leblanc's soda-preparation methods in the name of the revolution. Not only did Leblanc forfeit profits from his own production of soda (the St. Denis factory distributed its production to the state at no profit), but also he lost the rights to royalties from the use of his method to produce soda. He sank into poverty while fighting to regain control over his discovery.

Leblanc had performed early investigations on crystallization, and now, in 1802, he published his findings in his only book, *Cristallotechnie*. Also in that year, Napoleon returned the soda factory to Leblanc; however, in the absence of funds to make capital improvements, it proved untenable to run the factory. Leblanc hedged his bets on receiving ample compensation for his losses, but when his claim was decided in November 1805, the award fell far short of his projected need. Depressed by the outcome of this decision, and downtrodden in poverty, Leblanc committed suicide at St. Denis on or about January 16, 1806.

No records document Leblanc's inspiration for this discovery, as it does not follow logically from his previous investigations. In fact, it laid Leblanc open to claims against his propriety in this discovery. In 1810, Dizé published an article claiming a share of Leblanc's discovery. However, a special commission appointed by the Académie des Sciences convened in 1836 and concluded that Leblanc held sole claim to the discovery, based on the existing evidence.

Lebon, Philippe
(1767–1804)
French
Engineer, Chemist

Philippe Lebon was the first to develop a system for using "artificial" gas to fuel large-scale lighting and heating systems. He derived the gas from wood, heating sawdust until it emitted a smoky, pungent gas. He tried to entice the French government into incorporating his system for large-scale lighting projects (the streets of Paris had first been illuminated by lamplight in 1667), but the smell of his gas prevented him from securing any financial backing. Ironically, he was killed before he was able to implement his system to the extent he imagined it, but he paved the way for future development of artificial gas systems used to illuminate and heat on a large scale.

Lebon was born on May 29, 1767, in Bruchay, near Jonville, France. He studied science at Chalon-sur-Saône and then at the École des Ponts et Chaussées, a prestigious engineering school in Paris that prepared students for governmental service. He graduated into the rank of major in 1792 and served the state as a highway engineer in Angoulême, near Bordeaux. Soon thereafter, he returned to Paris to accept an appointment teaching mechanics at the École des Ponts et Chaussées.

Lebon first distinguished himself by improving the steam engine, an accomplishment that earned him a national prize of 2,000 francs. However, it was his work on "artificial" gas for lighting and heating that earned him lasting renown. In 1797, while at Bruchay, Lebon commenced experiments to produce gas from wood: he heated sawdust in a glass tube over a flame, producing a gas that burned with a strong smell and a cloud of smoke. Back in Paris, Lebon consulted several prominent scientists (including Antoine Fourcroy) in hopes of improving his system, which he believed could be used for lighting and heating on a large scale.

The idea of lighting and heating on a large scale was not original to Lebon, as lamps containing candles had illuminated the streets of Paris starting in 1667. This system was replaced in 1744 by oil lanterns, which were improved in 1786 by the addition of glass chimneys. However, Lebon was the first to conceive of synthesizing gas to use as a fuel for lighting spaces beyond the confines of individual private homes.

By 1799, Lebon felt confident enough in his discovery to patent it (as the "thermolampe," in reference to the heat it emitted in addition to light) and present his findings in a paper before the Institut de France. His attempts to secure a contract with the French government proved futile, as he had yet to solve the problem of the smell produced by the process. He next appealed directly to the public, leasing the Hotel Seignelay in Paris for several months in 1801 to install a demonstration by illuminating a fountain with a large version of his gas lamp. People flocked to see the spectacle, but unfortunately, the smell persisted, driving them away and preventing the public from supporting his system. Lebon continued to work on improvements, securing additional patents in 1801.

The opportunity to perfect his gas lighting and heating system was robbed from Lebon by an attacker who stabbed him to death on the Champs-Élysées on December 2, 1804. One myth alleges that Napoleon, whose collection of merkins (used to adorn his shaved pubis) was destroyed by a thermolamp, contracted the murder.

Concurrently but independently, William Murdoch was developing a lighting scheme similar to Lebon's. While working for MATTHEW BOULTON and JAMES WATT, Murdoch began experimenting in 1891 with the idea of deriving gas from coal through a process like the one used to convert coal to coke. He secured backing for his work after Watts's son Gregory returned from Paris, where he witnessed Lebon's gas-lamp exhibit. After further experimentation by Murdoch, Boulton and Watt built plants for gas production on a commercial basis, supplying industrial factories with light as of 1804.

That same year, the German scientist Frederic Winsor (who also witnessed Lebon's Paris gas-lamp exhibit) developed gas-lighting technology. His scheme differed in providing the gas by piping it in mains emanating from a centrally located power plant. He established the Gas Light and Coke Company in 1812 to provide London with gas. Lebon's gas-lighting lamp thus inspired these subsequent systems, which continued the work that he could not pursue from the grave.

⊠ Leclanché, Georges
(1839–1882)
French
Engineer, Energy and Power Industries

George Leclanché invented the Leclanché battery, an improvement over ALESSANDRO VOLTA's wet cell, and the direct precursor of current battery technology. Whereas Volta used copper or tin as the material for one of his electrodes, Leclanché substituted carbon as the cathode. Although Leclanché's battery was a "wet" cell, as the electrodes were submerged in an electrolytic solution, his design opened the door for the development of the "dry" cell technology that characterizes modern batteries. The vast majority of batteries currently in common use, powering devices from portable stereos to flashlights, are descendants of the Leclanché cell.

Leclanché was born in 1839 in Paris, France. His parents, Eugenie of Villenuve and Léopold Leclanché, sent him to England for his early schooling, then brought him back to France to attend the École Centrale des Arts et Manufactures. After he graduated in 1860, the Compagnie du Chemin de l'Est hired him as an engineer. Within six years of commencing his career, he had invented the battery that would secure his name in the history of science.

Alessandro Volta invented the first battery, or galvanic cell, in 1798. Dubbed a "Voltaic pile" after him, it consisted of copper (or tin) and zinc (or silver), separated by pasteboard or hide, submersed in an electrolyte of dilute acetic acid. The acid ate the zinc, producing electricity. This represented a significant scientific advancement, as it allowed for the harnessing of relatively large amounts of electricity. Two-thirds of a century later, Leclanché replaced Volta's copper with carbon, marking the first step toward the modern zinc-carbon battery.

In 1866, Leclanché patented his new type of cell, which consisted of a zinc anode and a carbon cathode submerged in a porous pot filled with an electrolytic solution of ammonium chloride. He also added a "depolarizer" of powdered manganese dioxide to the cathode. In the Leclanché battery, the zinc oxidized (or lost both its electrons) to become a positively charged ion that migrated away from the anode, which retained the electrons, creating a negative charge. This imbalance of negative charge induced the flow of electrical current to the cathode and through an external circuit, if connected. The electrons sacrificed by the zinc combined with the manganese dioxide and water, creating manganese oxide and negative hydroxide ions.

However, this was not the end of the chemical reaction, as the Leclanché battery featured a secondary reaction in which the negative hydroxide ions combined with positive ammonium ions, which formed when the ammonium chloride was dissolved in water to form the electrolyte. The result of this secondary reaction was ammonia and water. The cell, which carried about 1.5 volts (a measure of electricity named after Volta), "died" when the manganese dioxide was completely depleted. His original battery was too heavy and fragile for effective commercial distribution, requiring further design improvements.

The next year, Leclanché resigned his engineering position to dedicate himself to developing his battery design. The year after that, 1868, the Belgian Telegraphic Service adopted the Leclanché battery to power its systems, the beginning of widespread use of the battery (especially in telegraphy, signaling, and electric bell work), prompting Leclanché to establish a factory to produce his cells. One of Leclanché's improvements was to seal the system in such a way that the outside remained dry, despite the fact that the inside of the battery still relied on the "wet" cell design. In this sense, his design also acted as a forebear of modern battery design.

Other scientists also contributed to the advancement of battery technology. ROBERT BUNSEN, the inventor of the burner that bears his name, also designed a battery using carbon and zinc electrodes but retaining Volta's acid as an electrolyte (Bunsen substituted chromic acid for Volta's acetic acid). J. A. Thiebaut came up with the notion of encapsulating the negative electrode and the porous pot inside zinc, patenting this idea in 1881. Carl Gassner, a German scientist, was responsible for the first "dry" cell that succeeded commercially, thus ushering in the modern era of batteries.

Leclanché died on September 14, 1882, in Paris. He was relatively young, in his early forties, preventing him from making further improvements to his battery design. His brother, Maurice, assumed leadership of Leclanché's battery-production company.

Leibniz, Gottfried Wilhelm
(1646–1716)
German
Mathematician, Philosopher

Gottfried Wilhelm Leibniz is credited as the inventor of differential and integral calculus, though historically, questions arose about his precedent over Sir ISAAC NEWTON, who discovered differential and integral calculus independently. Current opinion favors the German mathematician over the Englishman. Leibniz also invented a mechanical calculator that represented a major improvement over BLAISE PASCAL's calculator design. Leibniz is also recognized for his contributions to philosophical thought.

Leibniz was born on July 1, 1646, in Leipzig, Germany. His mother was Katherina Schmuck, and his father, Friedrich Leibniz, was a professor of moral philosophy at the University of Leipzig who died when Leibniz was only six years old. The next year, he entered the Nicolai School and augmented his classical studies there by installing himself in his late father's extensive library, reading voraciously. In 1661, at the age of 15, he entered the University of Leipzig to study jurisprudence and philosophy. Two years later, he earned his bachelor's degree on the strength of his baccalaureate thesis, *De Principio Individui* ("On the Principle of the Individual").

Leibniz spent the summer term of 1663 at the University of Jena studying Euclidian geometry under Erhard Weigel and then returned to Leipzig to earn his master's degree in philosophy with a dissertation combining philosophy, law, and mathematics. The University of Leipzig refused to grant him a doctorate in law, purportedly due to his youth. Impatiently, Leibniz proceeded to the University of Altdorf, where he earned his doctor of law degree in 1667 for his thesis, *De Casibus Perplexis* ("On Perplexing Cases").

Leibniz turned down a chair at Altdorf, serving instead as secretary to the Nuremberg alchemical society until he met Baron Johann Christian von Boyneburg, who employed him as a secretary, librarian, lawyer, and adviser in Frankfurt starting in November 1667. During this time, he started work on his calculating machine, capable of multiplying, dividing, and extracting roots (by mechanically adding or subtracting repeatedly), representing a significant improvement over Blaise Pascal's calculator. Pierre de Carvaci, the royal librarian in Paris, requested a prototype in 1671, though Leibniz did not complete its construction until 1672.

That year, Leibniz and his patron traveled to Paris on a political mission, though Leibniz focused his attention on studying Cartesianism with CHRISTIAAN HUYGENS. In December 1672, Boyneburg died; bereft of employment, Leibniz accepted a political mission to London in January 1673, though again he focused more attention on his scientific career than on his assignment, as he gained induction into the Royal Society and met Isaac Barrow and Sir Isaac Newton.

Leibniz spent the years between 1673 and 1676 in Paris, vainly trying to secure a salary from the Académie des Sciences. In 1676, Johann Friedrich, the duke of Brunswick-Lüneberg, secured Leibniz's services as adviser, librarian, and genealogist. He also conducted scientific work on hydraulic presses, windmills, lamps, submarines, clocks, carriages, water pumps, and the binary number system.

In 1684, Leibniz published a paper explaining his differential calculus, *"Nova Methodus pro Maximus et Minimis"* ("New Method for the Greatest and the Least"), in the Leipzig journal *Acta Eruditorium* (its lack of proofs prompted Jacob Bernoulli to call it more of an enigma than an explanation). Two years later, he introduced his integral calculus in the same journal, complete with notation. It was not until the next year that Newton published his *Principia*; his calculus of fluxions, which he devised sometime between 1665 and 1671, was not published until an English edition appeared in 1736. This lag time between conception and publication complicated the issue of precedence.

In later years, Leibniz readily admitted Newton's influence on his own research on tangents and quadratures, the genesis of his discovery of differential and integral calculus. In 1699, Fatio de Duiller, a Swiss mathematician and fellow of the Royal Society, questioned the originality of Leibniz's discovery, and a dozen years later, the Royal Society endorsed Newton as the father of differential and integral calculus. However, evidence strongly suggests that Leibniz devised his calculus independent of Newton, despite the fact that he developed his calculus after meeting Newton. Leibniz considered himself and Newton "contemporaries in these discoveries."

In 1700, Leibniz helped found the Berlin Academy, which named him president for life. In 1712, Peter the Great of Russia appointed him a privy councilor, as did the Viennese court from 1712 through 1714 (though he had been seeking this position for years). However, these were minor victories for Leibniz, who sunk into relative obscurity later in his life. He developed a case of gout in about 1714 and died on November 14, 1716, in Hannover, Germany. Both the Royal Society and the Berlin Academy overlooked his death, and it was not until much later that his name was resurrected, and he was credited as the thinker who most clearly expressed the fundamentals of calculus.

Leopold, Aldo
(1886–1948)
American
Ecologist

Aldo Leopold is considered the father of wildlife ecology, largely on the strength of his book, *A Sand County Almanac, and Sketches Here and There*, published posthumously in 1949. Leopold was responsible for a sea change in the philosophical approach to the relationship between humans and nature, from a utilitarian approach based on maximizing the economic value of resources to a more holistic approach based on preserving the life-sustaining qualities of nature.

Leopold was born on January 11, 1886, in Burlington, Iowa, the eldest of four children born to Clara Starker and Carl Leopold. The landscape of his youth, on the bluffs of the Mississippi River where he tracked animals and hunted, played an important role in his developing respect for nature's value for sustaining all life, not simply for human gain. He attended the Sheffield Scientific School at Yale University, graduating in 1908. He conducted graduate work the next year at Yale's newly established School of Forestry, earning his master of forestry degree in 1909.

That year, the U.S. Forest Service hired Leopold as a ranger in the Arizona Territories (which included what is now New Mexico). The service rewarded his conscientious work with several prompt promotions—to deputy supervisor of Carson National Forest first, and then to supervisor of the forest. In 1913, he ascended to the post of assistant district forester in Albuquerque. The year before, on October 9, 1912, he had married Estella Bergere in her hometown of Santa Fe. Together, the couple had five children, who all became scientists like their father.

At this point, Leopold began his break with the conventional wisdom of conservation, which espoused the utilitarian philosophy, as voiced by the chief of the Forest Service (and fellow Yale alum) Gifford Pinchot, that nature existed for human exploitation. Leopold took the then-radical stance that nature held value in an undisturbed state by supporting the ecosystems that helped sustain human life on earth. He advanced these ideas in a 1915 game handbook he wrote for forest rangers.

Also in 1915, he established *The Pine Cone*, a newsletter for the Albuquerque game protection association he founded. The next year, the New Mexico Game Protection Association named him its secretary. In 1924, Leopold left the southwest to take up the post of associate director of the U.S. Forest Products Laboratory in Madison, Wisconsin. He left his mark on the departed region, though, as that year the Forest Service established the Gila Wilderness in New Mexico, the first federal wilderness reserve. Four years later, he left public service to work independently as a wildlife and forestry consultant.

Leopold's private work consisted primarily of conducting game surveys nationwide. This experience resulted in the writing of *Game Management*, which crystallized his wildlife ecology philosophy and proved his most influential book in his lifetime. Published in 1933, it coincided with his appointment to the professorship of wildlife management at the University of Wisconsin at Madison, a chair established specifically for him. The next year, President Franklin D. Roosevelt appointed Leopold to his Special Committee on Wild Life Restoration. That year also saw the passage of the 1934 Fish and Wildlife Coordination Act requiring environmental impact studies before development—credit for this law could be attributed directly to Leopold and his holistic ideology.

Leopold served his field administratively in the latter part of his career, as the Audubon Society appointed him as its director in 1935, the same year he founded the Wilderness Society. In 1947, the American Forestry Association elected him honorary vice president, the Ecological Society of America named him its president, and the Conservation Foundation appointed him to its advisory council.

Over the last 15 years of his life, Leopold recorded nature's changes in the landscape surrounding his farm on the Wisconsin River north of Madison and incorporated these observations into a series of poetic essays. Then, on April 21, 1948, Leopold died of a heart attack—appropriately, he was engaged with the forces of nature, fighting a brushfire on a neighbor's farm. After his death, his most famous book, *A Sand County Almanac*, was published. The book solidified a national movement toward the belief in sustainable conservationism and was influential in the passing of a string of important environmental federal legislation, including the 1969 National Environmental Policy Act, the Endangered Species Act of 1973, the Forest and Rangelands Renewable Resources Planning Act of 1974, the National Forest Management Act of 1976, and the Federal Land Policy and Management Act of 1976.

Lister, Joseph
(1827–1912)
English
Surgeon

Joseph Lister developed the antiseptic surgical techniques that greatly reduced the postoperative mortality rates of the day, which were in the 50 percent range. Although the surgical field was slow to acknowledge the efficacy of these techniques, they eventually revolutionized the practice of surgery, allowing wounds to heal without infection.

Lister was born on April 5, 1827, at Upton, in Essex, England. He was the fourth of seven children (and second of four sons) born to Isabella Harris, a former schoolteacher, and Joseph Jackson Lister, a physicist who earned his living as a vintner. When Lister was five years old, his father was inducted into the Royal Society in recognition of his construction of the achromatic microscope, and the elder Lister taught his son microscopy at an early age. In his youth, Lister attended private Quaker schools at Hitchin and then at Grove House in Tottenham.

In 1844, at the age of 17, Lister enrolled at University College, London (the only English college open to non-Anglicans). He earned his bachelor of arts degree within three years and remained at the university to study medicine. After overcoming a case of smallpox, he distinguished himself in his medical studies under the ophthalmic surgeon Wharton Jones and the physiologist William Sharpey, becoming house physician under W. H. Walshe and house surgeon under Sir John Erichsen. He earned his medical degree in 1852, when the Royal College of Surgeons admitted him into its fellowship.

Lister's first two published papers, "Observations on the Contractile Tissue of the Iris" and "Observations on the Muscular Tissue of the Skin," both appearing in 1853, established his reputation for excellent research internationally. Sharpey recommended Lister to professor of clinical surgery James Syme of the University of Edinburgh, where Lister became supernumerary house surgeon, then resident house surgeon, and then, in 1856, assistant surgeon at the Edinburgh Royal Infirmary. In April of that year, he married Syme's eldest daughter, Agnes; the couple had no children but maintained a long and happy marriage.

The Royal Society of London published a series of three papers by Lister in its *Philosophical Transactions* in 1858, resulting in his induction into the society's fellowship in 1860, at the age of 33. That year, the University of Glasgow appointed him regius professor of surgery, a position that carried no clinical duties. Within a year, however, he took charge of the surgical wards of the Royal Infirmary.

Despite Lister's vigilant supervision, the wards' postoperative mortality rates matched those of most hospitals at the time—about 50 percent. Surgical science had recently solved the problem of anesthetization, but it had not solved the problem of sepsis, which the German chemist Justus von Liebig proposed in 1839 to be combustion of body tissue when its moisture met oxygen. Lister was dubious of this explanation and considered sepsis to be a kind of decomposition of tissue.

In the early 1860s, LOUIS PASTEUR explained decay as a kind of fermentation caused by airborne organisms, a theory that supported and advanced Lister's own theory of sepsis. At about this same time, Lister learned of the antiparasitic action of carbolic acid, which destroyed entozoa (as well as the smell) in the sewage at Carlisle. Lister experimented with carbolic acid dressings, sprays, and putties through late 1866, and he humbly announced his results to the British Medical Association at its Dublin meeting in August 1867: his surgical wards at the Glasgow Royal Infirmary had remained free of sepsis for the last nine months, a development that defied the statistics of the time. Surgeons, stuck in their own dogma, resisted his antiseptic techniques until their efficacy became undeniable.

In 1869, Lister returned to Edinburgh to succeed his former mentor and father-in-law Syme (who had suffered a stroke) in the chair of clinical surgery. After eight happy years in his wife's hometown, Lister accepted the newly established chair of clinical surgery at King's College in London. The latter period of his career brought much recognition: he was knighted in 1883 and made a peer (Baron Lister of Lyme Regis) in 1897. From 1895 through 1900, he served as president of the Royal Society. Earlier, in 1891, he became chairman of the British Institute of Preventive Medicine (later known as the Lister Institute). Lister died on February 10, 1912, at Walmer, in Kent, England.

⊠ **Love, Susan**
(1948–)
American
Physician

Trained as a surgeon, Susan Love has devoted her career to the treatment of breast cancer. By the late 1980s, concerned that the usual avenues for breast cancer treatment were inadequate, Love began to write articles and a book for the general public on this issue. She also founded the National Breast Cancer Coalition, a lobbying group that pressed for greater funding for breast cancer research.

Born on February 9, 1948, in Little Silver, New Jersey, Susan Love is the daughter of Mr. and Mrs. James Love. Her father was a machinery salesman who was frequently transferred to new sales territories by the company he worked for. For this reason, Love went to middle and secondary schools in Puerto Rico and Mexico City. Love lives with her partner, Helen Cooksey, and has a daughter.

In 1966, Love enrolled in Notre Dame of Maryland, a small women's college run by the

Susan Love, a breast cancer surgeon, founded the National Breast Cancer Coalition. *(Courtesy of Susan M. Love)*

Sisters of Notre Dame. Love took premed courses at Notre Dame and for a short while became a novitiate nun. However, after six months, she quit and eventually transferred to Fordham University in New York City, where she finished her premed courses and earned a B.A. in 1970. After finishing her undergraduate degree, Love was admitted to the State University of New York's Downstate Medical College, located in Brooklyn, and graduated from there in 1974. She later completed a residency in surgery.

In 1980, Love entered private practice as a surgeon. Many of her cases were breast cancer patients, and Love soon realized that there was a role for her in this field. For several years, she worked as director of Beth Israel Hospital's breast clinic in New York City. In 1982, she became a surgeon specializing in cancer cases at the Dana Farber Cancer Institute's Breast Evaluation Center. In 1988, she moved to Boston to found the Faulkner Breast Center at Faulkner Hospital. The breast center was the first medical organization of its kind to have a cross-disciplinary, all-female staff. While at Faulkner, Love also worked as an assistant professor at the Harvard Medical School.

Love had developed tremendous experience with treating and counseling breast cancer patients, and she was unhappy with a number of approaches that were considered standard in this field. For instance, she felt that the typical treatment of surgery, radiation, and chemotherapy were used too often and that doctors did not do a good job laying out alternatives to this treatment to their patients. Also, she felt that there was far too little emotional support being given breast cancer patients by mostly male doctors.

To redress these problems, Love wrote a hugely popular book, aimed at a general audience of women, entitled *Dr. Susan Love's Breast Book* (1990). In her book, Love laid out alternative paths women could take if they discovered they had breast cancer. She suggested that women whose cancer had been detected early could opt for a removal of the cancerous node in their breast rather than a complete mastectomy, or complete removal of the breast, which was typically recommended by surgeons. The survival chances, Love noted, were about equal for both methods. Love has also questioned the need for mammograms, which are used to detect early breast cancers, for most women under the age of 50.

In 1992, Love moved to Los Angeles to direct the University of California at Los Angeles's (UCLA) Breast Center. She quit this job and briefly retired from medicine in 1996. However, she has returned to surgery as adjunct professor of surgery at the UCLA School of Medicine where she is working on a fiber-optic cable that can be threaded through the milk ducts in a woman's breast to look for early cancers.

Lucid, Shannon W.
(1943–)
American
Astronaut

Astronaut Shannon Lucid has set the record for the most flight hours in orbit—5,354 hours, or 223 days—of any woman in the world. This record made Lucid the most experienced American astronaut, male or female. In recognition of these feats, President Bill Clinton awarded Lucid with the Congressional Space Medal of Honor in December 1996, the first woman to receive it.

Shannon Wells was born on January 14, 1943, in Shanghai, China, where her parents were serving as missionaries. The family was imprisoned in a Japanese internment camp until a prisoner exchange freed them. They returned to the United States and settled in Bethany, Oklahoma, where she graduated from high school in 1960. Fascinated by Robert Goddard, the inventor of modern rocketry, she earned her pilot's license after high school in her first step toward her eventual career. She studied chemistry at the University of Oklahoma to earn her bachelor of science degree in 1963.

She remained at the university as a teaching assistant in the chemistry department until 1964, when she became a senior laboratory technician at the Oklahoma Medical Research Foundation. Next she served as a chemist at Kerr-McGee in Oklahoma City from 1966 until 1968. She then worked as a graduate associate in the Department of Biochemistry and Molecular Biology from 1969 until 1973. She conducted graduate study in biochemistry during this period, earning her master's degree in 1970 and her Ph.D. in 1973. She married Michael F. Lucid while still in school, and she delivered the second of her three children the day before an important examination, which she passed. After receiving her doctorate, she returned to the Oklahoma Medical Research Foundation as a research associate.

In January 1978, Lucid entered the first class of NASA's astronaut corps to admit women, and she became an astronaut in August 1979. She conducted numerous duties on ground before entering space. She first entered space aboard the space shuttle *Discovery* on a seven-day voyage in June 1985. She flew as a mission specialist twice aboard the shuttle *Atlantis,* in 1989 and in 1991, and once aboard the *Columbia,* in 1993. During these missions, she performed important tasks and conducted numerous experiments: she used the remote manipulator system to retrieve satellites; she activated the automated directional solidification furnace; she operated the shuttle solar backscatter ultraviolet instrument to map atmospheric ozone; she deployed the fifth tracking and data relay satellite; and she conducted neurovestibular, cardiovascular, cardiopulmonary, metabolic, and musculoskeletal medical experiments on herself as well as experiments on radiation measurements, polymer morphology, lightning, microgravity effects on plants, and ice crystal growth in space.

In March 1996, after a yearlong training regimen in Star City, Russia, she lifted off in the *Atlantis,* bound for the space station *Mir.* Onboard, she served as a board engineer 2, conducting experiments in the life sciences and the physical sciences as well as in diplomacy, as she coexisted with two Russian cosmonauts for 188 days. On this six-month mission, she traveled a total of 75.2 million miles, most of them at 17,300 miles per hour while 250 miles above the earth's surface.

Upon her return aboard the *Atlantis,* which was twice delayed, Lucid's own body served experimental purposes as scientists studied the effects on the human body of prolonged weightlessness from living in space. Amazingly, Lucid walked on her own two feet upon disembarking into the earth's gravitational pull. While in space, Lucid maintained her connection to earth by eating M&M candies, a story that made for interesting media coverage and probably increased sales of the chocolate confections.

Lumière, Auguste and Louis
(1862–1954; 1864–1948)
French
Cinematographers

Auguste and Louis Lumière invented the cinematograph, a significant improvement over THOMAS EDISON's kinetoscope. The Lumière apparatus acted as a camera, printer, and projector; this last component proved to be the key to their innovation, as it allowed for the public screening of films. Not only did the brothers film the first true motion pictures, but also they hosted the first public screening of cinematographic films, which titillated their audience and established the motion picture industry that would transform public perception in the 20th century. Throughout their career, the brothers made 1,425 film "shorts," or brief scenes.

Auguste-Marie-Louis-Nicolas Lumière was born on October 19, 1862, and his brother, Louis-Jean Lumière, was born on October 5, 1864. In 1870, their father, the painter-turned-photographer Antoine Lumière, moved his family from Besançon in eastern France to Lyon, beyond the threat of the Prussian army. There, he established a photographic studio in the town's center while sending his sons to La Martinière, a technical high school.

When the Lumière brothers graduated, they joined their father's business, contributing their youthful innovation. In 1880, Louis invented an improved dry-plate photo process, known as *Etiquette bleue,* which caught on immediately. Within two years, Antoine financed the purchase of a large plot of land in the Montplaiser section of Lyon, where the family business established a factory to manufacture the plates; by 1894, the company produced some 15 million plates per years.

That year, Antoine attended a demonstration of Thomas Edison's kinetoscope in Paris. His description of the contraption failed to impress his sons, who were convinced they could improve upon Edison's peephole viewer. Interestingly, Edison had neglected to include a mechanism for projection when he filed his patent in 1888, perhaps because he was preoccupied convincing GEORGE EASTMAN to create a celluloid film for him.

The Lumière brothers devoted themselves to improving Edison's motion-picture camera. They ended up inventing the cinematograph, which filmed at 16 frames per second and weighed significantly less than the kinetoscope. What is more, it functioned not only as a camera but also printed the film and then projected it on a screen large enough for public viewing.

Louis and Auguste set about to experiment with filming, using scenes from everyday life as their subjects. On March 19, 1895, they set up their cinematograph camera outside their photo plate factory on what is now known as the Rue du Premier Film. There, they made what is considered the first film in history, *"La Sortie des Ouvriers de l'Usine Lumière"* ("Workers Leaving the Lumière Factory"). Later in 1895, they filmed *"L' Arrivée d'un Train à la Ciotat"* ("The Arrival of a Train at la Ciotat"), as well as numerous other films.

On December 28, 1895, the Lumières hosted the first public film screening, consisting of 10 shorts, lasting about 20 minutes in all, in the basement lounge of the Grand Café on the Boulevard des Capucines in Paris. They had low expectations for the reception of their new technology, as they had simply filmed scenes that viewers would otherwise encounter in their daily lives. Much to their surprise, the audience screamed in fear of the approaching train, which they had filmed diagonally to appear as it were coming right at the screen. Suspension of disbelief reigned, as many in the crowd thought the train would run them over!

The Lumières made more than 40 shorts the next year, continuing to chronicle everyday scenes—a game of cards, a blacksmith at work, a mother nursing her baby, soldiers marching, and the like. They also experimented with humorous shorts, as well as newsreels and documentaries, including four films detailing the Lyon fire department. They even filmed the arrival of participants at the French Photographic Society Conference, screening the results 48 hours later to a large gathering.

While Auguste eventually left cinematography in order to conduct medical research, Louis continued to innovate improvements—he invented the photorama, which took panoramic shots, and, in 1907, a color-printing process that employed dyed starch grains. He even experimented with aerial shots before the advent of airplanes and later expanded into the realms of stereoscopy and three-dimensional filming. Louis died on June 6, 1948, in Bandol, France, and Auguste died on April 10, 1954, in Lyon.

M

Magellan, Ferdinand
(1480?–1521)
Portuguese
Explorer

Ferdinand Magellan led the first expedition that successfully circumnavigated the world, thereby proving beyond a shadow of a doubt that the earth is a globe. Tragically, Magellan did not make it back to the Spanish port from which he departed, as he died at the hands of native Filipinos whom he had mistreated. Despite the fact that he did not survive the circumnavigation himself, he is remembered as the author of this venture. His name also lives on in his geographic discovery of the Strait of Magellan, at the southernmost tip of South America.

Magellan was born Fernão de Magalhães in about 1480 in Oporto, Portugal. His parents, Alda de Mesquita and Rui de Magalhães, belonged to the Portuguese nobility. He served as a page to Queen Leonor in Lisbon in his youth, then went to sea in the fleet of Francisco de Almeida, the first Portuguese viceroy of the east, leaving Lisbon on March 25, 1505. Gaspar Correia reported that Magellan was wounded in a sea battle at Cannanore off the Malabar coast of India, though little is known of his early life as a sailor.

In November 1506, Magellan joined Nuno Vaz Pereira to establish a Portuguese outpost on the coast of Mozambique. Next, he took part in the infamous Battle of Diu in India on February 2 through 3, 1509. At Malacca, he saved the life of the Portuguese explorer Francisco Serrão, who repaid him later by providing information about the Molucca Islands. In late June 1511, he returned to Malacca in Albuquerque's fleet, which captured the territory within six weeks, thereby establishing Portuguese dominance in the region.

The year 1513 found Magellan in Morocco fighting at Azamor, when he sustained yet another wound, this one causing him to limp the rest of his life. Having sacrificed his body for his country, he requested a slight raise in pension (and in rank) in November 1514, but King Manuel refused. Magellen submitted his request again in early 1516, and yet again the king refused.

On October 20, 1517, Magellan landed in Seville, where he joined the Portuguese cosmographer Rui Faleiro on a journey to the royal court at Valladolid to offer their services to the Spanish King Charles I (later Emperor Charles V). When the king accepted their offer, Magellan renounced his nationality and hispanicized his name to Fernando de Magallanes.

Magellan and Faleiro had won over the king by proposing to find a western route to the Spice Islands, to discover if any of them existed on the Portuguese side of the Treaty of Tordesillas's line of demarcation, which in 1494 divided the world between Spanish and Portuguese hemispheres of influence. In 1517, Magellan married Beatriz Barbosa, who bore him a son, Rodrigo, before he departed. King Charles appointed Magellan and Faleiro joint captains general on March 22, 1518, granting them (and their heirs) rights to any lands they discovered, as well as one-twentieth of the voyage's net profits. Charles also appointed them to the Order of Santiago.

Charles outfitted the expedition with five ships—the *Trinidad, San Antonio, Concepción, Victoria,* and *Santiago*. Before they departed from Sanlúcar de Barrameda, the Portuguese tried to sabotage their ships, rendering them barely seaworthy. Nevertheless, on September 20, 1519, Magellan set sail, in command of some 270 men of nine nationalities, including his brother-in-law Duarte Barbosa, and João Serrão as commander of the *Santiago*, but without Faleiro, who had succumbed to insanity.

By October 3, Magellan had reached Brazil. After mistaking the St. Christopher River for the strait he was searching for, he retraced his route and continued south, reaching Port St. Julian on March 31, 1520. At midnight on Easter day, Spanish captains mutinied, though Magellan squelched the rebellion, ordering the execution of two of the captains—Luis de Mendoza, who was stabbed to death before his crew, and Gaspar de Queixada, who was decapitated and quartered.

The fleet wintered at St. Julian from March through August 1520; on August 24, they departed, leaving the *Santiago* wrecked at the mouth of the Santa Cruz River. On October 21, they rounded Cape of the Virgins, thereby discovering the Strait of Magellan. After spending a month traversing the strait (during which time they deserted the *San Antonio*), the three remaining ships entered the Sea of the South, which they called the Pacific Ocean after its calmness, on November 28.

Magellan navigated across the Pacific Ocean, spending 99 days without replenishing food or water; the dehydrated, scurvy-ridden crew resorted to eating rawhide. Finally, they landed on the island of Guam on March 6, 1521, departing again three days later, bound for the Philippines, where Magellan converted some 800 natives to Christianity. However, after he burned a village of natives who refused to convert, he ordered another village at Cebú on Mactan Island to send him provisions; when that village offered different terms from those he set, Magellan attacked them, and he was killed in the ensuing skirmish on April 27, 1521.

Magellan's fleet, reduced to a crew of merely 17 of those who originally set out, continued on without him, under the command of Elcano, originally captain of the *Concepción* and a participant in the St. Julian mutiny. They reached Spain on September 8, 1522, and King Charles granted Elcano a coat of arms with a globe bearing the inscription, *Primus circumdesisti me* ("You were the first to circumnavigate me"). However, history has ascribed credit to Magellan for leading the first circumnavigation of the earth.

Maiman, Theodore
(1927–)
American
Physicist

Theodore Maiman invented the laser, a high-intensity beam of radioactive light capable of performing a wide range of tasks, from measuring the distance from the earth to the moon with great accuracy to reading digital information on compact discs. After patenting the laser in 1960, Maiman proceeded to found several companies to develop different applications for the laser, thus fueling the proliferation of this technology that became ubiquitous by the end of the 20th century.

Theodore Harold Maiman was born on July 11, 1927, in Los Angeles, California. He followed in the footsteps of his father, an electrical engineer who invented a vibrating power supply system for cars, by entering the field of engineering. After serving in the U.S. Navy briefly in 1945, he matriculated at the University of Colorado, earning his tuition by working on electronics equipment and radios. He graduated in 1949 with a bachelor of science degree in engineering physics and then pursued graduate work at Stanford University. He earned his master of science degree in 1951, then he conducted doctoral work under Willis E. Lamb, the Nobel laureate in physics in 1955, the same year that Maiman received his Ph.D. in physics.

Maiman joined Hughes Research Laboratories (of Hughes Aircraft Company), where he rose to the rank of section head while conducting the research that resulted in his invention of the laser. In 1953, Charles H. Townes had set the groundwork for the laser with his invention of the maser, or microwave *amplification by stimulated emission of radiation*, which he had developed in collaboration with the Soviet scientists Nikolai Basov and Aleksandr Prokhorov. The maser created coherent, monochromatic microwaves; Maiman followed up on Townes's work by making several improvements to the maser design while also becoming interested in applying the same technological theory to optics.

Townes and Arthur L. Schawlow had proposed the theory of the laser, or *light amplification by stimulated emission of radiation*, in a published paper, and they actually commenced construction of a device for producing this coherent, high-intensity, monochromatic light. However, the team tried to propagate light in alkali vapor, a gas medium. Maiman realized that a solid medium might yield more promising results, a perspective supported by the work of Nicolas Bloembergen, Satoro Sugano, Yukito Tanabe, and I. Wieder. Maiman adapted their solid-medium approach to ruby, a corundum mineral laced with the impurity of chromium, but his experiments failed to yield success.

Maiman's breakthrough moment occurred when he realized the inaccuracy of previous calculations predicting the amplification necessary to cross the threshold for creating a high-intensity beam of light. Based on his own corrected calculations, Maiman designed and built a cylinder of synthetic ruby crystal, coating the ends with reflective silver that allowed light from a xenon lamp into the rod. This light bounced back and forth between the mirrored ends, one of which was semitransparent. When this light reached sufficient intensity, it broke through the reflective barrier, exiting the rod as a coherent, monochromatic beam of red light, at a wavelength of 6,943 angstroms.

In 1960, Maiman secured patent number 3,353,115 for his ruby laser, which emitted light in pulses; the next year, Ali Javan of the Bell Telephone Laboratories introduced the first continuous-beam laser. In 1962, Maiman left Hughes to found the Korad Corporation, which led the new field of laser technology by developing diverse applications. He served as Korad's president until Union Carbide acquired the company in 1968, at which point he established Maiman Associates, a firm that offered consulting services for lasers.

In 1972, Maiman founded yet another company, Laser Video Corporation, serving as its vice president for the next three years. The company applied laser technology to large-screen color video screens with laser displays. In 1976, TRW Electronics hired Maiman as its vice president for advanced technology. Throughout this time, the laser continued to find new and innovative uses, as a light scalpel in microsurgery, and as a scanning device for replaying music and retrieving information on compact discs. Maiman told his own version of the story of the development of the laser in his book, *The Laser Odyssey*, which gave a behind-the-scenes glance at the intrigues of scientific invention.

Maiman was amply recognized for the significance of his invention: in 1962, he received the Ballantine Medal from the Franklin Institute; in 1966, he won both the Buckley Solid State Physics prize from the American Physical Society and the Fanny and John Hertz Foundation Award for Applied Physical Science, which Lyndon B. Johnson presented to him at a White House ceremony. In 1984, the Inventors Hall of Fame inducted him into its membership. At the turn of the century, *Time* magazine named him to its list of 100 contributors of significant advances in "A Century of Science," and *Business Week* acknowledged him in its "100 Years of Innovation."

Marconi, Guglielmo
(1874–1937)
Italian
Physicist, Engineer

Although Guglielmo Marconi did not discover radio waves, he harnessed them to practical use by inventing and patenting devices for the transmission and reception of "wireless" communications, as they were then called. Marconi received the 1909 Nobel Prize, the first Italian so honored in physics, with Karl F. Braun, who increased the range of radio transmission fivefold.

Marconi was born on April 25, 1874, in Bologna, Italy. His father, Giuseppe Marconi, a wealthy landowner, had two older sons—Luigi from his first marriage and Alfonso from his second marriage to Annie Jameson, the daughter of the Irish whiskey distiller. On March 16, 1905, Guglielmo Marconi married Beatrice O'Brien of the Irish aristocracy, and together the couple had three daughters—Lucia, who died in infancy; Degna; and Goia Jolanda—and one son, Giulio. Their marriage ended in a 1927 annulment, and on June 12 of that year Marconi married Maria Bezzi-Scali of papal nobility. They had one daughter, Elettra.

Marconi's formal education was limited to study at the Instituto Cavallero in Florence and at the Technical Institute of Leghorn. More important was his private tutoring in physics by Professor Vincenzo Rosa. On his own Marconi began experimenting with HEINRICH RUDOLF HERTZ's "invisible rays" or "electric waves." In 1895, he succeeded in transmitting a signal the one-and-a-half-mile length of his father's estate, Villa Grifone. Unable to interest the Italian government in his innovation, Marconi traveled in February 1896 to London, where his cousin, Henry Jameson Davis, aided him in preparing a patent application that was accepted on July 2, 1897.

That year Marconi formed the Wireless Telegraph and Signal Co., Ltd., which he renamed Marconi's Wireless Telegraph Co., Ltd., in 1900. The previous year he had achieved the first international radio transmission of 85 miles across the English Channel from Chelmsford, England, to Wimereux, France. Scientific belief at the time held that radio waves traveled in straight lines, thus hampering long-distance transmission.

Marconi believed otherwise, that the waves would bend with the Earth's curvature, a hypothesis he set out to prove by setting up a transmitter at Chelmsford, England, and a receiver on Cape Cod, Massachusetts. When a storm destroyed the Cape Cod antenna, Marconi traveled to St. John's, Newfoundland, to fly a kite 400 feet high as an antenna. On December 12, 1901, the kite worked, receiving the transmission over the 1,800 miles from Chelmsford of the Morse code for the letter S.

Marconi subsequently attempted to increase the distance he could transmit radio signals by investing interest first in short wave, setting up a global short-wave system through the British government by 1927, and later in microwaves, which he began experiments with in 1932.

Besides the Nobel Prize, Marconi was honored with the title of marquis in 1929 and with the presidency of the Royal Italian Academy in 1930. His most treasured honor, though, was a gold tablet given to him by 600 survivors of the *Titanic*, who felt they were saved by his wireless technology. Marconi died on July 20, 1937, in Rome. In Britain telegraph and wireless offices marked the time of his funeral with two minutes of dead air in his honor.

Marcy, Geoffrey
(1954?–)
American
Astronomer

Geoffrey Marcy has discovered 19 out of 29 "extrasolar" planets (or planets outside of our solar system). He devised the "wobble" method of detecting Jupiter-sized planets (those 318 times the mass of the earth, the smallest "visible" by this technique) by looking for shifts in the motion of stars, indicating a gravitational pull on them that could signal the orbiting of a planet (or planets) calculable by simple Newtonian laws of physics. In 1999, he discovered a three-planet system corresponding to our own solar system. Marcy's work has ushered in a new era in astronomy, adapting an interdisciplinary approach to the search for planets with life outside our solar system.

Geoffrey W. Marcy was 14 years old when his parents bought him a telescope; his mother remembered him setting his alarm clock for two o'clock in the morning to get up and observe Saturn, his favorite planet. When he attended the University of California at Los Angeles, he decided to double-major in physics and astronomy, and he graduated summa cum laude and Phi Beta Kappa in 1976. He then pursued doctoral work in astronomy and astrophysics at the University of California at Santa Cruz, earning his Ph.D. in 1982.

Marcy conducted postdoctoral research as a Carnegie Fellow at the Carnegie Institution of Washington from 1982 through 1984. That year, he started his professional career as an associate professor of physics and astronomy at San Francisco State University, a school that was not distinguished for its astronomy program. Furthermore, Marcy felt uninspired by the mundane direction of astrophysical research at the time; nevertheless, he continued to conduct such research and write papers to earn tenure as a full professor.

While showering one day, Marcy experienced an epiphany: he determined to focus his research on the search for extraterrestrial life, a project that earned the derision and scorn of many colleagues. He suffered the "E.T." jokes (after the then-popular Steven Spielberg movie about creatures from space) good-naturedly while he attenuated his astronomical instruments to make them a thousand-times more sensitive than those of his fellow astronomers. Even with these adjustments, he could not see extrasolar planets telescopically, so he devised an ingenious means of detecting their existence.

The sun reacts to the gravitational pull of the nine planets orbiting it, shifting back and forth ever-so-slightly in ways calculable by straightforward Newtonian physics. Marcy dubbed this shifting "wobble" and applied this logic to his search for extrasolar planets by looking for shifts in the Doppler readings of 120 nearby stars in the Milky Way. Unfortunately, the Swiss astronomer Michel Mayor beat him to the punch, announcing in 1995 his discovery of a planet orbiting the star known as 51 Pegasi.

Marcy and his colleague R. Paul Butler hurriedly examined the data they had amassed over the previous seven years, and they realized, ironically, that they had the information on hand to announce the discovery themselves. However, they did find two more new planets in their data, announcing their discovery in January 1996.

The media flocked to them, as ABC News Hour named Marcy its "Person of the Week" that month, *Newsweek* named him one of its 100 Americans for the Next Century, and cover stories appeared in *Time, the Washington Post,* and the *New York Times* over the next few years. This coverage hailed a new era in astronomy with the advent of an interdisciplinary search for new planets and for evidence of life on them.

San Francisco State University named Marcy a Distinguished University Professor in 1997, while Marcy continued to survey 107 stars, undaunted by his celebrity. Over the next several years, he discovered two-thirds of the new planets. In January 1999, he announced a new planet around a star in the Serpens constellation, but this was just a teaser for his discovery announced three months later.

In April 1999, Marcy and Butler announced the detection of two more planets in addition to the one they had discovered earlier orbiting Upsilon Andromedae, a star 44 light-years from the earth and about three billion years old, or two-thirds the age of our sun. The pair was joined by another group of astronomers from the Harvard-Smithsonian Center for Astrophysics in Cambridge, Massachusetts, and the High Altitude Observatory in Boulder, Colorado, who made the same discovery simultaneously but independently, substantiating Marcy and Butler's results. The announcement captured the public imagination, as it confirmed the existence of another multiple-planet system orbiting a normal star, the first such conglomeration of planets analogous to our solar system.

On November 7, 1999, while awaiting the publication of their article, "Evidence for Multiple Companions to Upsilon Andromedae," in the December 1999 issue of *The Astrophysical Journal,* Marcy's team recorded a 1.7-percent decrease in Upsilon Andromedae's brightness, corresponding with their prediction that the largest of the three planets would cross between the star and the earth. This observation corroborated their "wobble effect" findings, removing the shadow of a doubt in their discovery.

The University of California at Berkeley did not require such evidence to appoint Marcy as a professor of astronomy and head of its newly established Center for Integrative Planetary Studies, an interdisciplinary program combining chemistry, biology, physics, and astronomy in the search for life beyond the solar system. The state of California named him its Scientist of the Year in 2000. However, instead of basking in the limelight, Marcy is busy promoting the use of the planned 2005 launch of the NASA Space Interferometry Mission to search for earth-sized planets, which cannot be detected from the earth with current technology, orbiting other stars.

⊠ Matzinger, Polly Celine Eveline
(1947–)
American
Immunologist

Polly Matzinger was a relative latecomer to science—she did not begin graduate work until she was 29—but when she finally decided that science could be an interesting career, she made up for lost time in a hurry. Working as a researcher in several labs in Europe and the United States, Matzinger developed a theory that the body's immune system responds to danger rather than what it detects as being foreign. This theory, called the danger model, challenges a longstanding model—the self-nonself model—about how the immune system works and has caused a great deal of controversy in the medical world.

Born on July 21, 1947, in La Seyne, France, Matzinger is the daughter of Simone and Hans Matzinger. Her father, who is Dutch, was a World War II resistance fighter, and her mother, who is French, is a former nun. Matzinger has never married. Her children are her four dogs—Annie, Lilly, Charlie, and Roy.

After finishing her secondary education in California in 1965, Matzinger worked at a series of odd jobs in that state while occasionally attending college. By the early 1970s, she was working as a cocktail waitress near the University of California at Davis. At her job, she met a group of science professors who often came to the bar after work for drinks and began engaging them in arguments about science topics. One of these scientists, Robert "Swampy" Schwab, encouraged Matzinger to return to the university to study science. Matzinger heeded Schwab's call and, in 1976, graduated with a B.S. in biology from the University of California at Irvine. After receiving her undergraduate degree, Matzinger enrolled in the graduate school at the University of California at San Diego. She earned a Ph.D. in biology from that institution in 1979.

For 10 years, Matzinger worked in Europe, first on a fellowship from the U.S. National Institutes of Health that sent her to Cambridge University's Department of Pathology and then as a researcher at the Basel Institute for Immunology in Switzerland. As a doctoral student, Matzinger had been puzzled about the existing theory—the self-nonself theory—that explained how the immune system worked. For instance, she reasoned, if T cells, which are a type of blood cell and a major component of the body's immune system, attack a foreign substance such as a virus or a skin graft, why does it not also attack equally foreign substances such as food in the digestive tract or a fetus in a mother's womb? The self-nonself theory had no answer for this question, yet it was still considered by most immunologists as the best explanation of the phenomena of immune response. With no encouragement from her peers, Matzinger set aside her doubts and tried to understand how the body made the self-nonself discrimination.

By the early 1990s, and now working at the U.S. National Institutes of Health (NIH)'s Laboratory of Cellular and Molecular Immunology, Matzinger decided to revisit this question. Prod-

Polly Matzinger's theories are changing the way scientists view the immune system and may liberate organ transplant recipients from the necessity of using dangerous drugs. *(Courtesy Polly Matzinger)*

ded by discussions with Ephraim Fuchs, who was questioning why the immune system did not react against cancers, Matzinger put together the elements of her theory of immune response. T cells are activated only if another type of cell, called dendritic cells, warns them that injury has occurred to other, formerly healthy cells. Whether a cell has originated from inside the body or is from outside (that is, foreign) is not relevant. What matters is that it is damaged.

Matzinger's theories have begun to be tested clinically. The most interesting results so far have been in the field of organ transplants. By treating an experimental group of rats that have received skin grafts with drugs that block the response of dendritic cells, doctors have enabled the rats to accept these grafts without traditional immunosuppressant drugs. By treating a group of monkeys with these same blocking drugs, another group of researchers has found a way to transplant kidneys in monkeys. The most immediately interesting new predictions are in cancer treatment. By injecting cancer with bacteria that contain a heat shock protein, some doctors have noted dramatic improvement in patients' conditions and signs that T cells have been activated to fight the cancer.

As of 2001, Matzinger continues to work at the NIH's Laboratory of Cellular and Molecular Immunology. For her insights about the immune system, she has been elected a lifetime honorary member of the Scandinavian Society of Immunology.

Mauchly, John William
(1907–1980)
American
Computer Engineer

John William Mauchly conceived of the first digital electronic computer, and, in collaboration with J. Presper Eckert, he designed and built a series of digital electronic computers: the Electronic Numerical Integrator and Computer (ENIAC), the Electronic Discrete Variable Automatic Computer (EDVAC), the Binary Automatic Computer (BINAC), and the Universal Automatic Computer (UNIVAC). Despite the fact that computer scientists overwhelmingly attribute the invention of the digital electronic computer to Mauchly, he lost an extended patent dispute over the issue, and therefore he has not received the recognition he deserves as one of the people primarily responsible for ushering in the digital era of the information age.

Mauchly was born on August 30, 1907, in Cincinnati, Ohio. His mother was Rachel Scheidemantel, and his father, Sebastian J. Mauchly, was an electrical engineer. When the Carnegie Institute hired his father to head its section of terrestrial electricity and magnetism in 1915, Mauchly's family moved to Washington, D.C. There, he attended McKinley Technical High School. In 1925, Mauchly commenced his study at Johns Hopkins University in the engineering department, but he soon transferred to physics.

A distinguished student, Mauchly entered Johns Hopkins's graduate program in physics in 1927, after only two undergraduate years. The university retained him as a research assistant for a year after granting his Ph.D. in physics in 1932. In 1933, Ursinus College in Collegeville, Pennsylvania (near Philadelphia), appointed Mauchly as head of its physics department. In the meanwhile, he had married Mary Augusta Walzl in 1930, and together the couple had two sons.

Mauchly's early interest in meteorology was counterbalanced by the tediousness of the computations required to compile accurate information for predicting weather patterns. Rather than abandon this field, he endeavored in the early 1930s to invent a machine for making calculations using vacuum tubes. He presented his idea to the American Association for the Advancement of Science at its December 1940 meeting, eliciting a response from Iowa State University's John Atanasoff, who was building an electronic computer. Mauchly visited Iowa in June 1941, only to find the prototype incomplete (Atanasoff never did finish it) and capable of calculating only linear equations.

With the advent of World War II, Mauchly studied electrical engineering that summer at the University of Pennsylvania's Moore School of Engineering, which held a contract with the U.S. Army for computing trajectory tables for artillery shells (the existing tables needed recalculation due to new battle conditions and weaponry). Mauchly soon realized that this task and his meteorological calculations were one and the same, prompting him to propose his vacuum-tube computer design for this project as well. In August 1942, he sent Moore administrator John Grist Brainerd a five-page memorandum, entitled "The Use of High-Speed Vacuum Tube Devices for Calculating," which resulted in the army extending its contract to cover Mauchly's proposal as of August 1943.

Mauchly collaborated closely with Presper Eckert, who acted as the chief engineer while Mauchly served as the principal consultant; other contributors included A. W. Burks, T. K. Sharpless, R. Shaw, and some 50 other scientists. They designed and constructed the ENIAC, the first digital electronic computer. It filled a 30-by-50-foot room and weighed 30 tons; it had 18,000 vacuum tubes, 70,000 resistors, and 5 million soldered joints; and it was capable of performing addition in 200 microseconds and multiplication in 300 microseconds. However, the computer did not go into operation until December 1945, too late to contribute to the war effort. The first trial took place at the U.S. Army's Aberdeen Proving Ground in Maryland, where it calculated ballistics tables and solved trajectory problems; subsequently, the ENIAC was employed to perform calculations for the construction of the hydrogen bomb.

Mauchly and Eckert resigned from the Moore School in protest over its patent policy, leaving behind their pioneering work on the EDVAC, which improved upon the ENIAC design by storing information for later retrieval (the ENIAC had to be reprogrammed after each calculation). They founded their own company, the Electronic Controls Company (later renamed the Eckert-Mauchly Computer Corporation in 1948), in order to apply for a patent (which was not granted for another 17 years!). Sadly, in September 1946, Mauchly's wife drowned while swimming in the Atlantic Ocean. He remarried on February 7, 1948, to Kathleen R. McNulty, a programmer on the ENIAC project. The couple eventually had four daughters and one son.

Mauchly and Eckert continued to innovate new applications for digital electronic computing. In 1947, the Northrop Aircraft Company hired them to design a small-scale binary computer, and they responded with the BINAC, which used magnetic tape instead of punch cards for inputting information, allowing for internal storage of the information. In the early 1950s, the company designed and built the UNIVAC, the first commercial digital electronic computer, which was tested by the U.S. Census Bureau in 1951.

By then, however, Mauchly and Eckert (both bad businessmen) had mismanaged the company, forcing them to sell (along with the patent claims) to Remington Rand in February 1950. Mauchly served as director of UNIVAC applications research for Remington Rand (which merged with the Sperry Corporation in 1955 to form Sperry Rand) from 1950 through 1959. That year, he resigned to establish his own consulting firm, Mauchly Associates. A decade later, in 1968, he founded another computer consulting firm, Dynatrend.

In February 1964, the U.S. Patent Office finally granted the ENIAC patent to Illinois Scientific Developments, a subsidiary of Sperry Rand, which sued the Honeywell Corporation three years later for patent infringement. In its defense, Honeywell claimed Atanasoff as the rightful inventor of the digital electronic computer. In his October 1973 decision, Judge Earl Larson of Minneapolis revoked Mauchly and Eckert's patent, in part granting priority to Atanasoff, who had not made such a claim until 1954, not of his own accord but on the advice of an IBM lawyer. Dejected, Mauchly returned to Sperry Rand in 1973.

Although the legal fracas prevented Mauchly from receiving proper recognition from the mainstream for his invention of digital electronic computer technology, he did gain such recognition in his field: he won the 1949 Howard N. Potts Medal, the 1961 John Potts Award, and the 1966

Harry Goode Award. A lifelong sufferer of hemorrhagic telangiectasia, afflicting him with both internal and external bleeding (such as bloody noses), Mauchly died during heart surgery on January 8, 1980, in Abington, Pennsylvania.

Maxwell, James Clerk
(1831–1879)
Scottish
Physicist

Regarded as one of the preeminent physicists of the 19th century, James Clerk Maxwell explored an array of scientific topics. His most renowned works were his unified theory of electromagnetism and his kinetic theory of gases. Building upon the groundbreaking research of Michael Faraday, Maxwell showed the connection between electricity and magnetism. Maxwell wrote four differential equations (known simply as Maxwell's equations) describing the transmission of electromagnetic waves. When taken together, these equations provided a complete description of how electric and magnetic fields were produced and how they were related. These equations influenced ALBERT EINSTEIN in his formulation of the theory of relativity. Maxwell's studies of gases and his theory that gases were composed of molecules in random motion led to the postulation of the Maxwell-Boltzmann distribution law, a statistical formula for determining a gas molecule's probable energy level.

Maxwell was born in Edinburgh on June 13, 1831, to John Clerk Maxwell and Frances Kay Maxwell. After his mother died when he was eight, his father, a lawyer, raised Maxwell with a profound interest in technical matters. Indeed, the elder Maxwell published a scientific paper and was a member of the Royal Society of Edinburgh. It was from his father that Maxwell inherited both his methodical manner and the family estate in Scotland, to which he retired to undertake the bulk of his scientific writing. He married Katherine Mary Dewar in 1858; the couple had no children.

Maxwell's talents were recognized early. At the age of 14 he published his first paper, an examination of the optical and geometric properties of ovals. He entered the University of Edinburgh in 1847, when he was 16, and obtained a mathematics degree from Trinity College, Cambridge, in 1854. At Cambridge William Hopkins influenced him.

Upon his graduation Maxwell held professorships of natural philosophy at Marischal College, Aberdeen (1856–1860), and King's College, London (1860–1865). He resigned in 1865 and retired to research and write at his family estate. In 1871, colleagues persuaded him

James Clerk Maxwell is best known for demonstrating the connection between electricity and magnetism. *(AIP Emilio Segrè Visual Archives)*

to return to the world of academia, and he accepted an appointment as the first Cavendish Professor of Experimental Physics at Cambridge. Over the course of his career Maxwell published four books and 100 papers on a variety of topics. His early work was devoted to such divergent problems as the mechanism of color vision and the composition of Saturn's rings. He concluded in 1849 that all colors were derived from the three primary colors—red, yellow, and blue—and in 1861 he produced the first color photograph. From 1855 to 1859 he showed that Saturn's rings were neither fluid nor solid, as had heretofore been supposed, but rather were composed of small bodies in orbit.

Maxwell's most significant work was conducted during the 1860s. It was then that he formulated an electromagnetic theory and a kinetic theory of gases. He illustrated that oscillating electric charges (that is, charges fluctuating at regular intervals between the highest and lowest values) would produce waves transmitted through the electromagnetic field. He concluded that light was an electromagnetic wave, as were infrared and ultraviolet radiation, and he postulated the existence of additional kinds of electromagnetic radiation. This theory was supported in 1888 when HEINRICH RUDOLF HERTZ discovered radio waves. In 1864 Maxwell published *Dynamical Theory of the Electric Field*, which contained his famous differential equations. One of the more influential aspects of these equations was that Maxwell established that the speed of a wave was independent of the velocity of its source.

Maxwell's research into gases built upon the existing idea that gas consisted of constantly moving molecules colliding with one another. In 1860 Maxwell (and Ludwig Eduard Boltzmann independently) employed statistical methods in the Maxwell-Boltzmann distribution to account for the wide variation in the velocities of the various molecules in gas. Maxwell divined that this variation resulted from temperature. He also proposed that heat from the motion of the gas molecules is stored in a gas, a theory that led to explanations of the viscosity and diffusion of gases.

Maxwell's untimely death of abdominal cancer in 1879 cut short his further investigation into electromagnetism. His innovations, nevertheless, were considered to be on a par with those of Sir ISAAC NEWTON and Albert Einstein.

McClintock, Barbara
(1902–1992)
American
Geneticist

Working in her cornfields while other geneticists investigated molecules, Barbara McClintock was ignored for decades because her discoveries ran as counter to the mainstream as her methods. In the end, however, she proved that, contrary to what almost everyone had thought, genes could move and control other genes. Organisms thus could partly shape their own evolution.

Barbara preferred her own company almost from her birth in Hartford, Connecticut, on June 16, 1902. As a baby she played happily by herself; as a teenager she liked to spend time simply "thinking about things." She grew up in Brooklyn, then a somewhat rural suburb of New York City, to which her father, a physician for Standard Oil, moved the family when she was six. Her parents, Thomas and Sara McClintock, encouraged independence in their four children by allowing them to skip school.

Barbara became determined to attend college, even though, as she said later, her mother feared that a college education would make her "a strange person, a person that didn't belong to society. . . . She was even afraid I might become a college professor." Thomas McClintock took Barbara's side, however, and in 1919 she enrolled in the College of Agriculture at

Barbara McClintock's pioneering studies of corn genetics demonstrated that genes could move and could control other genes, suggesting that living things may have a hand in their own evolution. *(Cold Spring Harbor Laboratory Archives)*

Cornell University, which offered free tuition to New York residents.

McClintock took a course in the relatively new science of genetics in her junior year, and by the time she graduated in 1923 she had decided to make genetics her career. As a graduate student in the university's botany department she studied Indian corn, a type of maize (corn) in which the kernels on each ear have different colors. The color pattern is inherited.

McClintock made her first important discovery while still a graduate student. Geneticists were realizing that inherited traits were determined by information contained in microscopic wormlike bodies called chromosomes. Each cell has a number of pairs of chromosomes, and each pair appears only slightly different from the others. McClintock was the first to work out a way to tell the 10 pairs of maize chromosomes apart.

McClintock earned her M.A. in botany in 1925 and her Ph.D. in 1927, after which Cornell hired her as an instructor. In 1931 she and another scientist, Harriet Creighton, carried out an experiment that firmly linked changes in chromosomes to changes in whole organisms, a link that some geneticists had still doubted. This experiment has been called "one of the truly great experiments of modern biology."

McClintock was earning a national reputation in genetics, but Cornell refused to promote her because she was a woman. Rather than remain an instructor forever, she resigned shortly after her landmark paper was published. For the next several years she led an academic gypsy life, living on grants and dividing her research time among three universities in different parts of the country. She did her commuting in an old Model A Ford, which she repaired herself whenever it broke down.

In 1936 the University of Missouri at Columbia, one of the institutions at which McClintock had done part-time research, gave her a full-time position as an assistant professor. While there she studied changes in chromosomes and inherited characteristics made by X rays, which damaged genetic material and greatly increased the number of mutations, or random changes, that occurred in it. This university, too, refused to treat her with the respect she felt she deserved, and she resigned in 1941.

McClintock was unsure what to do next until a friend told her about the genetics laboratory at Cold Spring Harbor on Long Island. Run by the Carnegie Institution of Washington, which the steel magnate and philanthropist Andrew Carnegie had founded, it had been the first genetics laboratory in the United States. McClintock moved to Cold Spring Harbor in 1942 and remained there for the rest of her life.

The discovery in the 1940s that the complex chemical deoxyribonucleic acid (DNA) was the carrier of most genetic information and the working out of DNA's chemical structure and method of reproduction in 1953 revolutionized genetics, turning attention away from whole organisms or even cells and toward molecules. Geneticists saw genes, now shown to be parts of

DNA molecules, as unalterable except by chance or the sort of damage that X rays produced. FRANCIS CRICK, the codiscoverer of DNA's structure, expressed what he called the "central dogma" of the new genetics: "Once 'information' has passed into protein [chemicals that carry out cell activities and express characteristics] it cannot get out again."

Barbara McClintock meanwhile went her own way, working with her unfashionable corn and "letting the material tell" her what was happening in its genes. Contrary to Crick's central dogma, she found genes that apparently could change both their own position on a chromosome and that of certain other genes, even moving from one chromosome to another. This movement, which she called transposition, appeared to be a controlled rather than random process. Furthermore, if a transposed gene landed next to another gene, it could turn on that gene (make it active, or capable of expressing the characteristic for which it carried the coded information) if it had been off, or vice versa. Genes that could control their own activity and that of other genes had not been recognized before. McClintock suspected that such genes and their movement played a vital part in organisms' development before birth.

Even more remarkable, some controlling genes appeared able to increase the rate at which mutations occurred in the cell. McClintock theorized that these genes might become active when an organism found itself in a stressful environment. Increasing the mutation rate increased the chances of a mutation that would help the organism's offspring survive. If a gene that increased its mutation rate could be turned on by something in the environment, then organisms and their environment could affect their own evolution, something no one had thought possible.

McClintock attempted to explain her findings at genetics meetings in the early 1950s, but her presentations were met with blank stares or even laughter. She offered ample evidence for her claims, but her conclusions were too different from the prevailing view to be accepted. The chilly reception "really knocked" McClintock, as she later told her biographer, Evelyn Fox Keller, and after a while she stopped trying to communicate her research. Most geneticists forgot, or never learned, who she was; one referred to her as "just an old bag who'd been hanging around Cold Spring Harbor for years." She did not let rejection stop her work, however. "If you know you're right, you don't care," she said later.

McClintock received a certain measure of recognition in the late 1960s. In 1967, the same year in which she officially retired (in fact, her work schedule continued unchanged), she received the Kimber Genetics Award from the National Academy of Sciences. She was awarded the National Medal of Science in 1970. Only in the late 1970s, however, did other geneticists' work begin to support hers in a major way. Researchers found transposable elements, or "jumping genes" as they became popularly known, in fruit flies and other organisms, including humans. The idea that some genes could control others was also proved.

The trickle of honors became a flood in the late 1970s and early 1980s. McClintock won eight awards in 1981 alone, the three most important of which—the MacArthur Laureate Award, the Lasker Award, and Israel's Wolf Prize—occurred in a single week. Then in October 1983, when she was 81 years old, she learned that she had won the greatest scientific award of all, a Nobel Prize. She was the first woman to win an unshared Nobel Prize in physiology or medicine.

These honors and their attendant publicity irritated McClintock more than they pleased her. She complained, "At my age I should be allowed to . . . have my fun," which meant doing her research in peace. McClintock continued to have her scientific "fun" among the corn plants almost until her death on September 2, 1992, just a few months after her 90th birthday.

Some people saw Barbara McClintock's solitary life as lonely and perhaps even sad, but she never viewed it that way. As she said shortly before she died, "I've had such a good time... I've had a very, very satisfying and interesting life."

Mead, Margaret
(1901–1978)
American
Anthropologist

The best-known anthropologist of the 20th century, Margaret Mead greatly expanded both the audience for and the topics investigated by anthropological studies. Mead was the first anthropologist to examine child-rearing practices and the role of women in other cultures. Unlike many of her contemporaries (who believed that genes determined human personality and roles), Mead held that culture was the primary shaper of human behavior. The author of hundreds of articles and books, Mead brought her message to a popular audience. Her most famous publications—including *Coming of Age in Samoa* and *And Keep Your Powder Dry*—were national best-sellers. For most of her career, Mead was associated with the American Museum of Natural History.

Born on December 16, 1901, in Philadelphia, Pennsylvania, Margaret Mead was the eldest of the five children of Edward and Emily Mead. Edward was an economics professor at the University of Pennsylvania's Wharton School of Business, while Emily was a sociologist. Mead's mother and grandmother (a child psychologist) cultivated the skills that she would later utilize as an anthropologist, teaching her to observe other children and take notes on their behavior.

Mead began her academic career at DePauw University in Indiana in 1919, but she transferred to Barnard College in 1920. There, she met prominent anthropologists Franz Boas and Ruth Benedict. Inspired by their work, she majored in anthropology, graduating in 1923. That same year, she married Luther Cressman. (Their marriage ended quickly.) Mead earned her master's degree in psychology from Columbia in 1924, and in 1925, she traveled to the island of Tau in Samoa to conduct fieldwork for her Ph.D. (which she would receive from Columbia in 1929). Her thesis explored whether adolescent girls in Samoa experienced the same anxieties and concerns as female American teenagers. She concluded that they did not and parlayed her research into her first book (published in 1928)—*Coming of Age in Samoa*—which proposed that culture (not genetic determinants) accounted for these differences.

After returning from Samoa in 1926, Mead was named assistant curator of ethnology at the American Museum of Natural History in New York. She would remain affiliated with this institution for her entire career, becoming associate curator in 1942 and curator in 1956. In 1928, Mead married New Zealand anthropologist Reo Fortune. That same year, the couple researched the Manus people in New Guinea. Their fieldwork provided material not only for several highly regarded scientific papers but also for Mead's books *Growing Up in New Guinea* (1930) and *Sex and Temperament in Three Primitive Societies* (1935). These popular works further explored the notion that cultural forces determined social behaviors (including gender roles).

Mead's marriage to Fortune ended in divorce, and in 1936, she married Gregory Bateson, another anthropologist. Their daughter, Mary Catherine, was born in 1939. (The couple would divorce in 1951.) In the late 1930s and early 1940s, Mead and Bateson studied the Balinese people. Again, Mead transformed her field notes into a book—*Balinese Character*—which was published in 1941. This text pioneered the extensive use of photographs in anthropological works. In 1942, Mead applied her methodology to American culture, producing the best-seller

Margaret Mead was the first anthropologist to examine child-rearing practices and the role of women in other cultures. *(National Library of Medicine, National Institutes of Health)*

And Keep Your Powder Dry: An Anthropologist Looks at America, which compared and contrasted American culture to seven others. From 1947 to 1951, she was a visiting lecturer at Columbia University's Teachers' College.

During the 1950s and 1960s, she lectured and wrote extensively about a range of topics, including education and family life as well as more academic topics. She also returned to the Pacific Islands, New Guinea, and Bali, recording changes in the cultures she had studied previously. In 1964, she officially retired from the Museum of Natural History, although she continued to maintain an office there until her death on November 15, 1978.

Mead is credited with widening the focus of anthropology. While the discipline had long neglected the study of women and children, Mead made these areas essential subjects of anthropological inquiry. Mead also popularized the often esoteric discipline with a mainstream audience. Her numerous accomplishments did not go unrecognized. She became the president of the Anthropological Association in 1960 and was elected to the National Academy of Sciences in 1975. After her death, the conclusions of *Coming of Age in Samoa* were called into question by Australian anthropologist Derek Freeman, who accused Mead of leading her subjects to the responses she wanted and of misunderstanding their teasing answers to her questions about sexuality. Whether or not these criticisms are true, they do not diminish Mead's considerable legacy.

Medawar, Peter Brian
(1915–1987)
Brazilian/English
Biologist, Medical Researcher

Peter Brian Medawar established the possibility of tissue transplantation when he grafted skin from one strain of mice onto another strain of mice. This discovery opened the door for other kinds of transplantation, eventually leading to organ transplantation in human beings. Medawar's discovery and the techniques that have developed from it have saved countless lives. For this work, he shared the 1960 Nobel Prize in physiology or medicine with Frank Macfarlane Burnet, who first proposed the theory that the immune system could develop recognition for nonnative tissue, a phenomenon that Medawar dubbed "actively acquired tolerance."

Medawar was born on February 28, 1915, in Rio de Janeiro, Brazil, where his father, Nicholas Medawar, a native of Lebanon and a naturalized citizen of England, was conducting business. His mother, Edith Muriel Dowling, was also English, so the family moved to England after World War I ended, when Medawar was still young. There, he attended Marlborough College for secondary school starting in 1928. Four years later, he matriculated in Magdalen College of Oxford University, where he studied zoology under John Young to earn his bachelor's degree in 1935.

After graduation, Medawar worked under Sir HOWARD WALTER FLOREY at Oxford's School of Pathology, an experience that opened his eyes to the medical application of biological methods. Later in 1935, Medawar returned to Magdalen as the Christopher Welch Scholar and a senior demonstrator. Jean Shinglewood Taylor, a fellow zoologist, worked beside him both at Magdalen and at Florey's lab, and the couple married in 1937. Together, they eventually had four children—two sons, Charles and Alexander, and two daughters, Caroline and Louise.

In 1938, Medawar won the Edward Chapman Research Prize and also passed the examination to become a fellow of Magdalen College. In 1939, Oxford granted him a master's degree. Up until this point, Medawar had focused his research on mathematical analyses of organisms' shape transformations, as well as on tissue culture and the regeneration of peripheral nerves. The advent of World War II turned Medawar's research to more practical applications, as the Medical Research Council requested that he investigate tissue transplantation and skin grafting to treat wounded soldiers.

Stationed at the Burns Unit of the Glasgow Royal Infirmary, Medawar addressed the question of skin-graft rejection, determining that the body develops immunity to donor skin (explaining the longer rejection time for a first graft, as compared to subsequent grafts of skin that the recipient body already recognizes as foreign). He also developed a kind of biological "glue"—made of a concentrated solution of fibrinogen, a blood-clotting protein—for affixing nerves together. In 1942, he was named the Rolleston Prizeman, and in 1944, he returned to Oxford, this time working at St. John's College as senior research fellow and university demonstrator in zoology and comparative anatomy.

Despite the fact that Medawar declined to accept the Ph.D. for which he had qualified at Oxford (lacking the funds to afford the title), the University of Birmingham appointed him as Mason Professor of Zoology in 1947. There, he collaborated with Rupert Everett Billingham (known as "Bill") and Leslie Brent on his most famous researches. In response to a challenge issued to Hugh Donald at the 1948 International Congress of Genetics in Stockholm, Medawar's team determined to devise a means of distinguishing monozygotic from dizygotic twin calves, expecting identical twins to accept grafts of skin that it recognized as its own, and fraternal twins to reject the skin as "foreign," as it developed from a slightly different immunological "template."

Interestingly, Medawar discovered that calves do not differentiate as expected: fraternal twins accept skin grafts as readily as identical twins. California Institute of Technology's Ray D. Owen similarly found that blood transfuses between fraternal and identical calf twins. For an explanation, Medawar turned to the work of the University of Melbourne's Frank Macfarlane Burnet, who believed that immunity continues to develop for a period after birth, thus the body can acquaint itself with external influences if introduced early enough in the growth process. In other words, Burnet's theory hypothesized that immunity can develop a "tolerance" for "foreign" tissue; Medawar and his team endeavored to prove this theory experimentally.

Medawar, Billingham, and Brent worked with two sets of mice, inoculating recipient mice embryos with the tissue of donor mice, then grafting donor-mice skin onto matured recipients. Mice usually reject skin grafts, but these recipients accepted the grafts, clearly because their immunological systems recognized the cells. In his announcement of these findings, published in a 1953 edition of the eminent British scientific journal *Nature*, Medawar dubbed this phenomenon "actively acquired tolerance." The team's results introduced the possibility of grafting not only skin but also other tissues; this was the first step toward organ transplantation in humans.

While conducting these studies, Medawar had served as a visiting researcher at Rockefeller Institute in New York in 1949, and two years later University College, London, named him Jodrell Professor of Zoology and Comparative Anatomy. In the wake of his discovery, he received numerous honors, including the 1958 Croonian lectureship, the 1959 Royal Medal of the Royal Society (which had inducted him into its fellowship a decade earlier), and the 1960 Nobel Prize in physiology or medicine (shared with Burnet; Medawar shared his prize money with his research partners, Billingham and Brent). In 1965, Medawar was knighted.

Three years earlier, in 1962, the National Institute for Medical Research named Medawar its director. In 1969, he suffered a stroke that severely limited his abilities to attend to his administrative responsibilities, forcing him to retire in 1971. However, he was still able to write, allowing him to continue to publish essays aimed at mainstream readers; best remembered of these was his 1964 essay "The Art of the Soluble" (in which he famously stated that "if politics is the art of the possible, research is surely the art of the soluble"); his 1979 book *Advice to a Young Scientist*; and his 1986 autobiography, *Memoir of a Thinking Radish*. Medawar suffered subsequent strokes in 1980 and in 1984; he died on October 2, 1987.

Meitner, Lise
(1878–1968)
Austrian/Swedish
Physicist

Lise Meitner played a central role in one of the most significant scientific discoveries of the 20th century. Together with her nephew, OTTO ROBERT FRISCH, Meitner determined that atomic nuclei could be split. This theory of nuclear fission revolutionized notions of atomic structure and enabled the invention of both nuclear power and the atomic bomb. Meitner was also the first female full professor of physics in Germany, though she experienced considerable discrimination because of her gender.

Born on November 7, 1878, in Vienna, Austria, Lise Meitner was the third of eight children born to Philipp and Hedwig Meitner. Her father was a wealthy lawyer, and her mother was a member of Vienna's elite social circle. Although Lise expressed an early interest in science, her parents discouraged her ambition. Her father sent her to study to be a French teacher at the Elevated High School for Girls in Vienna.

Meitner still wished to be a scientist, and she worked intensely with a private tutor to pass the competitive university entrance exams. In 1901, Meitner enrolled at the University of Vienna, where she studied physics. She received her Ph.D. in 1906, becoming only the second woman ever to earn a physics doctorate from that institution. Unable to find a faculty position after graduation, she remained at Vienna as an assistant in her adviser's laboratory.

In 1907, Meitner moved to the University of Berlin's Institute for Experimental Physics to study under MAX PLANCK—one of the progenitors of quantum physics. She persuaded a young chemist, Otto Hahn, to hire her as his assistant, and the pair began a fruitful collaboration, focusing initially on the behavior of beta rays (negatively charged particles emitted during the breakdown of radioactive atoms) as they passed through aluminum. In 1912, the duo moved to the nearby Kaiser-Wilhelm Institute after Hahn was hired to work in its new radioactivity department. Although World War I intermittently interrupted their research, by 1918 the duo had discovered what was then the second-heaviest element, which they named protoactinium (later shortened to protactinium).

Meitner's career was buoyed by this groundbreaking discovery. In 1918, she was named head of a new department of radioactivity physics at the Kaiser-Wilhelm Institute. In 1926, she was appointed to the faculty of the University of Berlin, becoming the first woman full professor of physics in Germany. She continued to investigate beta particles and reunited with Hahn in 1934 to determine what would occur when the heaviest natural elements were bombarded with neutrons. However, political turmoil caused Meitner to abandon her work temporarily. Jewish by birth (although baptized as an infant), Meitner experienced increasing anti-Semitism, and she fled to Sweden and a post at the Nobel Institute of Theoretical Physics in Stockholm in 1938.

In December of 1938, Hahn wrote Meitner a letter explaining a conundrum he had encountered in the laboratory. While bombarding uranium with neutrons, he did not produce a heavier substance, as expected. Instead, he created what appeared to be barium, an element much lighter than uranium. Meitner puzzled over this paradox with her nephew, Otto Frisch, who was visiting her in Sweden. They realized that Hahn's results could be explained if he had split the uranium nucleus—rather than simply adding or subtracting particles from it. This conclusion ran contrary to accepted notions, since most physicists believed it was impossible to split a nucleus. Meitner and Frisch reasoned that the electric charge of the heavy nucleus had offset the forces binding the nucleus together. They also determined that splitting a uranium nucleus would release a tremendous amount of energy.

The first female full professor of physics in Germany, Lise Meitner, together with her nephew Otto Frisch, discovered nuclear fission and named the process. *(AIP Emilio Segrè Visual Archives)*

(In the small quantities Hahn had used, the energy output was not apparent). In 1939, Frisch and Meitner published a paper explaining their discovery. Frisch named the process fission, likening the splitting of the atomic nucleus to the process of cell division.

Unwittingly, the two physicists had provided the theoretical framework for the atomic bomb, though they did not appreciate its full consequences until 1945, when the first atomic bomb was dropped. Meitner continued to live and work in Sweden, even after her official retirement in 1947. In 1968, she moved to London to be closer to Frisch. Although she did not share Hahn's Nobel Prize, she received numerous awards for her achievements—including the Max Planck Medal (1949) and the Enrico Fermi Award of the U. S. Atomic Energy Commission (1966). She died on October 27, 1968. When researchers discovered element 109 in 1982, they named it meitnerium in her honor. She never married nor had children.

Mendel, Johann Gregor
(1822–1884)
Austrian
Geneticist

Johann Gregor Mendel's pioneering plant hybridization experiments laid the foundation for the modern field of genetics. His careful analysis of research concerning the transmission of plant characteristics across generations provided the first mathematical basis for that science. His work led to the formulation of Mendel's laws of segregation and independent assortment, and it represented the first application of statistical analysis to the explanation of a biological phenomenon.

Mendel's intense interest in horticulture and plant hybridization was rooted in his childhood. Born on July 22, 1822, in Heinzendorf, Austria (now Hyncice in the Czech Republic), the young Mendel was raised helping his peasant farmer father, Anton, graft trees in the family orchard. Anton, who had served in the Napoleonic army, was dedicated to improving his crops by applying farming methods he had observed during his travels. Since the family lived on the border between German- and Czech-speaking regions, Mendel became fluent in both languages at an early age.

Mendel's parents and teachers noted his abilities and arranged for him to be educated at the Gymnasium in Troppau, where he studied from 1834 to 1840. After suffering a nervous breakdown brought on by stress, Mendel attended Olmütz University. In 1843, at the age of 21, he entered the Augustinian monastery in Brno. Although he felt he lacked a religious vocation, the monastery afforded him an excellent environment for his studies and freed him from financial worries. After a brief stint as a substitute teacher, he failed to pass the teachers' qualifying exam. His order sent him to the University of Vienna from 1851 to 1853. In addition to learning botany and chemistry there, Mendel studied under Franz Unger, a plant physiologist who taught him how to organize botanical experiments. Mendel then returned to Brno but again failed the teachers' exam.

Although he continued to serve as a substitute teacher until 1868, Mendel's true work took place in the monastery's garden. From 1856 to 1863 he conducted a massive study involving 28,000 edible pea plants. He had noted seven constant plant traits, including stem length (whether tall or short); flower position (whether axial or terminal); pod color (green or yellow); seed shape (round or wrinkled); flower color (purple or white); and the seed color (yellow or green). He then crossed plants bearing these distinct attributes. The plants were self-pollinated and then individually wrapped to prevent pollination by insects. After collecting the seeds, Mendel scrupulously studied the offspring.

His findings overturned the accepted heredity theory of the day when he proved that the seven characters did not amalgamate on crossing. That is, rather than producing offspring bearing a blend of these characteristics, the progeny manifested only one such trait (for example, either a tall stem or short stem rather than some combination). However, when two of these "hybrid" progeny were crossed, the product of that union could manifest the traits of either grandparent, indicating that those elements had remained distinct, even when one was latent. This insight became known after 1900 as Mendel's first law, or the law of segregation. Another discovery was dubbed Mendel's second law, or the law of independent assortment. This principle holds that the various characters transmitted from parent to offspring combine randomly, rather than in any grouped bunch. Mendel's experiments revealed that a given pea plant was statistically no more likely to have a tall stem and smooth pod form, for example, than a tall stem and a wrinkled pod form. Moreover, all of the various permutations of the seven pea traits occurred with equal frequency.

Although Mendel presented his conclusions in 1865 at a meeting of the Natural Sciences Society and published his results in 1866, the significance of his experiments was lost on his contemporaries. After he was appointed abbot of his monastery, his bureaucratic duties expanded, leaving him less time for research. It was not until the 20th century that his work became widely recognized. Then not only did Mendel's experiments help forge the science of genetics but they also bolstered CHARLES ROBERT DARWIN's theory of evolution. Mendel's demonstration of the mechanism of variability in plant species buttressed Darwin's overarching theory of natural selection by supplanting Darwin's erroneous notion of pangenesis as an explanatory principle. Mendel died on January 6, 1884, in Brünn and is buried in an Augustinian tomb in Brünn's central cemetery.

Mercator, Gerardus
(1512–1594)
Flemish
Cartographer, Geographer

Gerardus Mercator modernized cartography with the "Mercator projection," or the depiction of the world using straight lines for latitude and longitude. Although this representation necessarily distorted actual geography as it moved away from the equator toward the poles, it enabled sailors to navigate accurately with the aid of maps. This development acted as a boon to nautical navigation, an important step in the age of geographic discovery. Although it has proven impossible to completely reconcile Mercator's two-dimensional projection with three-dimensional reality, the Mercator projection remains in use by navigators more than four centuries after its invention.

Mercator was born Gerhard Kremer (or Gerard du Cremer) on March 5, 1512, in Rupelmonde, Flanders (now Roermonde, Holland). His father was a poor shoemaker, but he was raised by his uncle, Gisbert Mercator, a wealthy ecclesiastic who sent Mercator to the 's-Hertogenbosch school at the House of the Brethren of the Common Life. There, he received an education in philosophy and theology that customarily prepared students for the priesthood. In 1530, he entered the University of Louvain in Belgium, at which point he latinized his name to Gerardus Mercator. Two years later, he received a master of arts degree.

After graduation, Mercator traveled to Mechelen and Antwerp before returning to Louvain, where he apprenticed in engraving under Gemma Frisius, who also taught him mathematics and astronomy. Mercator also apprenticed in the making of scientific instruments under Gaspar a Myrica. In 1534, Mercator established a center for geographic study in Louvain, where he collaborated with Frisius and Myrica to construct his first terrestrial globe two years later. The next year, they collaborated on a celestial globe.

Also in 1537, Mercator published his first map—of Palestine—that he dedicated to Frans Craneveld, a councilor of Emperor Charles V, as was the customary return-in-kind for financial patronage. Other dedications included a 1540 map of Flanders, dedicated to Charles V (for whom he also made two sets of surveying instruments), and a 1541 terrestrial globe, dedicated to M. Nicolas Perrenot de Granvelle, privy councilor of Charles V.

In 1544, Mercator was imprisoned for heresy (he had experienced a religious crisis earlier in life, unable to reconcile the biblical conception of the universe with the Aristotelian one) and held for seven months. He was finally released due to insufficient evidence, after the University of Louvain intervened on his behalf. Soon thereafter, he moved to the Protestant territory of Cleves, where he became the court "cosmographer" for Duke William of Cleves two decades later. In the meanwhile, he constructed another celestial globe in 1551, this one dedicated to Prince-Bishop George of Austria, Bishop of Liège. The next year, he became a lecturer at the University of Duisburg.

In 1554, Mercator published the first modern map of Europe, for which he received an honorarium, presumably from those to whom he dedicated the work: Cardinal Archbishop of Mechelen and Bishop of Arras, Antoine Perrenot de Granvelle, and Phillip II. From 1559 through 1562, he taught mathematics at a grammar school in Duisburg. In 1564, the same year he took up the position of cosmographer under William of Cleves, he published a map of the British Isles.

In 1568 or 1569, Mercator published the work that made the most lasting influence: a world map featuring the "Mercator projection," or the depiction of longitudes and latitudes with straight lines. This map, which increased its distortion as it neared the poles, nevertheless allowed sailors to navigate accurately, as its angles of intersection remain constant across all meridians, as they do in reality. Only four copies of this map, which he dedicated to William, remain in existence. The Mercator projection remains an important nautical tool.

Also in 1569, Mercator commenced work on an "atlas," or a compendium of maps. Mercator coined this name after the Greek god who held the earth in his arms, an image that ultimately appeared on the cover of the atlas, which contained incredibly detailed and accurate maps of western and southern Europe. However, it was not until the year after Mercator died—on December 2, 1594, in Duisburg, Germany—that the atlas was published (in an unfinished state) by his son, Rumold, under the title "Atlas—or Cosmographic Meditations on the Structure of the World."

Mestral, George de
(1907–1990)
Swiss
Engineer, Textile Industry

As with many of the greatest inventions, George de Mestral stumbled upon his idea for Velcro quite by accident. Taking after the natural process by which plants spread their seeds, he innovated an ingenuous closure system based on the simple system of hooking into loops. The key was to compact thousands of hooks close together on one side of material, and compact thousands of loops on the other side of material, so that the two sides fastened together securely when brought together. Velcro mushroomed in popularity only after the expiration of de Mestral's patent, when cheap imitations flooded the market. However, de Mestral's global family of companies maintained control of the trademark and the best manufacturing processes, producing a product so superior that it maintained a foothold in the market.

De Mestral was born on June 17, 1907, in a village near Lausanne, Switzerland. As a child, he indulged two interests: nature and invention. By the age of 12, he had patented his first invention, the design for a toy plane. He worked odd jobs to finance his study at the École Polytechnique Fédérale de Lausanne. He graduated with a degree in electrical engineering and subsequently landed a job in a machine shop of a Swiss engineering company, though he continued work on his own inventions independently.

The story of de Mestral's invention of Velcro has evolved into almost mythical status: in the summer of 1948, he set out game-bird hunting in the Jura Mountains with his Irish pointer. Throughout the day, he cursed the cockleburs that stuck to him and his dog. Picking the burrs off upon returning home that evening, he was amazed at the tenacity with which the burrs clung to his trousers and jacket, as well as to his dog's coat.

De Mestral realized that nature engineered the plant to affix its seed pods to passing animals to migrate to new growing grounds. Curious about Mother Nature's design, he took some burrs inside to examine them under his microscope. He discovered that the surface of the burrs consisted of tiny hooks that fastened to the loops in woven fabric, as well as to the hairs of animal's coats. Inspired, de Mestral endeavored to replicate this design to create a better fastener than the zipper.

In 1951, de Mestral applied for a Swiss patent (which he received in 1955) for "velcro," the name he derived by linguistic mimicry of the hook and loop process, joining the "vel" from "velvet," a soft fabric, with the "cro" from "crochet," a handicraft utilizing a hooked tool. The next year, he quit his engineering job and took out a $150,000 loan to finance his continuing experimentation at perfecting a design and manufacturing procedure. He consulted fabric and cloth experts in Lyon, France, a mecca for textile design, about how to construct velcro. He happened upon a solution while waiting for a haircut at a local barbershop, where the shearing motion of the barber's scissors inspired him with a manufacturing design.

Throughout these several years of trial-and-error design, de Mestral's friend Alfred Gonet, head of Gonet & Co., had become interested in velcro and pledged his support in launching the novel product on the market. Still in 1952, de Mestral established Velcro S.A., a company dedicated to making its namesake. Over the ensuing years, de Mestral received patents for "the invention and fabrication of special napped piles of man-made material at least some of these loops having the means of hooking near their ends" in Switzerland, Germany, Great Britain, Sweden, Italy, Holland, Belgium, France, Canada, and the United States.

In 1957, de Mestral established the American Velcro Inc. (the name changed to Velcro USA Inc. in 1976) in Manchester, New Hampshire, where production began on shuttle looms. In 1958, de Mestral registered the Velcro name as a trademark in the United States as well as in other countries. By then, he had happened upon the best method for making Velcro, by sewing the nylon under infrared light.

Production of Velcro reached 60 million yards per year, but it remained expensive and was thus limited to industrial applications. Noted for its shear strength, or resistance to sideways forces, a piece of Velcro less than five square inches could hold a one-ton load. NASA used the wonder material on its space suits and in many other applications, such as on food pouches that stuck to the wall for easy access in zero-gravity conditions. This utilitarianism ran against the grain of 1960s fashion, preventing it from being employed in clothing design. The innovation of molded plastic hooks (in addition to the existing woven hooks) in 1967 set the stage for broader applications.

Ironically, it was not until after de Mestral's patent expired in 1978 that Velcro blossomed in popularity. The 1979 replacement of the shuttle loom with the needle loom also aided in the manufacturing boom. However, the flooding of the market in the 1980s with Velcro imitations cheaply manufactured in Taiwan and South Korea, and utilized on everything from kids' shoes to surfers' swim trunks, truly established the prevalence of this closure process. The official Velcro products benefited as well, as the upsurge in overall sales inevitably included them, and de Mestral's international Velcro companies began to advertise themselves as the "real McCoy," maintaining far superior quality standards.

De Mestral experienced the "Velcro decade" before dying on February 8, 1990. In 1999, he was inducted into the National Inventors Hall of Fame in honor of the product he invented that became as ubiquitous as its predecessor, the zipper.

Michaux, Pierre
(1813–1883)
French
Inventor, Transportation Industry

Pierre Michaux is considered one of the fathers of the bicycle, as he was the first to fit the front wheel with pedals. He was not the first to make a bicycle, but he was the first to market bicycles successfully. During the decade of the 1860s, bicycle fever swept France and subsequently invaded Britain and the United States. In fact, it was a former Michaux employee who introduced the bicycle industry to the United States.

Michaux was born in 1813. Little information is readily available about his life before he made his name as a bicycle maker, but various accounts place him as a cabinetmaker, locksmith, and carriage-repairer. He may have worked all three of these trades, as he applied the combined skills to build some of the first bicycles. The first treadle-driven bicycle was built in 1839 by the Scottish blacksmith Kirkpatrick Macmillan, who called it a velocipede. Three years later, the Frenchman Alexander Lefèvre built a similar rear-wheel treadle-driven bicycle, or *draisienne*, which he brought with him to California when he immigrated there in 1860 or 1861.

Also in 1861, Michaux built his first velocipede in his workshop near the Champs Élysées. Weighing an unwieldy 110 pounds, it consisted of a wooden frame mounted with a set of compression-spoked, iron-rimmed wooden wheels, which accounted for the bicycle's nickname, the "boneshaker." Michaux also introduced the innovation that marked the boneshaker as a novel design—front-wheel pedals.

Legend has it that when a hatter named Brunel brought his *draisienne* in for repair, Michaux conceived of the idea of adding a grindstone-like crank. He enlisted his 14-year-old son, Ernst, to construct such a design, though Henri, his other son, who was seven at the time, later denied that his brother had anything to do with the boneshaker's construction. Whatever the familial roles were in making the prototype, the bicycle also gained the name *Michauline* after its inventor.

By the next year, Michaux produced 142 boneshakers. He organized a company to manufacture the increasingly popular recreation vehicle, which could reach a top speed of six-and-a-half miles per hour. By 1865, Michaux et Cie was producing 400 boneshakers per year, as boneshakers were catching on as a fad.

In the midst of Michaux's success, he had managed to enrage one of his employees, Pierre Lallement, who immigrated to the United States that year, where he teamed up with James Carroll in Boston to produce his own rendition of the boneshaker. On November 20, 1866, the pair received the first patent ever issued for a pedal-driven bicycle, U.S. patent number 59,915. In an 1883 interview with Charles Pratt published in *Wheelman Illustrated*, Lallement claimed to have conceived of a pedal-driven bicycle in 1862—a year after Michaux had introduced the boneshaker, thus confirming the priority of his former employer.

In 1867, Michaux introduced a redesigned version of the boneshaker, making the front wheel larger than the back wheel. Replacing the wooden frame with a wrought-iron one reduced the weight to 59 pounds. He also added a leaf-spring saddle, lever shoe brakes, and an adjustable crank-length. He displayed this model at the World Exhibition in Paris, resulting in mushrooming orders—Napoleon III even ordered one, despite his invalidism.

In 1868, René and Aimé Olivier invested 100,000 francs in the company, allowing for the construction of a two-and-a-half-acre factory near the Arc de Triomphe that housed some 300 workers making three to five bikes a day. The next year, the Olivier brothers paid another 200,000 francs to buy ownership of the company from Michaux, who joined with L. G. Perreaux to build one of the first motorcycles by mounting a Perreaux steam engine on a boneshaker.

In 1870, the Franco-Prussian War broke out. Boneshaker production continued at an unprecedented rate, with 500 workers operating 57 forges in 1870. That year, however, Paris was besieged, and the factory was destroyed by cannon-fire. At the time, some 60 other bicycle makers flourished in Paris, with 15 more producers spread throughout the French provinces. Boneshakers had reached the height of their popularity and declined precipitously in sales thereafter.

Michaux's fortunes also declined, as he unsuccessfully sued the Oliviers and lost a countersuit as well, diminishing his holdings. Michaux died in poverty in 1883, having given rise to one of the most popular forms of recreational transportation ever.

Morgan, Thomas Hunt
(1866–1945)
American
Geneticist

Thomas Hunt Morgan helped establish the field of genetics by proposing the chromosome theory of heredity, which established the existence of genes on the chromosomes. For this work, he was awarded the 1933 Nobel Prize in physiology or medicine. More significantly, his theory led to a new understanding of the continuation of life and opened the door to highly contentious moral and ethical issues, as the knowledge of how life replicates itself treads on territory previously reserved for gods.

Morgan was born on September 25, 1866, in Lexington, Kentucky. His mother, Ellen Key Howard, was the granddaughter of Francis Scott Key, composer of the American national anthem, the "Star Spangled Banner." His father, Charlton Hunt Morgan, was the American consul at Messina, Sicily, in the early 1860s, when Giuseppe Garibaldi and his Red Shirts were attempting to unify Italy. Morgan went to the State College of Kentucky's prep school, then stayed on there (at what later became the University of Kentucky) for undergraduate work in zoology. He received his bachelor of science degree summa cum laude in 1886. That summer, he conducted research at the Boston Society of Natural History marine biological station in Annisquam, Massachusetts.

In the fall of 1886, Morgan commenced graduate school at Johns Hopkins University, studying comparative anatomy, embryology, and morphology under William Keith Brooks. For his doctoral dissertation, he explored the distinction of sea spiders as spiders, or arachnids, as opposed to crayfish, or crustaceans, determining a closer relationship to the former. He earned his Ph.D. in 1890 for this work. Hopkins retained Morgan for a year of postdoctoral research under a Bruce fellowship.

In 1891, Bryn Mawr College hired Morgan as an associate professor of zoology, a position he retained for more than a dozen years. While there, he spent 10 months between the years of 1894 and 1895 at the Naples zoological station, where he formed a professional relationship with Hans Driesch. In Europe, Morgan confronted the question of nature versus nurture in embryology—are individual characteristics inherited, or do they develop through the influence of the environment?

In 1903, Morgan published *Evolution and Adaptation*, a diatribe against CHARLES ROBERT DARWIN's theory of natural selection in favor of HUGO DE VRIES's theory of mutation, which relied more on chance and less on "conscious" selection. The next year, E. B. Wilson appointed Morgan as a professor of experimental zoology (and head of the department) at Columbia University. Also in 1904, he married Lilian Vaughan Sampson, a cytologist who studied under him at Bryn Mawr. Together, they had four children. Throughout his quarter of a century of service to Columbia, he took only one sabbatical, which he spent conducting genetics research at Stanford University from 1920 through 1921.

In 1908, Morgan commenced the research for which he became most famous. He started breeding fruit flies, or *Drosophilia melanogaster*, because of their short (three-week) breeding cycle (though it also proved fortuitous that they had only four chromosomes, as it turned out). He collaborated with undergraduates A. H. Sturtevant and Calvin B. Bridges and graduate student Hermann Joseph Muller. At first, Morgan attempted to verify De Vries's theory of mutation through exposure to radium, but when those experiments failed to yield promising results, he shifted his direction.

Up until this point in his career, Morgan had been skeptical of the genetic theories of JOHANN GREGOR MENDEL, the Austrian monk who tracked variations in subsequent generations of pea plants. However, Morgan became less dubious of Mendel when he noticed a variation in male fruit flies: their wild red eyes turned white in one generation, then returned to red the next generation (when mated with red-eyed females), but then, the white eyes resurfaced in some of the next generation, but only in males. This development suggested to Morgan that genetic inheritance might be linked to the X chromosome that determines the male sex.

The team of four continued to track variations, with Sturtevant devising a method for mapping them. Although genes are too small to see, their effects can be inferred by generational mutations, especially with the aid of the "crossing over" theory, which stated that the greater the distance between two genes on a chromosome, the more likely they would break to cause a genetic disruption (such as the white-eye phenomenon). By 1915, the group had gathered enough evidence to publish their findings in *The Mechanism of Mendelian Heredity*, which advanced the chromosome theory of heredity and established the field of genetics. The next year, Morgan published *A Critique of the Theory of Evolution*, a tempered version of his earlier book in which he did not attack Darwinism as much as he explained it in Mendelian terms.

Morgan devoted himself increasingly to administrative duties in the latter part of his career. In 1928, the California Institute of Technology appointed Morgan as director of its Laboratory of Biological Sciences. He served as president of the National Academy of Sciences from 1927 through 1931, and as president of the American Association for the Advancement of Science in 1930. In 1932, he presided over the Sixth International Conference on Genetics in Ithaca, New York.

In 1933, Morgan received the Nobel Prize in physiology or medicine for proposing his chromosome theory. Earlier, in 1924, he had won the Darwin medal, and later, in 1939, he won the Copley medal of the Royal Society. Morgan continued to work at Cal Tech until he died on December 4, 1945, in Pasadena, California.

N

Napier, John
(1550–1617)
Scottish
Mathematician

John Napier invented the logarithm, which he named after the Greek terms for "expression" (*logos*) and "number" (*arithmos*), in order to aid in computing the large numbers involved in astronomical calculations. He spent much of his lifetime involved in devising tables of logarithms that were revised near the time of his death to a more useful system using base 10 as its foundation. The existence of logarithms proved instrumental in Sir ISAAC NEWTON's statement of his theory of gravitation, and logarithms remain an integral component of modern mathematics.

John Napier was born in 1550 at Merchiston Castle near Edinburgh, Scotland, the first son of the seventh laird of Merchiston (and hence the eighth laird of Merchiston himself). His mother, Janet Bothwell, the sister of the Bishop of Orkney, was the first wife of Archibald Napier, a justice-deputy who was knighted in 1565 and appointed Master of the Mint in 1582. Little is known of Napier's early education, though he was probably tutored at home and may have attended Edinburgh High School. Soon after he went away to boarding school at St. Salvator's College of the University of St. Andrews in 1563, at the age of 13, his mother died. He continued to study at St. Andrews, but university records do not report his graduation; most likely, he finished his education on the Continent.

In 1571, Napier returned to attend his father's wedding to his second wife and arrange his own marriage to Elizabeth Stirling in 1572. The couple took up residence in a newly constructed castle on the Napier estate at Gartnes in 1574. They had two children, Archibald and Joanne, before his first wife died in 1579. Napier then married Agnes Chisholm of Cromlix, Perthshire, and together they had 10 children. Napier occupied himself with the running of his estate, distinguishing himself as an innovator: he pioneered the use of salt as a fertilizer, and in 1597, he patented a hydraulic screw and revolving axle for avoiding flooding in coal pits. He came to be known as the "Marvellous Merchiston" after he moved into Merchiston Castle upon the death of his father in 1608.

Napier made an even more prominent name for himself as an amateur theologian when he published *A Plaine Discovery of the Whole Revelation of St. John* in 1593, a virulently anti-Papist tract. The book went through several editions in Dutch, French, and German translations, establishing Napier as one of the most vociferous proponents of Protestantism. Throughout his life, though, Napier exercised his passion for astronomy and mathematics, devising formulas for solving spherical triangles—a mnemonic and two formulas known as "Napier's analogies." He also introduced the use of the decimal point to signify fractions, a practice adopted by mathematicians that remains in use to this day, though Napier is not often remembered as its inventor.

Napier is best remembered for his introduction of logarithms, a system of simplifying the calculation of large numbers that he developed in order to avoid the "slippery errors" endemic to astronomical computations. He conceived of the notion early in his lifetime and spent several decades compiling tables of logarithms. He finally explained his invention in 1614, when he published *Mirifici logarithmorum canonis descriptio*, which included his table of logarithms. The work had an immediate influence on mathematicians, and it generated an even broader influence after Samuel Wright published his father's English translation of Napier's Latin original in 1616, after Edward Wright had died (Napier approved of the translation).

In the summer of 1615, Henry Briggs, the Gresham professor of geometry at the University of London, traveled four days to visit Merchiston and consult with Napier. When Briggs suggested that the logarithm table would be more useful if founded on base 10, Napier revealed that he had already come to the same conclusion but lacked the energy, as he was suffering from gout and was too sick to recalculate the table. Briggs volunteered to do so himself, publishing the results in *Logarithmorum chilias prima*. Interestingly, Joost Bürgi, a watchmaker and astronomical instrument-maker, independently developed a system of logarithms at about the same time as Napier, but he did not publish his results until 1620, after Napier had established himself as the inventor of logarithms.

The year that Napier died, he published *Rabdolgia* (or "numeration by little rods"), in which he described his invention of a mechanism consisting of ivory rods (resembling bones, hence their name of "Napier's bones") marked with numbers that aligned into tables for reading the results of multiplication and division and deriving square and cube roots. Napier died on April 4, 1617, at Merchiston. Two years after his death, in 1619, his second son and literary executor, Robert, published *Constructio*, a text gathering his writings into the clearest explanation of his logarithms.

Newcomen, Thomas
(1663–1729)
English
Inventor, Energy and Power Industries

Thomas Newcomen invented the atmospheric steam engine, an innovation that greatly increased the efficiency of pumping water from flooded mines, its main use. This development helped fuel the burgeoning Industrial Revolution, as raw materials (such as coal and other minerals) became more available and in much larger quantities. Although Newcomen created a highly original design, THOMAS SAVERY worded his existing steam pump patent so broadly as to block any subsequent innovations, no matter how different, thus preventing Newcomen from financially benefiting as much as was his due. Furthermore, JAMES WATT receives the most credit for inventing the steam engine, when in reality he simply added a secondary condenser to Newcomen's design. Newcomen thus deserves much more credit than he has received for helping to spawn the Industrial Revolution.

Newcomen was born at Dartmouth, in Devon, England, in 1663, and was christened there on February 24 of that year. His father was Elias Newcomen, a merchant whose family had been in the aristocracy until losing its property under King Henry VIII's reign. Little is known of Newcomen's early life and education, though it is believed that the Nonconformist John Favell educated him. It is also believed (but not known for certain) that he apprenticed to an ironmonger in Exeter. It is known that Newcomen was a practicing Baptist who acted as a lay preacher, delivering fervent sermons.

In about 1685, Newcomen apparently returned to Dartmouth to establish his own blacksmith and ironmonger business with his assistant, the plumber John Calley. In 1707, he was successful enough to rent a large house between Higher and Lower Streets in Dartmouth, where he set up his shop. By then, he was almost certainly working on his own steam engine design, most likely in response to the obvious need exhibited by his main clients, those in the mining industry, whose tools he fashioned and repaired.

The main impediment to more efficient mining was the constant flooding of the mine shafts and the limited draining technology, which consisted of horse-drawn ropes pulling hand-filled buckets up and out of the shafts, an incredibly inefficient system. In 1698, Thomas Savery had invented a steam pump for clearing flooded mines, and he had secured a broadly worded patent (covering "all vessels or engines for raising water or occasioning motion to any sort of millworks by the impellent force of fire") that effectively prevented others from making any innovations to steam engine technology.

Despite this deterrent, Newcomen experimented throughout the early part of the 18th century, tinkering with different steam engine designs. He installed his first steam engine in 1712 at a South Staffordshire colliery near Dudley Castle in Wolverhampton. This engine exhibited a distinct advantage over the Savery design, in that Newcomen did not rely on steam pressure itself but rather converted it into atmospheric pressure by means of a piston-driven cylinder.

The vertical cylinder measured 21 inches in diameter and almost eight feet in length. A boiler filled the cylinder with steam, which was then released by an automatic valve, lowering the piston by vacuum pressure. The piston was mounted on a wooden beam, and its other end was attached to a pump, such that when the vacuum pressure pulled the piston down, the pump pulled water up. Newcomen's first engine operated at 12 strokes per minute to lift 10 gallons of water as far as 156 feet. The engine operated at less than one percent efficiency, or about five-and-a-half horsepower, which nevertheless represented a significant increase over the efficiency of existing methods.

Newcomen's engine burned huge amounts of coal, as new steam had to be produced for each stroke, thus tailoring his engine for use at collieries, where coal was "free." The brass cylinder (iron could not be founded accurately enough) made the engine quite expensive, costing about 1,000 pounds in 1714. However, Newcomen never realized significant profits, as he had to pay royalties to Savery (despite the fact that his design was radically different from Savery's), whose patent was extended until 1733 by the Fire Engine Act of 1698. To waylay this expense, Newcomen also rented out his engines at a rate of about seven pounds a week.

By the time that Newcomen died, on August 5, 1729, in London, at least 100 of his engines were in operation internationally—in Britain, France, Belgium, Holland, Sweden, Hungary, Germany, and Austria. Ironically, Newcomen has never received the recognition he deserved, as credit for inventing the steam engine is often accorded to James Watt, who merely improved upon Newcomen's design in 1765 by adding an extra chamber for condensing the steam, thus avoiding the constant temperature fluctuations in the cylinder itself. However, only about 100 Watt engines ever sold, due to their expense, while the more simplistic Newcomen engines were selling in the thousands by this time.

Newton, Sir Isaac
(1642–1727)
English
Mathematician, Physicist

Sir Isaac Newton's discoveries are considered some of the most influential and important in the history of science. He made profound contributions to the fields of physics, mathematics, and optics. His formulation of three laws of motion laid the foundation for classical mechanics, explained planetary motion, and provided the foundation for his general theory of gravitation. In the realm of mathematics he invented the calculus (independently of Leibniz) and conceived the binomial theorem. His research in optics led him to propose his theory of light—that white light was a mixture of various colors in the spectrum.

Sir Isaac Newton's numerous contributions to the fields of physics, mathematics, and optics included his theory of light, which proposed that white light was a mixture of various colors in the spectrum. *(AIP Emilio Segrè Visual Archives)*

Newton was born prematurely on December 25, 1642. His father, who owned the manor of Woolsthorpe in Lincolnshire, England, had died three months before he was born. His mother, Hannah Ayscough Newton, remarried when Newton was three and left him in the care of his grandmother at Woolsthorpe. Upon the death in 1653 of her second husband, the Reverend Barnabas Smith, she returned to Woolsthorpe. Although the family owned quite a bit of land, they were neither wealthy nor noble.

After Newton had attended the King's School in Grantham for some time, his mother withdrew him with the intention of making him a farmer. He showed little skill or interest in this vocation, however. With the encouragement of his uncle, Newton was prepared for the university instead. In 1661 he was admitted to Trinity College, Cambridge, where he earned his bachelor of arts degree in 1665.

Newton was forced to return to his home in Woolsthorpe in 1665 when Cambridge closed because of an outbreak of plague. His time away from the university proved to be fruitful. Newton himself later termed this period his *annus mirabilis* (miraculous year), because he began to ponder the forces behind natural phenomena he witnessed and to research the subjects that would later make him famous—gravity, optics, and light. When he returned to Cambridge in 1667, he was elected a fellow at Trinity, and in 1669 he was named Lucasian Professor of Mathematics at the age of 26.

Newton's work in mechanics and gravity earned him most fame. In 1687 he published his seminal work, *Mathematical Principles of Natural Philosophy*—known simply as the *Principia*—which discussed his three laws of motion. The first law posited that a body at rest or in uniform motion will continue in that state unless a force is applied; the second defined force as equaling the mass of a body multiplied by its acceleration; the third stated that if a body exerted a force (action) on another, there would be an equal but opposite force (reaction) on the first body. These laws provided the basis for Newton's law of gravitation, which proposed that any particle of matter in the universe attracts any other with a force determined by the product of their masses but decreases by the square of their distance apart. These theories, elaborated in the *Principia*, explained a diverse range of natural phenomena—from the motion of the planets and their moons to the tides—and became essential building blocks for classical mechanics and future scientific endeavors.

Newton's contributions to mathematics and optics were no less significant. In 1665 he first laid out the binomial theorem, which was a general formula for writing any power of a binomial without multiplying out. He invented the calculus in 1669, although he did not publish his findings until 1674. He published *Opticks*, in which he described his theory of light and included other mathematical research, in 1704.

Newton's career was not devoted exclusively to science. From 1689 to 1690 and from 1701 to 1702 he served as a member of Parliament for the university. He was appointed warden of the Mint in 1696 and master of the Mint in 1699. After resigning his professorship at Cambridge in 1701, he was knighted for his services to the Treasury in 1705. After this time he dedicated little time to science, although he did remain president of the Royal Society until his death. His legacy as a scientist remains unsurpassed. Newton died in London in 1727 at the age of 85.

Niepce, Joseph
(1765–1833)
French
Chemist

Joseph-Nicéphore Niepce created the first photograph, which he called a "heliograph" (after the Greek for "sun"), a landscape that took eight hours to expose. As his death grew imminent, he reluctantly shared his secret with Louis Daguerre, who advanced the process by introducing mercury-vapor development. Early photographs came to be known as "daguerrotypes," though Niepce was properly the father of photography, which became an important tool for historical documentation, artistic expression, and, ultimately, personal recreation.

Joseph Niepce was born in 1765 in Chalon-sur-Saône, France. He hailed from a wealthy family, which proved vital to the support of his later experimentation. By 1789, he had been appointed to a professorship at an Oratorian *collège* in Angers. Instead of pursuing this academic career, however, he joined the French army as a staff officer stationed in Italy. He returned to France in 1794, marrying and securing a position as administrator of the district of Nice in early 1795. However, he retained this post only a few months, until his brother Claude, also a retiree from the army, wooed him away to pursue scientific research.

The Niepce brothers returned to the family seat of Chalon-sur-Saône and their estate at St. Loup de Varenne to conduct experiments on the *pyréolophore*, a prototypical internal combustion engine using lycopodium powder as fuel. They secured patent number 405 from the Institut National de la Propriété Industrielle on August 3, 1807, and conducted powerboat demonstrations for Lazare Carnot and Berthollet, who conveyed encouraging reviews to the Académie des Sciences. They then tried to convert their engine's fuel consumption of costly lycopodium powder to pulverized coal or petroleum.

Niepce diversified his research, entering his design for a hydraulic ram in a state-sponsored competition that same year, but it failed to win the prize and replace the existing technology providing water to Versailles from the Seine. In response to yet another government competition, the brothers attempted to find a dye to replace indigo, which was in short supply in 1813, but again they failed to win the prize.

They also experimented with sugar-extraction from beets, starch-extraction from pumpkins, and fiber-production from various plants.

Also in 1813, Niepce became interested in lithography, the art and science of transferring images. He made improvements to the process pioneered by Alois Senefelder, employing tin plates in place of stones. Lacking artistic skills, Niepce relied on his son's drawings as the originals for transfer. In 1814, the French army drafted his son to fight at Waterloo, depriving Niepce of his artist.

Niepce turned to nature as his artist, using existing scenes as his "original," which he would transfer by modifying lithographic techniques. By May 5, 1816, he was able to report to his brother in England that he had produced a negative of the view from his workshop, and two weeks later, in a letter dated May 28, he reported the production of four prints, which he transferred to paper imbued with silver chloride and fixed with nitric acid. In March 1817, he began experimenting with Judean bitumen as his fixing agent, borrowing this asphalt from lithography. Having depleted his sizable trust in trying to perfect the *pyréolophore*, Niepce secured support from the Société d'Encouragement pour l'Industrie National in 1817.

A decade later, in June or July 1827, Niepce created the first true photograph, what he called a "heliograph," or a positive image. He set up a camera obscura before a scene of a building, a tree, and a barn, leaving it there for the eight hours it required to expose the pewter-coated plate, with bitumen of Judea hardening on the light areas, and dark areas remaining soluble in water, washing off later.

Louis Daguerre, hearing of Niepce's experiments, sent him a letter of inquiry in January 1826, and another a year later. Niepce was reluctant to share his secret, but by June 1827, he relented and sent Daguerre a heliograph. By October 1829, he consented to cooperate with Daguerre, and on December 14, 1829, they signed an agreement. In 1831, they commenced experiments with silver iodide for developing.

Two years later, Niepce suffered an attack of apoplexy and died two days later, on July 5, 1833, at St. Loup de Varenne, France. Daguerre continued their experiments, eventually innovating a mercury vapor development process that proved effective (though dangerous). Niepce's nephew, Claude-Félix-Abel Niepce de Saint-Victor, also advanced the photographic process, becoming the first to use albumen and creating photographic engravings on steel.

Nipkow, Paul Gottlieb
(1860–1940)
German
Electrical Engineer, Communications Industry

Paul Gottlieb Nipkow, while still a student working on his engineering doctorate, proposed the first theory of electromechanical television broadcasting. The signature innovation of his system was the Nipkow scanning disk, a platter pocked with holes through which light cast by an image passed. The element selenium, with its recently discovered photoconductive properties, transformed this light into electronic pulses of varying degrees, based on the intensity of light issuing from the image. A receiver, with a synchronously spinning disk, captured the transmitted image and reconstituted it for projection. All subsequent electromechanical television systems used the Nipkow disk, either in actuality or in theory, but the advent of the all-electronic television system eclipsed electromechanical ones.

Nipkow was born on August 22, 1860, in Lauenburg, Pomerania, which was then in Germany and is now Lebork, Poland. Little is generally known about his youth. It was not until 1883 that Nipkow entered posterity when, as a 23-year-old doctoral student in engineering in Berlin, he proposed his pioneering theory. The next year, he filed and received patent number 30,105, covering his *electrisches teleskop,* or "electric telescope," best known for its Nipkow disk.

Nipkow's system relied on British engineer George May's 1873 discovery of the photoconductive properties of the metallic form of the element selenium, which not only conducted light energy but also transformed it into electrical energy, which led to the development of the photoelectric cell. Nipkow applied the photoconductivity of selenium to capture images by discerning variations in light and shadow.

The key to Nipkow's system, however, was the scanning disk that bears his name. The surface of the disk was punched with 24 square holes that functioned as apertures—spiraling the edge of the disk—which were placed between the brightly illuminated scene and the selenium cell. When the disk rotated, at 600 revolutions per minute, the varying degrees of light from the scene filtered through each hole, generating successive vertical lines of the resolution on the cell; each revolution of the disk thus scanned the entire image, amounting to one frame of the television picture.

The selenium cell captured the dissected image in its discrete units, which it analyzed by transforming it into an electrical impulse that it then transmitted to a receiver. The receiver reversed the process, with a spinning disk rotating in synchronicity with the transmitting disk. The receiver reconstituted the image from the sequential lines of resolution, with light and shadow corresponding to the intensity of the signal voltage. The raster, or resolved image, was then projected onto a screen by means of the Faraday effect, thus completing the transmission.

No evidence exists that Nipkow actually built a prototype of this system, but if he had, the clarity of the image would have been severely limited by the small number of resolution lines created by the disk. As well, selenium's relative insensitivity to shades of light and dark would have limited the image's clarity.

Regardless of whether he built a prototype or not, he certainly could not market the invention, and when the patent lapsed, he lacked the funds to pay for the extension fees for renewal, so he let it expire. He spent the rest of his career as an engineer at a company that manufactured railway-signaling equipment.

The 1907 invention of the amplification tube overcame some of the technical limitations of Nipkow's television system, which was modified to send images over the wire via cable telegraphy from London to Paris. Selenium was later replaced by more light-sensitive elements, further improving the clarity and resolution of the system.

Both major innovators of electromechanical television systems—PHILO FARNSWORTH in the United States and JOHN LOGIE BAIRD in Britain—relied on the Nipkow scanning disk (or variations thereof) in their designs. In recognition of the fact that the technological roots of television resided in Nipkow's invention, the newly established German Television Society named Nipkow its honorary president in 1934.

Three years later, however, the fully electronic television broadcasting system, based on ALAN ARCHIBALD CAMPBELL-SWINTON's earlier theories proposing a television system that forsook all mechanical components, supplanted electromechanical systems based on Nipkow technology as the standard for television broadcasting in Britain, and subsequently throughout the world. Nipkow died two days after his 80th birthday, on August 24, 1940, in Berlin.

Nobel, Alfred Bernhard
(1833–1896)
Swedish
Industrial Chemist

Alfred Nobel invented dynamite and amassed a huge fortune on this and other inventions. He put this wealth to good use by endowing a foundation that after his death awarded annual prizes to the people who made the greatest contributions to humankind in one of several areas: physics, chemistry, physiology or medicine, literature, and peace. Established in 1901, the Nobel Prize quickly became (and remains) the most prestigious award in the world.

Alfred Bernhard Nobel was born on October 21, 1833, in Stockholm, Sweden. His father, Immanuel Nobel, was an inventor and engineer who went bankrupt when barges carrying bridge and building construction supplies were lost. While his father traveled to Finland and Russia to recoup financially, his mother, Andrietta Ahnsell, established and ran a grocery store to support her family. Nobel attended St. Jakob's Higher Apologist School in Stockholm until 1842, when his father had amassed enough wealth selling the naval mines he invented (wooden casks filled with gunpowder submerged just beneath the water's surface) to the Russian military to summon his family to St. Petersburg.

Nobel continued his education in St. Petersburg under private tutors, learning fluency in four foreign languages (Russian, French, English, and German). His father disapproved of his love of literature and poetry, instead encouraging his innate interest in chemistry and physics. In 1850, his father sent him on a two-year continental and transatlantic chemical engineering tour, studying in the United States, Sweden, Germany, and France. In Paris, his favorite city, he worked in T. J. Pelouze's chemistry laboratory. There, he met the Italian chemist Ascanio Sobrero, who had invented nitroglycerine three years earlier by mixing glycerin with sulfuric and nitric acid to produce a highly volatile and explosive liquid (it produces 1,200 times its volume in gas upon explosion).

Upon his return to St. Petersburg in 1852, Nobel worked in his family's business, which provided munitions to the Russian military. At the end of the Crimean War in 1856, however, the munitions market collapsed and Nobel's father again went bankrupt, prompting him to move the family back to Sweden in 1859. There, Nobel and his father worked independently on explosives studies. In 1862, the elder Nobel devised a means of producing nitroglycerine industrially, and the next year, Nobel himself invented a mercury fulminate detonator. Tragedy struck in 1864, when the family's nitroglycerine factory exploded, killing Nobel's brother Emil, as well as several others.

Undeterred, Nobel continued to experiment with nitroglycerine, though Stockholm officials forced him to conduct these studies outside city limits, on a barge anchored in Lake Mälaren. Within three years, he had devised a means of stabilizing nitroglycerine by adding the silica kieselguhr, a porous powder that turned the liquid into a paste that could be shaped. In 1867, he patented this invention (in Sweden, Britain, and the United States) as "dynamite," a product that proved extremely successful commercially. Over the course of his lifetime, he established manufacturing plants and laboratories in 90 different locations in 20 different countries.

The volatility of "guhr dynamite," as it was called, prompted Nobel to continue researching, striving to improve upon his invention. By 1875, he had created blasting gelatin, or gelignite, which added a colloidal solution of gun cotton (or cotton fiber nitrated with a mixture of nitric and sulfuric acids) to nitroglycerin, making it more stable as well as more explosive. He also improved upon blasting powder by mixing nitrocellulose as well as camphor and other additives with nitroglycerin to create ballistite, which was nearly smokeless, in 1887. He continued to conduct research on explosives, as well as other commercial applications, such as synthetic rubber and leather and artificial silk. For these and other inventions, he held 355 patents. He also conducted less market-oriented research in electrochemistry, optics, biology, and physiology at his laboratories in Hamburg, Germany; Ardeer, Scotland; Karlskoga, Sweden; Paris and Sevran, France; and San Remo, Italy.

When Nobel died on December 10, 1896, in San Remo, Italy, his relatives were surprised to discover that he had willed his $9 million fortune to endow a foundation for awarding annual prizes in physics, chemistry, physiology or medicine, literature, and peace to "those who, during the preceding year, shall have conferred the greatest benefit on mankind." In 1969, the Nobel Foundation added the category of economic science to the medal list. The responsibility for establishing the Nobel Foundation fell upon the engineers Ragnar Sohlman and Rudolf Lilljequist. The first Nobel prizes were awarded in 1901. In 1905, the Nobel Peace Prize went to Countess Bertha von Suttner, a peace activist and author of the classic book *Lay Down Your Arms*, who had worked briefly as Nobel's assistant. It was she who probably planted the seed of the idea for establishing prizes that honored great contributions to humankind. Nobel was honored in return in 1958, when element number 102 was named nobelium after him.

Oppenheimer, J. Robert
(1904–1967)
American
Physicist

J. Robert Oppenheimer directed the Manhattan Project, the collaboration of scientists on the design and building of the first thermonuclear weapon. Although his work had seemed like a moral imperative in the race against the Nazis to build an atomic bomb, Oppenheimer reconsidered the ethical status of his work after the detonation of the atomic bombs above Hiroshima and Nagasaki killed millions of civilians. Oppenheimer understood his responsibility for ushering in the nuclear age, and after World War II, he sought to mitigate this negativity by devoting his efforts to limiting the proliferation of nuclear weapons. However, this stance earned him political retribution, as his patriotism was publicly challenged in the McCarthy era of anticommunist hysteria.

Oppenheimer was born on April 22, 1904, in New York City. His mother, Ella Friedman, was a painter, and his father, Julius Oppenheimer, was an émigré from Germany who established a successful textile-importing business in the United States. When Oppenheimer was 11 years old, he became the youngest member of the New York Mineralogical Society, where he presented a paper as a 12-year-old. Upon his graduation from New York's Ethical Culture School, he toured Europe, where he contracted dysentery, requiring a year's rehabilitation.

In 1922, Oppenheimer matriculated in Harvard College, where he studied physics under Percy Bridgman. He graduated summa cum laude in 1925, completing the four-year liberal arts curriculum in a mere three years. He spent the next four years in Europe studying theoretical physics with the preeminent physicists of the day: he worked with Ernest Rutherford, Werner Heisenberg, and Paul Dirac at the Cavendish Laboratory in Cambridge, England; with Max Born at the University of Göttingen (where he earned his doctorate in 1927); with Paul Ehrenfest in Leiden, Holland; and with Wolfgang Pauli in Zurich, Switzerland. Oppenheimer established himself at the forefront of the burgeoning field of quantum physics, developing the Born-Oppenheimer method of addressing quantum mechanics at the molecular level (as opposed to the atomic level). He also conducted research on *bremsstrahlung*, or the continuous emission of radiation.

Upon his return to the United States, Oppenheimer took up a national research fellowship, and then, in 1929, he landed dual appointments as assistant professor at both the California Institute of Technology and the University of California at Berkeley, teaching alternating semesters at each. He ascended to associate professor in 1931 and to full professor in 1936.

In 1930, Oppenheimer predicted the existence of the positron, three years before his pupil Carl Anderson actually discovered this antimatter equivalent of the electron, or the antielectron, which scientists presumed to be the proton before Oppenheimer's hypothesis. He also conducted research on the "cascade process," whereby cosmic radiation breaks down into mesons, the newly discovered atomic components. In 1935, he discovered that the addition of a proton to a neutron to create a deuteron allowed for greater acceleration than a neutron could achieve alone. Throughout this decade, Oppenheimer not only established himself as an important theorist and experimentalist but also as a respected teacher. In 1940, Oppenheimer married Katherine Harrison.

J. Robert Oppenheimer directed the Manhattan Project, which designed and built the first thermonuclear weapon. *(Mrs. J. Robert Oppenheimer, courtesy AIP Emilio Segrè Visual Archives)*

With the advent of World War II, Oppenheimer not only recognized the possibility of harnessing nuclear power in an atomic bomb (as did many other physicists) but he also foresaw the necessity of orchestrating a combined scientific effort conducted in strict secrecy in order for the Allies to achieve this objective before the Nazis. Colonel Leslie Groves of the U.S. Army thus appointed him director of the Manhattan Project, based in remote Los Alamos, New Mexico (at Oppenheimer's suggestion), where the physicist gathered a crack team of leading scientists to collaborate on the design, building, and testing of an atomic bomb.

In May 1945, Oppenheimer conferred with ENRICO FERMI, Ernest Orlando Lawrence, and Arthur Compton to formulate the scientific opinion regarding the use of the atomic bomb (instead of an Allied invasion of Japan) to end the war. Judging that a demonstration of the bomb's destructive capability on an uninhabited island would not convince the Japanese to surrender, the panel voted to use the bomb on a "military target" also populated by civilians, a decision Oppenheimer later regretted. On July 16, 1945, he witnessed the detonation of the first test bomb at Alamogordo, New Mexico, an experience that moved him to quote the *Bhagavad-Gita*: "I am become Death, the destroyer of worlds." On August 6, 1945, the first uranium bomb exploded over the Japanese city of Hiroshima, fulfilling Oppenheimer's work.

After World War II, Oppenheimer continued to influence policy regarding nuclear weapons by cowriting the Acheson-Lilienthal Report that condemned the proliferation of nuclear weapons, in favor of strict international limitations. As chairman of the general advisory committee to the Atomic Energy Commission, a post he took up in 1946, he promoted similar policies curbing the use of nuclear technology for weaponry.

In October 1947, Princeton University's Institute for Advanced Studies appointed Oppenheimer as its director, a position that called on many of the same organizational skills and foresight as the Manhattan Project, and he met with similar success. Under his tenure, the institute transformed into the world's foremost purveyor of theoretical physics.

Oppenheimer's moral opposition to the proliferation of nuclear weapons clashed with the political expediency of developing a hydrogen bomb, a conflict that resulted in his being blacklisted. He was stripped of his security clearance in December 1953, and his patriotism was publicly challenged in congressional hearings. His service on the Manhattan Project confirmed his status as a "loyal citizen," but his contacts with communists in the 1930s were labeled as "defects of character," blocking the reinstatement of his security clearance.

A decade later, in 1963, Oppenheimer received the Enrico Fermi Award from President Lyndon Johnson, an apologetic gesture by the government that had insulted his patriotism. Three years later, he retired from his directorship of the Institute for Advanced Studies, though he continued to serve as a Princeton professor. A lifelong smoker, Oppenheimer battled throat cancer until he died at his home in Princeton, New Jersey, on February 18, 1967.

Ørsted, Hans Christian
(1777–1851)
Danish
Physicist

Hans Christian Ørsted devised a simple but ingenious experiment to demonstrate electromagnetism, as electricity and magnetism were believed to be two separate phenomena at the time. Following up on the fact the electrically charged wires emitted both heat and light, he reasoned that they must emit magnetism as well, so he connected a wire to a battery and ran it by a magnetic compass, diverting its needle

Hans Christian Ørsted's work established the existence of electromagnetism. *(Original watercolor by Day, courtesy AIP Emilio Segrè Visual Archives, Knapp Collection)*

perpendicularly. This demonstration of electromagnetism created a paradigm shift in physics, establishing the existence of electromagnetism and thereby paving the way for the discovery of radio waves and the statement of the theory of relativity, among other significant developments in physics.

Ørsted was born on August 14, 1777, at Rudkøbing in Langeland, Denmark. His mother was Karen Hermansen, and his father, Søren Christian Ørsted, was an apothecary. A German wigmaker and his wife raised Ørsted and his younger brother, as his parents believed that their professional commitments prevented them from raising the boys properly. At the age of 11, Ørsted returned to Denmark to work in his father's pharmacy.

In 1794, Ørsted earned honors on the qualifying examination (despite lacking a school education) for entrance into the University of Copenhagen. He studied pharmacy as well as natural philosophy, astronomy, chemistry, and math, earning his pharmaceutical degree with high honors in 1797. He continued with doctoral studies in natural philosophy, writing his dissertation on Kantian philosophy, interpreting it to say that science amounted to more than simple empirical observation of Nature; further, human perception found structure in Nature, which it expressed in the form of scientific laws. Ørsted earned his doctorate in 1799, then remained at the university as chemical assistant to the medical faculty.

From 1801 through 1803, Ørsted made a scientific tour of the Continent, visiting laboratories in Germany, Holland, and France. After failing to land a professorship in physics upon his return to Denmark, he commenced public lecturing, earning enough of a reputation for the University of Copenhagen to hire him as an extraordinary professor of physics in 1806. As early as 1812, he expressed in his *View of Chemical Laws* his belief in the necessary interrelationship between electricity and magnetism, more a philosophical than a scientific conviction, as he believed in the inherent interconvertibility of all forces.

In April 1820, Ørsted devised a means of demonstrating his theory: he ran a conductive wire, connected to a battery and thus electrically charged, past a magnetic compass. Before he could conduct the experiment, his students filed into the room for his lecture, so he decided to demonstrate the effect publicly. Although his class remained unimpressed by the slight motion of the compass needle at a perpendicular to the wire, Ørsted recognized this as proof of electromagnetism. He communicated his findings in the major European scientific journals in July of that year, creating a sea change in scientific activity. Ironically, the Italian jurist Gian Domenico Romagnosi had discovered the same effect in 1802, but the scientific community ignored his announcement.

Ørsted left it up to other scientists to find applications of the electromagnetic effect. ANDRÉ-MARIE AMPÈRE witnessed Dominique Arago's demonstration of the Ørsted effect at the Académie des Sciences on September 11, 1820, and the next year Ampère returned with a more advanced demonstration: that electrical currents traveling the same direction through parallel wires attract each other but repel each other when traveling in opposite directions. Ørsted turned his attention to chemistry, discovering piperine (the pungent component of pepper) later that year.

In 1822, Ørsted used a piezometer to ascertain a fairly accurate value for the compressibility of water, and the next year, he compared voltaic cells to thermoelectric junctions, noting that the latter created relatively high currents with relatively low potential differences. In 1824, he founded the Danish Society for the Promotion of Natural Science, and in 1825, he isolated metallic aluminum, a development overshadowed by the 1828 preparation of pure aluminum.

In 1829, the Polytechnic Institute of Copenhagen appointed Ørsted as its director, a position he retained for the remainder of his career. He indulged his interest in philosophy in his later career, publishing a nonscientific text, *The Soul in Nature*, in 1850. Ørsted died on March 9, 1851, in Copenhagen. In 1908, the Danish Society for the Promotion of Natural Science established the Ørsted Medal, awarded to Danish physical scientists for outstanding contributions to the field. In 1932, the scientific community adopted the term "oersted" to measure magnetic intensity in physical units.

Ortelius, Abraham
(1527–1598)
Belgian
Cartographer, Geographer

Before Abraham Ortelius published his *Theatrum Orbis Terrarum*, the first comprehensive atlas of the world, the geographic-minded had to sort through individual maps of varying sizes and scales. Ortelius, more an editor than an originator, imposed order on the previously chaotic field of cartography by collating uniformly sized maps with accompanying text, all bound in one book, enabling the 16th-century mind to conceive of its entire world. Ortelius is thus considered the "father of modern cartography."

Ortelius (or Oertel) was born on April 4 or 14, 1527, in Antwerp, Brabant (now Belgium). He considered himself "Belgo-Germanus," as his Catholic family originally hailed from Augsburg, Germany. His father, a prosperous merchant, died when Ortelius was eight years old; to help support his family, he began selling maps and geographic charts. At the age of 20, he was admitted to Antwerp's guild of St. Luke as a map illuminator, a craft and trade involving the purchase, coloring, mounting, and resale of maps.

The more valuable the original map, the more lucrative Ortelius's refinished version, so he traveled throughout Europe, visiting England, Italy, and Germany (especially the annual fairs in Frankfurt and Leipzig) in search of the finest maps he could find. Often, he was accompanied by GERARDUS MERCATOR, the great cartographer who later aided navigation with his innovation of the Mercator projection. It was Mercator who encouraged his friend to enter cartography himself, resulting in Ortelius's production of an eight-leaved map of the world in 1564, some five years before Mercator produced his own renowned map of the world. Ortelius continued to draw his own maps, publishing a two-sheet map of Egypt in 1565, a two-sheet map of Asia in 1567, and a six-sheet map of Spain in 1570.

There is some speculation that Mercator, who coined the term *atlas* after the earth-bearing god, also conceived the idea of uniform map-sizing, and there is further conjecture that he delayed the publishing of his own atlas, begun in 1569 but ultimately published posthumously in

1595, to allow his friend Ortelius priority, though no evidence exists to confirm either supposition. Whatever the case, in May 1570, Ortelius published what would become his magnum opus, the *Theatrum Orbis Terrarum*, a compendium of maps charting the entire known world.

The *Theatrum*, which was published by Egedius Coppens Diesth and printed by Christoph Plantin (or Christoffel Plantijn), consisted of 53 sheets containing 70 uniformly sized maps engraved individually on copper plates, with explanatory text on the back of the folio sheets, all bound in book format. It opened with an allegorical title page depicting the five known continents as native goddesses, as well as a dedication to King of Spain and Netherlands, Philip II (who later appointed Ortelius as his *geographus regius*, or Royal Geographer). Mercator contributed a letter of recommendation to the text, and in return Ortelius derived at least eight plates of the *Theatrum* derived from Mercator's 1569 world map.

Ortelius sought out the most current and reliable maps available, respecting the original cartographers by representing their maps unchanged (except for sizing) and acknowledging their authorship, amending only unsigned maps as he saw fit. Ortelius also included a *Catalogus Auctorum*, or index of authors, listing not just the 33 whose maps he reprinted but all 87 known cartographers of the time. This list served as an invaluable tool for geographic historians, as these lesser-known figures would have faded from the historical record if not for their inclusion in this lasting reference.

Ortelius considered geography the eye of history, or the *historiae oculus*, and therefore included written contextualization of each map. This component, along with his standardized sizing, distinguished his work from his numerous competitors' and ultimately created its success. The *Theatrum* had gone through 31 editions, including translations into seven languages, before the copyright expired in 1625. Throughout these reprintings, he updated and added new material: by 1593, the maps numbered 137, and by 1595, the *Catalogus Authorum* included 183 bibliographic references. By 1612, the atlas had mushroomed to 167 maps, and 7,300 copies had been printed.

By the end of the 1500s, Ortelius was considered the "Ptolemy of his century," both for his cartographic preeminence and for his freeing cartography from the influence of Ptolemy. Ortelius never married, and he died on June 28 or July 4, 1598, in Antwerp.

Otis, Elisha Graves
(1811–1861)
American
Engineer, Transportation Industry

Elisha Otis invented the mechanism that instilled public confidence in the elevator: a safety feature that kept the elevator from falling if its cable broke. While elevators existed before Otis's invention, they were shunned as unpredictably dangerous; after the advent of his safety elevator, elevators became standard equipment in tall buildings. This development coincided with advances in materials and building techniques, resulting in the explosive growth of skyscrapers in urban areas, a phenomenon that transformed the demographic makeup of cities.

Elisha Graves Otis was born on August 3, 1811, in Halifax, Vermont, the youngest of six children. His father was a farmer and justice of the peace who also served in the Vermont state legislature for four terms. Otis attended school in Halifax until he was 19, when he headed west to Troy, New York, where he worked as a builder.

Plagued by bad health and bad luck, Otis bounced from job to job over the next decade and a half. During this period, he worked in the haulage business, then set up his own mill; when this business failed, he converted the mill into a factory where he constructed wagons and carriages. His failing health forced him out of his own business, and he rebounded by becoming a master mechanic at the O. Tingley & Company bedstead factory in 1845.

Otis remained at this position for the next three years, innovating a safety brake for railroad car engineers and improving the design of the turbine wheel. He struck out on his own again as soon as he had saved enough to establish his own machine shop; however, bad luck visited him again, as Albany city officials rerouted a stream that had turned the water turbine powering his shop.

In 1852, Otis returned to industry work as a master mechanic at yet another bedstead factory—Maize & Burns, in Yonkers, New York. One of the owners, Josiah Maize, charged Otis with the task of building a hoist to lift heavy equipment to the newly established factory's upper levels. Aware of the failure rate of lifts at the time, Otis determined to improve upon the design of the elevator to make it safer.

Otis fitted the elevator shaft with toothed guide rails on each side of the elevator cage, to which he fitted a steel wagon spring. He strung the hoist's rope to the middle of the spring, such that it maintained tension; if the rope broke, the slackening of tension would release the spring, which would catch like a ratchet on the teeth of the side rail, stopping the cage in its place. He drummed up sufficient interest in his mechanism to take over the factory when Maize & Burns went out of business, filling the three orders he had received.

It was not until the 1854 Crystal Palace Exposition in New York that business picked up. In a move of marketing genius, Otis put his life on the line to demonstrate his safety elevator: he entered the elevator alone, ascended halfway up the shaft (which was left exposed for public viewing), then ordered the operator on the ground to cut the rope, which held the elevator aloft, with an ax. With a gasp, the crowd watched as Otis's invention sprung into action, activating the ratchet to catch the elevator before it fell.

Orders followed, and the public soon trusted their lives to the Otis safety elevator, as the first public passenger elevator, powered by steam, was installed at the E. V. Haughwout and Company china company in New York City in 1857. In 1860, Otis secured patent number 31,128, covering his elevator-mechanism improvements, including a double-oscillating machine for lifting.

Otis died on April 8, 1861, but his sons, Charles and Norton, continued the family business, naming it Otis Brothers & Co. in 1867. By 1873, the company had produced more than 2,000 elevators. In 1878, the brothers introduced the first hydraulic-powered passenger elevator, and in 1889, the company introduced the first direct-connected, electric elevator with gears. In 1898, the brothers joined forces with 14 other elevator companies to create the Otis Elevator Company, and in 1903, the company introduced the gearless traction electric elevator, which became the standard in the skyscraper era. Otis was inducted into the National Inventors Hall of Fame in 1988.

Otto, Nikolaus August
(1832–1891)
German
Engineer

Nikolaus August Otto designed and built the first gas engine that effectively replaced the steam engine. His four-stroke internal combustion engine revolutionized power production, introducing a vastly more efficient system. The Otto engine, as it was called, powered automobiles, motorcycles, and motorboats and served (and continues to serve) as the prototype for internal combustion engines. The Otto engine thus sparked a vast transformation in transportation, allowing people to travel longer distances faster and more independently than ever before. However, the Otto engine also introduced the large-scale burning of fossil fuels into the environment, creating unimaginable atmospheric consequences.

Otto was born on June 10 (or 14), 1832, at Holzhausen, in Nassau, Germany. The son of a farmer, he dropped out of high school at the age of 16 to work in a grocery store. Over the next dozen years, he worked various jobs: in a merchant's office, in Frankfurt as a clerk, and as a traveling salesman based out of Cologne, hawking sugar, tea, and kitchenware to grocers on the Belgian-French border. There, he became acquainted with Etienne Lenoir's gas engine.

Otto recognized the inefficiencies of Lenoir's design—its high fuel consumption (100 cubic feet of gas per horsepower per hour), its problematic ignition system, its overheating, and its need of constant oil lubrication. To solve these problems, Otto designed a carburetor that processed liquid fuel, vastly improving the engine's efficiency, as well as making other improvements. In 1861, he patented his two-stroke gas engine, known as the "silent Otto." Three years later, in 1864, he teamed up with Karlsruhe Polytechnic graduate Eugen Langen to found a company for manufacturing these engines, and he also collaborated with Franz Reuleaux, a former fellow student.

Otto and Langen exhibited their engines at the 1867 World's Fair in Paris. Competing against 14 different gas engines, Otto's design won the gold medal. Encouraged by their success, N. A. Otto & Cie. constructed a factory (*gasmotorenfabrik*) at Deutz, near Cologne, in 1869. In 1872, Otto hired GOTTLIEB WILHELM DAIMLER (who went on to build one of the first motor vehicles) as its technical director and his partner, Wilhelm Maybach, as head of its design office.

While Otto concentrated on launching the business, he left Langen, Daimler, and Maybach to the task of making engineering improvements to the engine design. However, Otto did not abandon his own engineering efforts forever. In May 1876, he built what became known as the "Otto engine," the first four-stroke piston cycle internal combustion engine. He secured patent number 365,701 on it, and it soon became the first practical and successful replacement of the steam engine.

The Otto engine ran on a compressed mixture of gas and air: the piston's "induction" stroke drew this mix into the cylinder; the "compression" stroke pressurized the fuel; the "power" stroke ignited the aerated gas with a spark provided by a plug, representing the "internal combustion"; the "exhaust" stroke released the charred by-products, preparing the system to repeat the four-step process again indefinitely. Thirty thousand units of this engine, which had a vertical design, sold within its first decade. Otto then refined a horizontal design, improving performance.

In 1862, unbeknownst to Otto, the French engineer Alphonse Beau de Rochas had patented the four-stroke cycle, which he had described in an obscure journal. History supports the fact that Otto had arrived at his design independent of Rochas; nonetheless, Otto's patent was revoked in 1886. In the annals of history, Otto has retained precedence, and his name is associated with the invention of the first four-stroke internal combustion engine, as he effectively introduced his design to the world.

Daimler and KARL FRIEDRICH BENZ both installed Otto engines on their prototypical motor vehicles and continued to use Otto's design to power their automobiles. RUDOLF DIESEL also based his design for the diesel engine on the Otto cycle, and diesel engines became the standard in the railroad industry, as well as in the nautical industry. Otto died in Cologne on January 26, 1891. His company continued to produce engines and continues to exist as Klockner-Humbolt-Deutz AG, the largest manufacturer of air-cooled diesel engines in the world. Otto was inducted into the National Inventors Hall of Fame in 1981.

P

Papin, Denis
(1647–1712)
French
Inventor, Energy and Power Industries

Denis Papin invented an atmospheric steam engine, a similar design to THOMAS NEWCOMEN's atmospheric steam engine, except that it contained a single cylinder for both boiling and condensing the steam. Papin's design ultimately proved unworkable in reality, but it did represent one of the first examples of a piston-driven steam engine, the machine that fueled the Industrial Revolution by de-flooding coal mines.

Papin was born on August 22, 1647, in Blois, France. His parents, Magdeleine Pineau and Denys Papin, were Huguenots, a sect later barred from practicing their religion freely in France. Although Papin himself became a Calvinist, he received his early education at a Jesuit school until 1661, when he matriculated in the University of Angers, where he studied medicine. He received his medical degree in 1669.

After graduation, Papin moved to Paris, where he served as an assistant to CHRISTIAAN HUYGENS while living in the Dutch physicist's apartments in the Royal Library. Together, they conducted experiments on an air pump that Papin constructed, publishing their findings in 1674. The next year, Papin moved to London, where he collaborated with Robert Boyle on further air pump investigations. The pair published their results in 1680 in *A Continuation of New Experiments*, which described Papin's invention of the pressure cooker.

The year before, in May 1679, Robert Hooke arranged for Papin to demonstrate his autoclave, or "steam digester," before the Royal Society. The device captured the pressure generated by the condensation of steam by sealing the system, which significantly raised the boiling point of water. To avoid explosions, Papin designed a steam release valve that activated before the pressure reached a dangerous level, an innovation that proved important to later steam engine designs. Hooke subsequently employed Papin as a low-paid letter-writer (at two shillings per letter) for the Royal Society, which inducted him into its fellowship in 1680.

That year, Papin returned to Paris to continue his collaboration with Huygens, and in 1681, he moved to Venice to serve as the director of experiments at Ambrose Sarotti's Accademia publicca di scienze. Unfortunately, Sarotti intended the experiments to entertain the society's members, not to advance scientific knowledge; Papin retained the position for only three years, returning to London in 1684 to serve as the

Royal Society's curator of experiments. This similar position had its own set of drawbacks, as it paid a mere 30 pounds a year. It did allow him to conduct his own experiments, though, so he continued to investigate hydraulics and pneumatics, publishing his findings in the *Philosophical Transactions* of the Royal Society.

In 1687, Papin moved to Germany, where the University of Marburg appointed him as a professor of mathematics. Ensconced in a community of Huguenots, he married and had many children. It was here that Papin confronted the problem of raising water to fill a canal between Kassel and Karlshaven. At first, Papin proposed a solution he'd devised with Hugyens (though he presented the approach as his own in a letter to GOTTFRIED WILHELM LEIBNIZ)—to raise a piston pump by exploding gunpowder. Unfortunately, this technique left 20 percent residual air in the piston chamber, prompting Papin to devise a more efficient mechanism for powering the piston.

Papin returned to the pressure he had encountered in his earlier experiments with steam to design a steam engine. In 1690, he published *De novis quibusdam machinis,* a tract containing his description of a steam engine with a single cylinder that acted as both boiler and condenser, generating atmospheric pressure to drive a piston, but this design proved too impractical to implement.

In 1705, Leibniz sent Papin a letter that included a sketch of THOMAS SAVERY's steam pump design. In 1707, Papin published *Ars Nova ad Aquam Ignis Adminiculo Efficacissime Elevandam* ("The New Art of Pumping Water by Using Steam"). The paper presented a retooled version (based on Savery's design) of his steam engine, one that proved even less efficient than his original piston design, which was not as successful as Thomas Newcomen's atmospheric steam engine. This was a similar but workable design for an atmospheric steam engine Newcomen arrived at independent of Papin.

Also in 1707, Papin abandoned his patronage in the court of the Landgrave of Hesse (a position he had secured in 1696) to return to London, where he built a man-powered paddleboat. However, he could not secure a post at the Royal Academy, and he fell into poverty and obscurity. Papin's last existing letter was dated January 23, 1712. It is presumed that he died later that year in London.

Pappus
(300?–350?)
Greek
Mathematician

Pappus of Alexandria contributed to the history of science more as a filter than as an originator of his own novel ideas. In his *Collection*, he discussed the entire body of Greek geometry in great detail, thereby recording for posterity the only existing accounts of many texts that did not survive from antiquity. However, Pappus was not merely a transcriber, as he also introduced his own improvements and extensions of the original ideas.

No details of Pappus's life have survived—the dates of his birth and death are thus not recorded and can only be interpolated from clues in existing evidence. The *Suda Lexicon*, a source of questionable reliability, places him as a contemporary of Theon of Alexandria during the reign of Theodosius I, from 379 through 395 A.D. Elsewhere, Theon himself places Pappus as a contemporary of Diocletian, dating from 284 to 305 A.D. The most reliable reference pinpointing Pappus in time is the solar eclipse of October 18, 320, which Pappus mentions in his commentary on Ptolemy with the conviction of one who had witnessed it personally with the eyes of a mature scientist.

Knowledge of Pappus derives almost exclusively from his own writings, many of which have survived only in translation. His notoriety and lasting significance rests on his *Synagoge*, more commonly known as his *Collection*—as this name suggests, it gathered together works that may have been composed separately and later collated. It is organized into eight parts, though evidence within the text itself suggests that he planned, and may have written, at least 12 parts. Compilation of the *Collection* occurred soon after 320 at the earliest, or as late as about 340. The text passed from antiquity to modernity in a single 12th-century manuscript—*Codex Vaticanus Graecus 218*.

The significance of the *Collection* resides primarily in its documentation of the work of previous scholars, much of which has been lost in its original form. Pappus intended his *Collection* as a workbook more than an encyclopedia, so he assumed that his readers would have at hand the works to which he referred. This approach helps historians when Pappus discusses preceding works in detail, allowing them to reconstruct the originals; however, it also hinders this reconstruction when his discussion skips over details he considered redundant with the original in front of the reader.

The faithfulness with which Pappus discusses known works inspires historians with confidence in the accuracy of his discussions of unknown texts. Among the works the *Collection* rescues from oblivion are Euclid's *Porisms* and Apollonius's *Cutting Off of an Area, Determinate Section, Tangencies,* and *Plane Loci*. However, Pappus does not simply record the work of his predecessors—he also extends and expands upon it, offering his original analysis.

In the first section of Book IV, Pappus offers an even broader generalization of the Pythagorean theorem than does Euclid. In the third section, he transcribes several solutions to squaring the circle, including ARCHIMEDES' use of a spiral and Nicomedes' use of a conchoid, but his own use of a quadratrix differs significantly from that of Archimedes. Book VII contains perhaps his most influential original work, dubbed "Pappus's problem" in 1637 by René Descartes, who reconciled the problem of conceiving the product of more than three straight lines geometrically by means of his new Cartesian algebraic symbols. In this respect, Pappus can be considered inspirational to the development of modern analytic geometry.

Besides contributing to the history of geometry in his *Collection*, Pappus also recorded work in astronomy, devoting Book VI to the astronomical work of Theodosius, Autolycus, Aristarchus, and Euclid. Outside of his *Collections*, he wrote surviving commentaries on Ptolemy's *Syntaxis* or *Almagest*, and on Euclid's *Elements*, as well as a lost commentary on Euclid's *Data*. Other lost works, according to existing evidence, might include *Description of the World*, a geographical essay, as well as a work on music and another on hydrostatics, which is suspected to have included a description of the instrument he invented for measuring the volume of a liquid.

Although Pappus did not leave much evidence of his own life, we do know that Pappus had at least one son, Hermodorus, to whom he dedicated one of his works. What little Pappus did leave to the historical record makes up for its limited size with its expansive significance, as his works salvaged much knowledge of ancient geometry from the dustbin of history.

Parsons, Sir Charles Algernon
(1854–1931)
English
Mechanical Engineer

Sir Charles Parsons invented the steam turbine engine, significantly improving the efficiency of the existing steam engine design by breaking down the process of steam expansion into many separate steps, allowing for much more power derivation. Parsons initially used the engine for electricity generation, but for most of his career, he concentrated on designing turbine-driven engines for ships, greatly increasing the speed of seafaring transportation.

Charles Algernon Parsons was born on June 13, 1854, in London. He was the sixth and youngest son born to the Countess and the Third Earl of Rosse, who was the president of the Royal Society. Sir Robert Ball, a scientist, tutored Parsons in Ireland until the age of 17, in 1871, when he matriculated in Trinity College in Dublin. After two years there, during which time he won a mathematics prize, he transferred to St. John's College at Cambridge, graduating with first-class honors in mathematics in 1877.

Parsons then made an unexpected career choice for a young man of his social station: he paid 500 pounds to obtain a "premium" apprenticeship in engineering at the Elswick works of William G. Armstrong in Newcastle upon Tyne. There, he designed an "epicycloidal" engine, which he had first conceived while at Cambridge. Kitsen's Company of Leeds manufactured 40 of these engines under his supervision. While in Leeds, he impressed Katherine Bethell with his handiwork in needlepoint, and the couple married in January 1883.

The next year, Parsons joined the ship equipment-manufacturing firm of Clarke, Chapman and Company as a junior partner and director of its electrical section. He applied himself to the problem of electrical generation, which was exceedingly inefficient at the time—steam engines wasted much of their potential by allowing steam to expand in one step. Parsons resurrected an idea two millennia old, first conceived in basic form by HERO OF ALEXANDRIA—the turbine, which broke down its operation into incremental steps.

Parsons designed a turbine that allowed steam to expand gradually, in 15 stages, which greatly increased the efficiency of the steam engine. A rotor turned blades housed in a cylinder such that the space between blades enclosed a pressurized space where the steam could expand; each successive blade-bound space had lower pressure, eliciting increasing degrees of expansion. The energy created not only exited the system as usable energy but also helped fuel the system, turning the rotor. He patented this steam turbine engine in April 1884, and he also designed a dynamo, or a mechanism for converting mechanical energy to electrical energy by electromagnetic induction, specifically tailored to his steam turbine.

In 1889, Parsons enlisted several friends to help found the C. A. Parsons Company at Heaton, near Newcastle, to build steam turbine engines using an inferior design until 1894, when Parsons secured the patent rights from Clark, Chapman. The year before, he had established yet another business, the Marine Steam Turbine Company at Wallsend on Tyne; Parsons's main goal was to use steam turbine technology toward marine applications. First thing, Parsons designed a vessel with a length of 104 feet and a maximum width, or beam, of nine feet, and constructed of very light steel, to demonstrate the use of a steam turbine engine to power a boat. The *Turbinia*, as the vessel was appropriately dubbed, was launched on August 2, 1894, achieving a speed of 20 knots.

Convinced he could achieve much greater speeds, Parsons built a glass tank in which to observe propellers turning in water. He noticed the effects of "cavitation," or a vacuum that formed behind the propellers, which moved too quickly for the water to adhere to them. He replaced the *Turbinia*'s single turbine/propeller system with a triple turbine/propeller system, increasing the prototype's speed to 34 knots, the equivalent of 40 miles per hour (an unheard-of speed on the seas).

In a bold move, Parsons "debuted" the *Turbinia* by arriving uninvited at Queen Victoria's Diamond Jubilee Fleet Review at Spithead on June 26, 1897, weaving in and out of the slowly parading boats (Parsons, who was at the helm, almost clipped a French yacht). Even the British Admiralty's patrol boats could not catch the speedy *Turbinia*, effectively advertising its superiority over existing technology.

Orders for steam turbine engines flooded in from around the world, prompting Parsons to establish yet another company, the Parsons Marine Steam Company Limited, specifically for the manufacture of vessels powered by *Turbinia*-based steam turbine engines.

In 1898, the Admiralty ordered the torpedo boat HMS *Viper* and the next year added an order for the HMS *Cobra*. In 1901, both ships crashed within a month of each other, though through no fault of Parsons's design or construction. Nevertheless, the loss of 77 lives in the latter crash, including some of Parsons's workers, haunted him for the rest of his life. That same year, the *King Edward* became the first turbine-driven passenger ship to sail the seas, and in 1905, the turbine-driven passenger ships *Victorian* and the *Virginian* began crossing the Atlantic Ocean. Two years later, the turbine-driven *Luisitania*, capable of sailing at 26 knots, won the Blue Riband for the fastest Atlantic crossing.

In honor of the advances he made in marine transportation, Parsons was knighted in 1911, and he received the Order of Merit in 1927. Four years later, while on a cruise in the West Indies, Parsons became ill while in Jamaica, and died on February 11, 1931, aboard the *Duchess of Richmond*.

⊠ Pascal, Blaise
(1623–1662)
French
Mathematician, Physicist

Blaise Pascal exerted a wide-ranging influence on society, contributing to mathematics, meteorology, and theology. In mathematics, he proposed Pascal's theorem, which introduced the mystic hexagram. Pascal's work on conics anticipated the development of projective geometry, and he collaborated with Pierre de Fermat to introduce the theory of probability, inventing Pascal's triangle to aid in calculating possible combinations. Pascal's late work on the theory of indivisibles led to the creation of integral calculus. In the face of intense skepticism, he confirmed the existence of an absolute vacuum, which led to the use of the barometer as a meteorological instrument. Finally, his best-known book, *Pensées*, fused the discipline of scientific thinking with the contemplation of spiritualism.

Pascal was born on June 19, 1623, in Clermont-Ferrand, France. He was the third child (and only son) born to Antoinette Begon, who died when Pascal was three years old, and Étienne Pascal, a judge who home-schooled Pascal, though he forbade the study of mathematics.

Blaise Pascal's influence ranged across disciplines, including mathematics, meteorology, and theology. *(AIP Emilio Segrè Visual Archives, E. Scott Barr Collection)*

Étienne's injunction only excited Pascal's curiosity, as he discovered geometry on his own, arriving at his own proof that the angles of a triangle equal those of two right angles by folding a triangle inscribed within a circle. Étienne relented and gave his son a copy of Euclid's *Elements*, which Pascal devoured.

In 1631, the family moved to Paris, where Pascal began attending, at the age of 16, meetings of "Académie Parisienne" at the Convent of Place Royale directed by Father Marin Mersenne. Pascal became fascinated with the work of one member of the discussions, Girard Desargues, who had just published a profound but inscrutable mathematical work, *Experimental Project Aiming to Describe What Happens When the Cone Comes in Contact with a Plane*. Though young, Pascal understood the work, and what was more, he could translate the ideas into more comprehensible language. In June 1639, he presented a paper to the academy in which he not only explained some complex theorems of projective geometry in terms more comprehensible than Desargues, but he also extended Desargues's principles to introduce his own "mystic hexagram," also known as Pascal's theorem, which stated that the points of intersection of opposite sides of a hexagon inscribed in a conic will lie on the same line.

In December 1639, Pascal and his family moved to Rouen, as his father was appointed tax collector for Upper Normandy. In February 1640, Pascal published his first work, "Essay on Conic Sections"—the lucidity of his explanation of Desargues's work on projective geometry earned him a reputation among mathematicians. The pamphlet also outlined a complete work on conics, which he never published, though GOTTFRIED WILHELM LEIBNIZ preserved accounts of Pascal's manuscript in letters.

After observing the difficulties his father experienced computing taxes, Pascal resolved to design a calculating machine for Étienne. Between 1641 and 1645, he said that he built "more than fifty models, all different" before perfecting his design. This represented the first practical calculator, for although W. Schickard had designed one in 1623, it only reached the prototype stage. Pascal manufactured and marketed his "Pascaline," as the calculator was called, but its high price discouraged its popularity. Of remaining significance is his 18-page pamphlet describing the machine, and the seven existing Pascalines, one of which he gave to Queen Christina of Sweden in 1652.

Pascal next entered the debate over the existence of a vacuum; the conventional scientific wisdom sided with Aristotle's theory of *horror vacui*, that "nature abhors a vacuum." In January and February 1647, Pascal repeated experiments originally performed by Evangelista Torricelli in 1643. Pascal replaced Torricelli's mercury, which air pressure raised about 30 inches in a sealed glass tube, with water and wine, which rose some 39 *feet* in tubes tied to ship masts. Convinced that vacuums exist, he published *New Experiments Concerning Vacuums* in October 1647, though Descartes pronounced after his famous visit to Pascal on September 23 and 24 of that year that the younger scientist had "too much vacuum in his head."

A year later, on September 19, 1648, Pascal's brother-in-law, Périer, confirmed Pascal's prediction that air pressure decreases at higher altitudes by repeating Torricelli's mercury experiment atop Puy de Dôme, which finally proved the existence of an absolute vacuum. This discovery led to the development of the barometer, a fundamental instrument in weather forecasting. Although some questioned Pascal's priority in claiming responsibility for the development of the barometer, nevertheless Pascal's vacuum experiments represented a vital step in the establishment of the barometer as an instrument for measuring air pressure.

Pascal proceeded to investigate pressure in liquids and gases, resulting in his formulation of Pascal's law of pressure, published in the 1653

Treatise on the Equilibrium of Liquids. The law holds that pressure exerts its force equally in all directions in a fluid or gas. Then, in five letters exchanged with Fermat during the summer of 1654, Pascal set the foundation for the later development of his theory of probability, based on the question of the division of stakes in an interrupted game of dice. He invented the so-called Pascal's triangle to facilitate his combinational analysis.

Later in 1654, Pascal had a driving accident on a bridge over the River Seine, resulting in his "night of fire" religious experience. This incident inspired him to write his best-known work, *Pensées*, a diversion from scientific work into spiritualism. However, he did not completely abandon science thereafter, as he developed his theory of indivisibles, which led to the creation of integral calculus. Pascal's health continued to decline, as a malignancy that originated in a stomach ulcer spread to his brain; he died in Paris on August 19, 1662. That year, the city of Paris inaugurated the public transportation system designed by Pascal.

⊠ Pasteur, Louis
(1822–1895)
French
Chemist, Microbiologist

Louis Pasteur is considered one of the most important scientists in the history of the field. His work led to the establishment of the new fields of stereochemistry and microbiology. He developed the process of pasteurization, which has become indispensable in the preparation of safe foods. Later in his career, he theorized the idea of attenuating microbial diseases as a means of immunization, and he developed the first rabies vaccination. Most significantly, though, he combined many of the elements of these researches to propose the germ theory of disease, which revolutionized the practice of medicine.

Pasteur was born on December 27, 1822, in Dôle, in the Jura region of France. His mother was Jeanne-Etiennette Roqui, and his father, Jean-Joseph Pasteur, was a tanner. Pasteur had two older siblings—a brother who died within months of his birth, and a sister named Jeanne-Antoinne (called Virginie), he also had two younger sisters—Joséphine (who died of consumption at 25 and Emilie (who died of epilepsy at 26). The family moved to Arbois, where he attended the local primary school (though his father largely educated him at home). At the age of 16, he traveled to Paris to continue his education at the Collège Saint-Louis, though the homesick youngster soon returned to his family.

In 1839, Pasteur entered the Royal Collège of Franche-Comté at Besançon, where he fashioned himself as a portraitist (an artistic perspective informed all his lifework) while conducting general studies in *philosophie*. He earned his bachelor of letters degree at Besançon on August 29, 1840, and prepared to take his *baccalauréat* examination the next year. He failed in his first attempt, requiring him to remain in Besançon for a second year of advanced mathematics studies. He passed his second *baccalauréat* examination (with a grade of "mediocre" in chemistry) to earn his bachelor of science degree in 1842. The next year, he commenced doctoral studies at the École Normale Supérieure in Paris, earning his doctorate in 1847.

That year, Pasteur was appointed professor of physics by the Dijon Lycée, where he commenced his first important studies, on the molecular asymmetry of tartaric acid: he discovered that the two sides of the molecule rotate plane polarized light in opposite directions, and that living microorganisms can absorb one side but not its antipode. He presented these findings to the Paris Academy of Sciences in 1848, sowing the seeds that would establish a new field of study—stereochemistry.

In 1849, the University of Strasbourg appointed Pasteur as a chemistry professor; that same year, he married Marie Laurent, the university rector's daughter. The couple had five children together, three of whom passed away in childhood. In 1857, the École Normale Supérieure named Pasteur director of scientific studies, a post he relinquished a decade later to relieve himself from administrative duties. In 1863, the school created a chair for him that combined chemistry, physics, geology, and the fine arts. Also in that year, Lille University appointed him as dean of its new science faculty.

Pasteur conducted his next round of significant research at Lille. At the request of a distiller named Bigo, he investigated fermentation, discovering differentiation between yeast cells in good wine (which were spherical globules) and in wine that had gone bad (which were elongated molecules). He thus ascertained that different types of fermentation result from different types of yeast. Pasteur also discovered fermentation to be an anaerobic phenomenon. In an attempt to prevent wines from spoiling, he devised a means of killing the yeast cells after they had done their job by gently heating the fermented wine to 55 degrees Celsius. This process came to be known as "pasteurization," as it was applied to many other foods to prevent spoilage.

Other important studies included Pasteur's final disproof of the theory of spontaneous generation, as well as his identification from 1865 through 1868 of a parasite infecting silkworms. He extended these studies to develop the germ theory of disease, which held that microbial germs spread disease, and that sterile conditions could reduce the spread of disease. This theory proved of extreme significance to the medical field.

In 1867, Emperor Napoleon III granted Pasteur a pension sufficient to support his scientific studies without requiring him to perform bureaucratic duties. The next year, he suffered a stroke that partially paralyzed him, though he recovered well enough to continue his work. Over the next two decades, his work helped establish the field of microbiology; in the second decade, he discovered three bacteria that caused illness: staphylococcus, streptococcus, and pneumococcus. In 1881, he turned his attention to creating vaccines by heating pathogenic microorganisms to attenuate their virulence enough to trigger a response from the immune system without eliciting the disease itself. He applied this method successfully to anthrax bacili in sheep, as well as creating a vaccine for fowl to combat chicken cholera.

The crowning achievement of Pasteur's already distinguished career came when Pasteur developed a vaccine against rabies. He started this work in 1882, and by July 6, 1885, he saved the youngster Joseph Meister from rabies. He presented his findings to the French Academy of Sciences on March 1, 1886; as a result, the Pasteur Institute for rabies and infectious disease studies was approved by the French president in 1887, and the next year it opened. Pasteur remained there as director for the rest of his career, expanding the institute's influence by opening satellites internationally (the first, in Saigon, Vietnam, opened in 1891). Pasteur died in Paris on September 28, 1895.

Pavlov, Ivan Petrovich
(1849–1936)
Russian
Physiologist

Although Ivan Petrovich Pavlov earned the Nobel Prize in 1904 for his pioneering studies of the digestive system, it was his research on conditioned reflexes, conducted after 1904, that won him international fame. This research contributed to the development of a physiologically oriented school of psychology that studied the significance of conditioned reflexes on learning and behavior. The expression "Pavlov's dog" refers to the famous experiments in which Pavlov taught a dog to salivate at the sound of a bell by associating the bell with the sight of food.

Pavlov's intense devotion to his work was probably inspired by his father, Pyotr Dmitrievich Pavlov, an influential priest in the town of Ryazan in central Russia, where Pavlov was born on September 26, 1849. A reader and scholar himself, Pyotr taught his son to read all good books at least twice so that he would understand them better. Varvara Ivanova, Pavlov's mother, also came from a family of clergy.

Pavlov studied, in accordance with his family's wishes, at the Ecclesiastical High School and the Ecclesiastical Seminary in Ryazan. While at the seminary, he encountered the work of Dmitri Pisarev, who taught many of CHARLES ROBERT DARWIN's evolutionist theories. Not long after, he left the seminary to enter the University of St. Petersburg, where he studied physiology and chemistry. Pavlov received his M.D. from the Imperial Medical Academy in St. Petersburg in 1879 and completed his dissertation, for which he received a gold medal, in 1883. Between 1884 and 1886 he studied in Leipzig, Germany, under the direction of Carl Ludwig, a cardiovascular physiologist, and Rudolf Heidenhain, a gastrointestinal physiologist. Another mentor, Sergei Botkin, asked Pavlov to run an experimental physiological laboratory. It was under Botkin's guidance that Pavlov first developed his interest in "nervism," the pathological influence of the central nervous system on reflexes.

Pavlov's first independent research when he returned to Russia, conducted between 1888 and 1890, was on the physiology of the circulatory system. Working on unanesthetized dogs, he would introduce a catheter into the femoral artery and record the influence of various pharmacological and emotional stimuli on the dog's blood pressure. This work earned him his first professorship at the Military Medical Academy, then in 1895 Pavlov became the chair of the physiology department at the St. Petersburg Institute for Experimental Medicine, where he spent most of his career.

From 1890 to 1900, and to a lesser degree until about 1930, Pavlov studied the physiology of the digestive system. In one famous experiment, he slit the gullet of a dog he had just fed, causing the food to drop out before it reached the dog's stomach. He was able to show that the sight, smell, and swallowing of food were enough to start secretion of digestive juices. For this work he received the Nobel Prize in 1904.

Pavlov next developed the laws of conditioned reflex by focusing on neural influences in digestion. He noted that his dogs would sometimes salivate at the sight of a lab assistant who often fed them. Through a series of repeated experiments, he demonstrated that if a bell is rung each time a dog is given food, the dog eventually develops a "conditioned" reflex to salivate at the sound of the bell, even when food is not present. Pavlov's conditioned reflex theory became the subject of much research in the fields of psychiatry, psychology, and education.

In addition to his influential career in science, Pavlov was a political activist who frequently spoke out against communism and protested the actions of Soviet officials. In 1881 he married Seraphima Vasilievna Karchevskaya, a friend of the writer Fyodor Dostoyevsky. They eventually had four sons and a daughter. In addition to the Nobel Prize, Pavlov was awarded the Order of the Legion of Honor of France and the Corey Medal of the Royal Society of London. He remains one of the fathers of modern science.

Pennington, Mary Engle
(1872–1952)
American
Chemist, Food Technology

Mary Engle Pennington was sometimes called the "Ice Lady," in reference to her expertise in refrigeration and frozen foods. She designed the first refrigerated boxcars for railroad transportation of perishables, while also developing the standards governing railway refrigeration until the 1940s. Earlier in her career, she had implemented standards for milk inspection that were subsequently adopted by health boards throughout the United States.

Pennington was born on October 8, 1872, in Nashville, Tennessee. Her mother was Sarah B. Molony, and her father was Henry Pennington, a label manufacturer who piqued his daughter's interest in food and botany by gardening together. She graduated from high school in 1890 and then attended the Towne Scientific School at the University of Pennsylvania to study chemistry and biology. The university did not grant bachelor's degrees to women, only certificates of proficiency, which she earned in 1892; but in an ironic twist, she continued with doctoral work in chemistry (with minors in zoology and botany) and obtained her Ph.D. in 1895 through a bureaucratic loophole allowing women to earn doctorates in "extraordinary cases."

Pennington remained at the University of Pennsylvania for two years (from 1895 through 1897) on postdoctoral fellowships in chemical botany; she next spent a postdoctoral year (from 1897 through 1898) at Yale University, conducting research in physiological chemistry. She then returned to Philadelphia to serve as a lecturer and director of the chemical laboratory at the Women's Medical College of Pennsylvania until 1906. Simultaneously, she worked as a researcher at the University of Pennsylvania until 1901, when she established her own laboratory, the Philadelphia Chemical Laboratory, to provide bacteriological analyses for physicians.

In 1904, the Philadelphia Health Department hired Pennington as director of its bacteriological laboratory. She focused her efforts at this post on reducing bacterial contamination of dairy products, especially ice cream. During her three-year tenure there, she simultaneously performed some consulting work for the Bureau of Chemistry of the United States Department of Agriculture. The head of this bureau, Harvey W. Wiley, was also a family friend who encouraged her to apply for a newly created government position overseeing compliance with the Pure Food and Drug Act of 1906. Aware of the inherent sexism she would encounter, she veiled her sex by applying as "M. E. Pennington." Her qualifications earned her a position in 1907 as a bacteriological chemist, much to the consternation of the Food Research Laboratory, which did not expect a woman to fill the post.

Within a year, however, the excellence of Pennington's work revealed that her gender was immaterial to her performance, so she was appointed as head of the lab. Over the next decade, she spearheaded multiple advancements in the storage and transportation of perishable foods: she developed milk inspection standards that were later adopted by many health departments throughout the United States; designed egg packaging that reduced breakage; and improved methods for freezing and storing fish safely. During this period, she also invented refrigerated train cars, designing insulation systems to retain the cold from ice beds. President Herbert Hoover conferred on her the Notable Service Medal for her work on railway refrigerator cars during World War I.

In 1919, Pennington left her position as a civil servant for a post in industry as the director of research and development at the American Balsa Company, an insulating materials manufacturer. She remained there only three years, though, before launching her own consulting firm in New York City in 1922. Her firm specialized in dispensing information and advice on the processing and storage of frozen foods, both at the industrial and the household level. In 1933, she coauthored the book *Eggs*.

In 1940, the American Chemical Society presented Pennington the Garvan Medal, awarded annually to the outstanding woman chemist in the United States. She served as vice president of the American Institute of Refrigeration and was the first woman member of the

American Society of Refrigerating Engineers. She was also the first woman inducted into the Poultry Historical Society's Hall of Fame, in recognition of her development of slaughtering methods that extended the freshness of poultry. Pennington continued to work until the end of her life; she died in New York City on December 27, 1952.

Perkin, William Henry
(1838–1907)
English
Organic Chemist, Textile Industry

William Henry Perkin discovered the first synthetic dye at the age of 18, quite by accident. He had been trying to synthesize quinine, an antidote to the fatal disease of malaria, but yielded only a reddish sludge that turned out to be an excellent dye—aniline purple, more popularly known as mauve. With the advent of synthetic dyes, an industry grew around the manufacturing of coal-tar dyes. Perkin can rightly be considered the father of this industry segment.

Perkin was born on March 12, 1838, in Shadwell, South London, England. His father was George Fowler Perkin, a builder and contractor who wanted his son to become an architect. However, Perkin became enamored with science at the age of 12 or 13, when a friend demonstrated some chemical experiments on crystalline substances. In 1851, Perkin matriculated in the City of London School, where his master, Thomas Hall, encouraged his interest in chemistry. Hall convinced Perkin, who was 15 years old at the time to pursue undergraduate study at the Royal College of Science, where he became assistant to the eminent scientist August Wilhelm von Hofmann in his second year. In the meanwhile, Perkin had been pursuing scientific research on his own, setting up a laboratory in his home.

Perkin's first experiments involved dyes, and in collaboration with Arthur H. Church, he derived one of the first azo dyes from naphthalene, "nitrosonaphthalene," which the pair immediately patented. Interestingly, Perkin turned his attention away from dyes for his next experiments, in which he attempted to synthesize the antimalarial substance quinine at Hofmann's suggestion. So during Easter vacation in 1856, Perkin tried to re-create quinine's chemical formula by mixing chromic acid (from potash) first with toludine, then with aniline. Both experiments yielded a dark sludge, representing unpromising results to most chemists, who looked for clear crystalline solutions.

Ever curious, Perkin continued to experiment with the sludge, seeing what the addition of alcohol would extract from it. The aniline solution turned strikingly purple, a result Perkin considered promising—not as a discovery of quinine, but potentially as a new dye. Silk soaked up the dye, Perkin discovered, prompting him to send a sample to Pullars of Perth, Scotland. The dyers responded enthusiastically, inspiring him to patent his dye-preparation method. Against the advice of his mentor, Hofmann, Perkin borrowed money from his father to finance the establishment of a dyestuff works on the banks of the Grand Union Canal in Greenford Green, West London, in June 1857. Two years later, he married Jemima Harriet Lissett, who bore him two sons—William Henry Jr. and Arthur George, both of whom followed in their father's footsteps into chemistry. Lissett died in 1862, and in 1866, Perkin married Alexandrine Caroline Mollwo, and together the couple had one son and four daughters.

Perkin did not simply rest on his laurels, however: in 1869, he developed a commercially viable process for synthesizing alizarin—the year before, Carl Graebe and Karl Lieberman had discovered this red dye to replace madder-root dye, but their process proved prohibitively expensive. However, Perkin submitted his patent application a mere day after the German chemical manufacturers BASF. Despite lacking a patent, Perkins & Sons manufactured a ton of alizarin by the end of the year, and by 1871, it produced 220 tons a year.

In 1874, Perkin sold Perkin & Sons to Brooke, Simpson and Spiller, using the proceeds to secure retirement funds ample enough to support research in pure science for the rest of his life. Although Perkin placed little stock in public recognition, he received copious amounts of it, garnering the 1879 Royal medal and Davy medal a decade later, both from the Royal Society, which had inducted him into its fellowship in 1866. He also received the 1888 Longstaff medal and the 1890 Albert medal of the Royal Society of Chemical Industry.

The 50th anniversary of Perkin's discovery of aniline purple (1906) brought him honor upon honor: the German Chemical Society granted him its Hofmann medal, the French Chemical Society its LAVOISIER medal. He also received the Perkin medal, named after him by the American section of the Society of Chemical Industry. As well, he was knighted. However, under the strain of such notoriety, Perkin developed pneumonia and died on July 14, 1907, in Sudbury, England.

Piaget, Jean
(1896–1980)
Swiss
Psychologist

Jean Piaget proposed a theory of developmental psychology that respected the natural cognitive states of childhood instead of assessing childhood development in adult terms and thereby judging it lacking. Piaget's "genetic epistemology," or timetable of natural development, continues to wield a profound impact upon education by encouraging teachers to meet students at their own level rather than forcing children to strive for levels of understanding that they are cognitively incapable of attaining.

Piaget was born on August 9, 1896, in Neuchâtel, Switzerland, the eldest child of Rebecca Jackson, a strict Calvinist, and Arthur Piaget, a professor of medieval literature at the University of Neuchâtel. Piaget attended Neuchâtel Latin High School, where, in 1907 at the age of 10, he wrote his first published scientific paper—a short notice on an albino sparrow he had sighted. At the age of 15, he published a paper on mollusks that gained him international recognition.

Piaget continued to study zoology at the University of Neuchâtel, where he earned his doctorate in 1918. He then became interested in psychology, and he studied psychoanalysis for a semester under Carl Gustav Jung at the University of Zurich. He moved to Paris, where he spent two years working at the École de la rue de la Grange-aux-Belles, a boys' school founded by ALFRED BINET, creator of the intelligence test. While administering true/false tests, he noticed that children consistently make similar "mistakes," leading him to consider whether the fault was not with the children but with the tests.

Piaget began exploring the idiosyncratic logic of childhood by asking children open-ended but probing questions, such as "What makes the wind?"

"The trees," five-year-old Julia replied.

"How do you know?" Piaget asked, to which Julia replied, "I saw them waving their arms."

"How does that make the wind?" Piaget inquired, and Julia answered, "Like this," waving her hand in front of her face. "Only they are bigger. And there are lots of trees."

"What makes the wind on the ocean?" Piaget asked, introducing adult logic into the conversation, to which Julia answered, "It blows there from the land. No. It's the waves. . .".

Instead of invalidating childhood logic by prioritizing adult logic over it, Piaget respected the cognitive state of childhood. "Children have real understanding only of that which they invent themselves, and each time that we try to teach them something too quickly, we keep them from reinventing it themselves," he explained. Such innovative thinking earned him the directorship of Geneva's Institut J. J. Rousseau in 1921. In 1923, he married Valentine Châtenay, and together the couple had three children—Jacqueline, Lucienne, and Laurent, who grew up as the casual subjects of their father's psychological observations.

Over the next five decades, Piaget held a series of impressive and broad-ranging appointments, many of them concurrent: he returned to the University of Neuchâtel in 1926 as a professor of philosophy, remaining there until 1929; he then held an array of positions at Geneva University—professor of child psychology and history of scientific thought from 1929 through 1939, professor of sociology from 1938 through 1952, and professor of experimental psychology 1940 through 1971; he also served as director of the Institut Universitaire des Sciences de l'Education in Geneva from 1933 through 1971 and director of the International Bureau of Education in Geneva from 1929 through 1967; finally, he also returned to Paris as a professor of genetic psychology at the Sorbonne from 1952 through 1963. In 1955, Piaget secured funding from the Rockefeller Foundation and the Swiss National Foundation for Scientific Research to found the International Center of Genetic Epistemology at Geneva University, serving as its director until his 1971 retirement.

Piaget's theory of childhood psychological development included four stages: sensorimoter, from birth through the age of two, during which time infants learn about objects through tactile contact; preoperational, lasting from ages two through seven, when children begin to represent objects with language; concrete operational, the stage from ages seven through 12, when children learn numbers, and distinguish between present and past, as well as between similar and different; and finally, the formal operational stage from 12 years on, when children develop adult logic and mathematical abilities.

Piaget disseminated his theories in an astounding number of publications, numbering more than 60 books and monographs—many of them written in the technical vernacular of psychology. His best-known books translated into English include *The Language and Thought of the Child*, published in 1923; *Judgment and Reasoning in the Child*, published the next year; and *The Origins of Intelligence in Children*, published in 1948. Piaget died on September 16, 1980, in Geneva.

Planck, Max
(1858–1947)
German
Physicist

As the originator of quantum theory, Max Planck revolutionized physics. His work provided the foundation for a great many discoveries in modern physics. Planck's primary achievement, which ultimately won him the Nobel Prize, was his radical reworking of the conception of energy. Classical physics postulated, and indeed the entire scientific community believed, that energy was always transmitted in a continuous form, such as a wave. Planck's research led him to conclude that energy could exist in discrete units, which were later given the name quanta. In further elucidating this theory he discovered a universal constant in nature—of equal magnitude to other universal constants such as the speed of light—which came to be known as Planck's constant.

Planck was born in Kiel, Germany, on April 23, 1858, to Johann von Planck and Emma Patzig Planck. Max was his parents' fourth child and also had two step-siblings from his father's previous marriage. When Planck was nine, the family moved from Kiel to Munich.

After graduating from the Königliche Maximillian Gymnasium in Munich in 1874, Planck entered the University of Munich. His

Max Planck is recognized as the creator of quantum theory. *(AIP Emilio Segrè Visual Archives)*

education was interrupted by a severe illness in 1875 that forced him to withdraw. In 1877, he completed two semesters at the University of Berlin, as he continued to recuperate. He returned to Munich in 1878 and passed the state examination for higher-level teaching in mathematics and physics that year. He wrote his dissertation on the second law of thermodynamics and received his Ph.D. from the University of Munich in 1879.

Unfortunately his personal life was filled with tragedy. He married his childhood sweetheart, Marie Merck, in 1885, and the couple had three children, but after his wife died in 1909, both his daughters died while giving birth. His son was then killed during World War I. Planck eventually remarried, but his son from this union was executed in 1944 for plotting to overthrow Hitler. Planck's home was destroyed during an air raid in 1945, just two years before his death.

Prior to attaining his first university appointment in 1885 at the University of Kiel, Planck was a Privatdozent at the University of Munich. He was offered an assistant professorship at the University of Berlin in 1888; he remained there until 1926. He turned his attention to the topic of blackbody radiation in 1896. A blackbody is an ideal body or surface that absorbs all radiant energy when heated and gives off all frequencies of radiation when cooled. Physicists had tried to formulate a mathematical law that applied to the blackbody's radiation of heat, but no scientist could adequately explain blackbody radiation across the entire spectrum of frequencies. Planck solved this problem in 1900 with his groundbreaking theory of the quantum (Latin for "how much"). He proposed that energy moved in a stream of discrete units. Planck determined that the energy of each quantum is equal to the frequency of the radiation multiplied by the universal constant he discovered. The numerical value of Planck's constant is 6.62×10^{-27}.

Planck then decided to investigate the application of his quantum theory to the emerging field of relativity. He retired from the University of Berlin in 1926 and four years later became the president of the Kaiser Wilhelm Society (subsequently renamed the Max Planck Society) in Berlin. He increasingly dedicated his energies to philosophical and spiritual questions. In 1935, he published *Die Physik im Kampf um die Weltanschauung*, which explored the connections among science, art, and religion. His posthumously published *Philosophy of Physics* addressed the importance of finding general themes in physics.

Planck's contribution to the field of physics was immense. Not only did he launch a new subfield, quantum mechanics, but his work also made future endeavors, such as the discovery of atomic energy, possible. It provided a foundation for the theories of Werner Karl Heisenberg, Wolfgang Pauli, and later ALBERT EINSTEIN. Planck was awarded the Nobel Prize in physics in 1918 and achieved a status as one of the greatest scientific minds of his era. He died in 1947. On his headstone was engraved the mathematical expression of Planck's constant.

Playfair, Lyon
(1818–1898)
English
Chemist

Sir Lyon Playfair is best known not for his own greatest achievement—the discovery of nitroprussides—but for his promotion of another scientist's invention. In the mid-19th century, Sir CHARLES WHEATSTONE developed a promising new cipher, but it was Playfair who proved its best advocate, endorsing its adoption as a strategic weapon in the Boer War. The cipher came to be known as "Playfair," and it continued in use through World War I and into World War II.

Playfair was born on May 21, 1818, in Chunar, India. Relatives in Scotland raised Playfair while his father, Dr. George Playfair, served as inspector general of hospitals in Bengal and as a medical officer with the East India Company. In 1835, Playfair matriculated at the Andersonian Institution in Glasgow, where he studied chemistry under Thomas Graham as a medical student. He transferred to the University of Edinburgh, where he received his medical degree, but a severe case of eczema prevented him from pursuing a career in medicine. After failing in his attempt at a business career in India, he returned to London to serve as assistant to his former mentor, Graham.

From 1839 through 1841, Playfair worked under Justus Liebig in Giessen, Germany, where he conducted studies on myristic acid and caryophyllene. However, it was his translation of Liebig's *Die organische Chemie in ihre Anwendung auf Agricultur und Physiologie* that secured his reputation in the field of chemistry. He worked in the industry briefly, serving as chemical manager at Primrose Mill, near Clitheroe, Lancashire.

In 1842, the University of Toronto offered Playfair a chair in chemistry; however, Britain's prime minister, Sir Robert Peel, intervened to offer Playfair a post as chemist to the Geological Society and a chair of chemistry at the newly established School of Mines in order to retain Playfair in England. There was one catch—these positions were not open at the time, so Playfair spent the next three years as a professor of chemistry at the Royal Institution in Manchester, where he collaborated with JAMES PRESCOTT JOULE investigating the atomic volume of substances in solution and with ROBERT WILHELM BUNSEN analyzing blast-furnace gases.

Playfair held this dual appointment for more than a decade, during which time he performed numerous other functions—he served on the Royal Commissions on the Health of Towns and on the Irish Potato Famine, and as Prince Albert's chief adviser on the Great Exhibition of 1851 in London. From 1850 through 1858, he served as inspector of schools of science in London. During this period, he also discovered nitroprussides, his greatest contribution to chemistry.

Playfair's most significant contribution to society fell outside of his own discipline. In 1854, his neighbor, Sir Charles Wheatstone, invented a cipher that proved much more effective than simple substitution and even polyalphabetic (or Vigènere substitution), the prevailing types of existing ciphers. Wheatstone's cipher came to be known as "Playfair" after the man who promoted the system most enthusiastically. Playfair advocated the cipher on the strength of its simplicity compounded by its impenetrability. Nevertheless, the cipher met some resistance—when Wheatstone pointed out that elementary schoolboys could learn the cipher in 15 minutes, a governmental undersecretary replied, "That is very possible, but you could never teach it to attachés."

Despite such complaints, the British services adopted the Playfair cipher in the Boer War and continued to use it through World War I. More sophisticated ciphers replaced it by the time of World War II, though it remained a backup cipher for emergency situations. The cipher hinged on the choice of a keyword, arranged in some order, such as in a square, in columns, or in a spiral, with the remaining letters of the alphabet following the keyword in their customary order. The message itself is broken down into two-letter pairs, then letters are substituted based on their position in the keyword arrangement.

In 1858, Playfair accepted the University of Edinburgh's chair of chemistry, though he failed to produce any significant work in his decade-long tenure there. He did distinguish himself, however, through his public service to the government thereafter. In 1868, he won a seat in Parliament representing the Liberal Party and the Universities of Edinburgh and St. Andrews, and he promoted the cause of universal compulsory education. He served as postmaster general for a year starting in 1873 and chaired the House of Commons' ways and means committee while also serving as deputy speaker from 1880 through 1883.

That year, Playfair retired. In honor of his years of civil service and his contributions to society, he was knighted. Playfair died on May 29, 1898, in London, England. However, his influence continued to exert itself whenever the cipher bearing his name was used. Its most prominent employment occurred in World War II, when Lt. John F. Kennedy sent an emergency message using the Playfair cipher from Plum Pudding Island, where he washed ashore after a Japanese cruiser sank the PT-109 under his command. His encrypted message evaded detection by the Japanese forces swarming the Solomon Islands, resulting in his (and his surviving crew's) rescue by Allied coast-watchers.

Pliny the Elder
(23?–79)
Roman
Naturalist

If it were not for the work of Pliny the Elder, much early science history would not have been recorded. In his 37-volume book, *Historia Naturalis*, Pliny not only transcribed the existing knowledge of "natural history" (as science was then known) from Greek and, to a much lesser extent, Roman scholarship, but he also added his own observations. *Historia Naturalis* represents the first time that an individual author sought to encompass the entirety of scientific understanding. Pliny the Younger called his uncle's book "a diffuse and learned work, no less rich in variety than nature itself."

Pliny, or Gaius Plinius Secundus, was born in about the year A.D. 23 in Como. It appears that he came from a family of means that relocated to Rome by the time he was 12. After receiving an education in literature, law, and oratory, he joined the military at the age of 23. During one of his first military assignments, as commander of a cavalry squadron stationed on the Rhine, he wrote a monograph on the use of the lance on horseback in warfare.

While continuing his career in the militia, Pliny followed this literary contribution with several more—including a history of Roman campaigns in Germany and a biography of his commander, Pomponius Secundus. In about A.D. 57 or 58, he retired from public life in the military (perhaps due to differences with the new emperor, Nero) and dedicated himself over the next decade to writing, and possibly to practicing law. He wrote a grammar and a work on oration; however, none of his early works have survived.

In A.D. 69, when Vespasian inherited the throne from Nero, Pliny reentered public life in service of the state, visiting the new emperor daily as an unofficial adviser while in Rome; while in the provinces of Germany, Spain, and perhaps even Tunisia, he served as an overseer of finances. His literary production increased throughout this period, as he worked on a history of the years 44 through 71, published posthumously but lost to posterity.

The only work surviving work by Pliny, the 37-volume *Historia Naturalis*, amply establishes his significance in history. Dedicated to Vespasian's son Titus and published in A.D. 77, two years before his own death, the tome covers the entirety of natural history—astronomy, meteorology, geography, physiology, zoology, botany, geology, and mineralogy.

Pliny almost certainly began gathering information for his natural history years before he started composing it. He was a scrupulous note-taker, going so far as to dictate notes to his secretary while eating. He included not only his own original observations but also a staggeringly comprehensive catalog of existing Greek and Roman scientific information. "Things must be recorded because they have been recorded" was his motto, and in his preface he states that he referenced more than 100 principal authors for some 20,000 facts. However, the index contained in book one, which is not even exhaustive, cites 473 authors for 34,707 facts.

Pliny considered his job of compiling a complete natural history both a duty as well as a mark of distinction. "It is godlike for man to help man," he states in book two, which he dedicates to astronomy, recording the works of Hipparchus and Eratosthenes as well as adding original research of his own. To compile a catalog of stars, he devised a classification system that lasted more than 17 centuries. He also posited one of the earliest hypotheses on the nature of gravity, and later, in book 36, he recorded one of the first examples of art history.

Pliny imposed a logical order on *Historia Naturalis*, discussing different fields of natural history in successive books: he devoted books three through six to geography and ethnography; book seven to human physiology; books eight and nine to marine animals; book 10 to birds; book 11 to insects; books 12 through 19 to botany, agriculture, and horticulture; books 20 through 27 to medicine and drugs; books 28 through 32 to medical zoology; and books 32 through 37 to mineralogy.

No evidence suggests that Pliny married, and in fact he designated his nephew, Pliny the Younger, a prominent biographer in his own right, as his sole heir. His last commission placed him in command of the fleet at Misenum, embarking from the northwest edge of the Bay of Naples when he observed a strange bank of clouds. His scientific curiosity got the best of him, and he landed at Stabiae only to discover that the clouds issued from the erupting Mount Vesuvius. Poisonous gases overwhelmed him, and Pliny died on August 25, 79.

Polhem, Christopher
(1661–1751)
Swedish
Mechanical Engineer

Known as the "father of Swedish mechanics" and as the "Archimedes of the North," Christopher Polhem promoted the use of waterpower to replace human labor, establishing a factory to employ this technique for manufacturing tools of his own invention. He also believed in the division of labor, some two centuries before this concept gained common acceptance as the most efficient means of production.

Polhem was born on December 18, 1661, in Visby, on Gotland Island in Sweden. His father, the merchant Wolf Christopher Polhamma, died when Polhem was eight years old. Polhem's stepfather sent him to an uncle in Stockholm, where he received his education at the Deutsche Rechenschule. However, when his uncle died, Polhem had to abandon school for work, which he found near Uppsala, as an estate bailiff for the Biorenklou family. He also honed his carpentry and mechanical skills at the lathe and the forge, which would prove important later, as he made all his own instruments.

In 1687, Polhem entered the University of Uppsala, where he studied astronomy and mathematics. He also applied his mechanical skills repairing clocks—two astronomical pendulum clocks and a dilapidated clock in a medieval cathedral. The latter job, which took more than a year to complete, established his reputation, earning him a 500-daler salary from the king of Sweden to support continued mechanical experimentation and work.

Polhem's reputation extended beyond Sweden's borders with the installation of the Blankstot hoist at the Stora Kopparberg mine in 1693, an engineering feat. He spent the next several years, from 1694 through 1696, entertaining invitations to tour factories, mills, and mines in Germany, Holland, England, and France, where he soaked up various diverse mechanical methods and instrument designs.

Upon his return to Sweden in 1697, Polhem established the Laboratorium Mecanicum, a mechanical laboratory. There, he invented new mechanical techniques, including brick-, barrel-, and wheel-making, bell-casting, organ building, and such oddities as a tap "to prevent serving maids from sneaking wine from the cask." He also designed hoisting devices, mine pumps, and the Polhemsknut, a flexible shaft coupling.

The next year, the state of Sweden appointed Polhem as its director of mining engineering. In this position, he drew up plans to power all the machines at the Falu mine with three large, remote waterwheels; however, he never completed this project. Later in his career, he planned to build a canal between Göteborg and the Baltic Sea, but he never completed this project either. However, he did design and oversee the construction of two sluices, the first in Stockholm, and the second, named the Polhem sluice, in Trollhättan.

Polhem strongly believed in the replacement of human labor with waterpower, and he applied this theory in the founding of a waterpowered tool-manufacturing factory in Stjärnsund at Husby in southern Dalarna in 1700. He designed all the production machinery, developing numerous innovations that improved efficiency. The factory produced a wide variety of tools and implements, such as knives, scissors, pans, bowls, and trays, as well as Polhem locks (or burglarproof locks) and Stjärnsund pendulum clocks. Water powered all the mechanical processes—cutting gears for the clocks, shaping hammers for plates, and mill-rolling sheet metal. Also in 1700, he was appointed master of construction for the Falu mine, where he installed a Karl XII hoist.

In 1707, the elector of Hannover invited Polhem to visit mines of the Harz district, where he continued to absorb new information, such as mining methods. A decade later, in 1716, Frederick I of Sweden knighted him in recognition of his contributions to Swedish society. He also contributed outside of Swedish society, building a minting machine for George I of England.

Polhem wrote incessantly, producing some 20,000 pages of manuscripts. His 1729 book, *A Brief Account of the Most Famous Inventions*, recorded his essential work. Polhem died on August 30, 1751, in Tingstäde, Sweden. However, his legacy lived on into the 21st century, as a team of Swedish computer engineers honored his memory for 3Dwm who named some of their software after Polhem in recognition of his significance as an inventor. He is also enshrined at the Historical Museum of Gotland, which devotes an entire room to Polhem's technical inventions.

Polo, Marco
(1254–1324)
Venetian
Explorer

Ever since the turn of the 13th century, when Marco Polo published accounts of his travels in the far east in the book *Divisament dou Monde*, or *Description of the World*, the validity of his stories have been called into question. Incredulous readers dubbed his book *Il Milione*, or *The Million Lies*, as they could not believe Polo's descriptions of a world far more advanced than their own. As recently as 1995, British librarian Frances Wood published a book entitled *Did Marco Polo Go to China?* Although scholarship has revealed some of Polo's more fantastic depictions to be mere fabrications, much of what he reported has been borne out as true, such as his description of howling sand dunes (which turned out to be the wind whipping across sand-covered granite). Some skeptics point out that he could have gathered accounts of the far east from Arab or Persian travelers. No matter whether he visited China himself or not, his accounts exposed Europe to the far east for the first time, creating an impression that prevailed until the 18th century and even persists in the present.

Polo was born in about 1254, in Venice (or possibly on the island of Curzola, his family's trading outpost off the coast of Dalmatia). When Polo was six years old, his father and uncle, Niccolo and Maffeo, embarked on a journey eastward to China, reaching the capital city of Beijing in 1266, two years after the Mongolian emperor Kublai Khan had established the Yuan dynasty. Intrigued by Christianity, Khan dispatched the Polo brothers with a golden tablet ensuring safe passage, as well as a letter in Turkish to Pope Clement IV, requesting him to send back oil from the Holy Sepulchre in Jerusalem and a hundred scholars to teach Khan of Western science and religion. The Polo brothers returned to Venice in April 1269.

By 1271, with preparations for the return trip complete, 17-year-old Marco Polo accompanied his father and uncle first through Acre (now Akron, Israel) to Jerusalem to procure the holy oil and back to Acre, then continued overland through Armenia, Afghanistan, Hormuz, and across the Gobi desert. After traveling some 5,600 miles, Polo arrived at Shang-tu, Kublai Khan's summer residence in May 1275 and subsequently traveled to the capital in Beijing.

Polo won Khan's favor, securing political assignments carrying him to Burma and India, as well as appointment as prefect of Yangchow for three years (though Chinese records do not mention Polo). Chinese culture impressed Polo with its paper currency ("With these pieces of paper they can buy anything and pay for anything. And I can tell you that the papers that reckon as ten bezants do not weigh one"), coal ("stones that burn like logs"), and postal service (with second, first, and imperial class).

Wood questions why Polo failed to mention the Great Wall, tea ceremonies, and foot binding. Historian John Larner counters that the Great Wall had crumbled by the 13th century only to be built back up in the 16th century, that the popularity of tea ceremonies had not yet migrated from the south, and that the practice of foot binding was limited "to upper-class ladies . . . confined to their houses."

In 1292, Khan dispatched Polo with the Mongol princess, Lady Kokachin, bound for marriage to the Persian prince, Arghun, Khan's grand-nephew. Polo arrived in Persia, after two years of arduous sea travel, only to learn that the prince had died (the princess married his son), as had Kublai Khan (releasing Polo from the obligation to return). Polo continued on to Venice, arriving home in the winter of 1295.

In 1298, war broke out with the Genoese, who captured the galley commanded by Polo and imprisoned him. Polo dictated the story of his eastern travels to his cell-mate, the romance author Rusticello of Pisa, who wrote up this account and published it. The book sold extremely well, thereby establishing Europeans' geographic and cultural understanding of the far east.

After his return to Venice in the summer of 1299, Polo married Donata Badoer, who bore him three daughters. Legend has it that on his deathbed, Polo replied to the priest who pleaded for a retraction of his "lies" by reiterating, "I have not told the half of what I saw." Polo died in Venice on January 9, 1324, and was buried at the Church of San Lorenzo. After his death, geographers and historians substantiated much of the information transcribed in his book, thus partially resurrecting his reputation. It is known that Christopher Columbus owned a copy of Polo's book, which influenced his decision to sail westward in search of the far east.

Poncelet, Jean-Victor
(1788–1867)
French
Mathematician, Engineer

Some two centuries after BLAISE PASCAL proposed the theory of projective geometry, Jean-Victor Poncelet established this new geometry as a veritable field. Poncelet himself first posited his notions while incarcerated as a prisoner of war, but professional resistance to his theories and the intervention of his military and academic career in mechanics intervened, delaying his final exposition of projective geometry by half a century.

Poncelet was born on July 1, 1788, at Metz, in Lorraine, France. It was not until after a family in Saint-Avold had raised him to the age of 16 that his father, the wealthy landowner and advocate of the Metz Parliament Claude Poncelet, acknowledged paternity of Poncelet by Anne-Marie Perrein. At that point, in 1804, Poncelet returned to his birthplace to study at the Metz *lycée*. On the strength of his performance there, he graduated in October 1807 to the École Polytechnique in Paris, where he studied under the likes of ANDRÉ-MARIE AMPÈRE and Gaspard Monge, becoming a protégé of the latter. After two years of study and one year recuperating from illness, Poncelet entered the corps of military engineers, studying at the École d'Application at Metz from September 1810 to February 1812.

In June 1812, Lieutenant Poncelet was assigned to the engineering general staff of Napoleon's disastrous Russian campaign. In the confusion of the retreat in the Battle of Krasnoï on November 18, 1812, French forces left Poncelet for dead, but in fact Russian forces had captured him and held him captive at a prison camp on the Volga River in Saratov. He remained a prisoner of war there until June 1814, and he returned to France in September of that year.

Instead of wasting away in jail, Poncelet spent the year and a half recounting the foundations of geometry from memory, deprived as he was of textbooks, as a stepping-stone for his leap into the uncharted territory of projective geometry, advancing beyond the realm reached by Pascal and his predecessor, Girard Desargues. Poncelet recorded his theories in his "cahiers de Saratov," using these notes to reconstruct his theories later. It took more than half a century before the pioneering work he performed there was published in its entirety.

The engineering corps appointed Poncelet to the rank of captain upon his return to France, assigning him to fieldwork back in Metz. There, he designed a variable counterweight drawbridge in 1820, publishing a description of this innovation in 1822. However, his engineering duties did not prevent him from continuing his work in projective geometry.

Also in 1822, Poncelet published *Traité des propriétés projectives des figures*, applying his notion of the "ideal chord" and his "principle of continuity" to transform geometry from a disparate form into one capable of sustaining generalizations. As with Pascal and Desargues before him, he considered the specific case of conics, extrapolating them to infinity and to imaginary points. His presentation of these ideas to the Académie des Sciences two years earlier had met with severe criticism, due more to misunderstanding on the audience's part than to mistaken reasoning on his part; nevertheless, the unenthusiastic reception of his theory discouraged him from elucidating it in its entirety, thus postponing the introduction of projective geometry by four more decades.

In May 1824, the École d'Application de l'Artillerie et du Génie appointed Poncelet professor of mechanics applied to machines, displacing his attention from geometry to engineering. That summer, he doubled the efficiency of existing waterwheels by designing a system with curved paddles. The Académie des Sciences awarded him with a mechanics prize for this innovation in 1825. His focus on engineering supplanted his geometry work, causing him to delay the publication of writings that formulated the principle of duality; in the meanwhile, Joseph Gergonne proposed similar ideas, giving him priority in the establishment of this principle. A close historical look reveals that Poncelet's memoir of 1824, not published until 1829, had anticipated Gergonne by several years.

Poncelet devoted himself to his teaching of applied and industrial mechanics for the remainder of his career. As well, he served on the municipal council of Metz and as secretary of the Conseil Général of the Moselle in 1830. He moved to Paris in 1834, serving in 1837 as professor of physical and experimental mechanics on the Faculty of Science of Paris. Five years later, he married Louise Palmyre Gaudin.

Although Poncelet was slated to retire from the engineering corps in 1848, he continued to work for two more years due to the Revolution of 1848, which resulted in his appointment as professor of mechanics at the Collège de France. In addition, he was promoted to the rank of brigadier general and made commandant of the École Polytechnique, a duty he filled by reorganizing the school's curriculum. Upon his retirement in 1850, he took up the task of compiling a survey of the advances in industrial machinery thus far in the 19th century for presentation at the Universal Expositions of London in 1851 and of Paris in 1855.

Finally, in 1857, Poncelet cleared his slate sufficiently to return to his work on projective geometry, collecting his work on the subject for publication. With the appearance of the two volumes of *Applications d'analyse et de géométrie* in 1862 and 1864, Poncelet finally enunciated his theory of projective geometry in a complete form, allowing for its broad implications to be absorbed and synthesized with existing knowledge.

Poncelet published two more volumes on projective geometry in the form of a second edition of his seminal 1822 text, *Traité des propriétés projectives de figures*, thus solidifying the establishment of this new field. The history of mathematics marks this as the beginning of a modern geometry. Poncelet died in Paris on December 22, 1867.

Popov, Alexander Stepanovich
(1859–1906)
Russian
Physicist, Communications Industry

The Soviet Union celebrated "Radio Day" every May 7, in honor of Alexander Popov, considered the "Father of Radio." Evidence suggests that Popov and GUGLIELMO MARCONI may have developed telegraphy concurrently, but lack of documentation weakens the case for Popov. Regardless of actual priority, Marconi certainly established telegraphy as a significant means of communication commercially. However, Popov's contributions to the development of telegraphy cannot be ignored.

Alexander Stepanovich Popov was born on March 16, 1859, in Turinsk, a mining village in the Ural Mountains (now Krasnoturinsk, Sverdlovsk oblast), one of seven children. His father was a priest who sent him to the free seminary in Perm (formerly Molotov). However, Popov's interest in physics, mathematics, and engineering diverted him from entering the priesthood. Instead, he entered St. Petersburg University in 1877 to study under the Faculty of Physics and Mathematics. He supported his education by working at the Elektroteknik artel, the first electric power plant in Russia. He defended his doctoral dissertation in 1882 and graduated with distinction the next year.

Also in 1883, Popov accepted an assistantship at the Russian Navy's Torpedo School in Kronstadt on Kotlin Island in the Gulf of Finland, where he conducted investigations into magnetic and electrical phenomena. In 1888, the school promoted him to an instructorship. That same year, HEINRICH RUDOLF HERTZ demonstrated a detection of electromagnetic waves, confirming JAMES CLERK MAXWELL's 1865 prediction of their existence. In 1894, Oliver Lodge invented the "coherer," a device for detecting electromagnetic waves based on Edouard Branly's 1890 observation that the proximity of an electrical charge decreased the electrical resistance of fine metal powders.

Popov improved Lodge's coherer by installing a tapping mechanism that restored the metal powders to equilibrium after each detection, thus allowing the device "to note separate, successive discharges of an oscillatory character," as Popov noted in 1895. And whereas neither Branly nor Lodge conceived of practical applications for their detection devices, Popov recognized the possibility of detecting lightning at a distance, thus increasing the reliability of weather prediction.

On May 7, 1895, Popov presented a lecture entitled "On the Relation of Metallic Powders to Electrical Oscillations" to the Russian Physical and Chemical Society in St. Petersburg. By connecting the coherer to a wire antenna on one end and a ground wire on the other end, Popov was able to receive electromagnetic waves from distant lightning bolts. He made further demonstrations of his "storm indicator" later that summer at the Institute of Forestry in St. Petersburg, detecting lighting charges as far as 20 miles away.

In January 1896, he published "An Apparatus for Detecting and Recording Electrical Oscillations" in the *Journal of the Russian Physical and Chemical Society*. He ended the article with a prediction of the development of radiotelegraphic technology: "In conclusion, I may express the hope that my apparatus, when further perfected, may be used for the transmission of signals to a distance by means of rapid electric vibrations if only a source of such vibrations can be found possessing sufficient energy."

On March 24, 1896, Popov reportedly demonstrated just such a telegraphic transmission by sending "rapid electric vibrations" in the form of Morse Code some 800 feet from one building at St. Petersburg University, where the Russian Physical and Chemical Society was again meeting, to another building, where the society's president, F. F. Petrushevsky, transcribed the message onto the blackboard: "Heinrich Hertz." Unfortunately, no written record survived of this historic demonstration, and it was not until 30 years later that witnesses attested to this occurrence.

On June 2, 1896, Marconi filed a patent for his invention of telegraphy, which he demonstrated that July to William Preece, the chief engineer of the British Post Office. Although Popov disputed Marconi's claim to priority, the Italian inventor and entrepreneur was clearly responsible for promoting the new technology and thereby establishing its significance as an efficient means of communicating. Most experts accord Marconi with responsibility for inventing telegraphy.

In 1901, Popov was appointed a professor at the Electrotechnical Institute in St. Petersburg, which promoted him to its directorship in September 1905. However, when the governor of St. Petersburg ordered him as director to repress student protests against tsarism, Popov refused due to his belief in freedom of political expression. However, he fell ill on January 10, 1906, from the resulting stress, and he died of a brain hemorrhage three days later, in St. Petersburg, Russia.

In 1908, the physical section of the Russian Physical and Chemical Society convened a commission of competence, which concluded that Popov "was justified as being recognized as the inventor of the wireless telegraph." However, this pronouncement failed to shift popular opinion from holding Marconi's priority.

Quimby, Edith H.
(1891–1982)
American
Biophysicist

Edith H. Quimby developed diagnostic and therapeutic uses for X rays, radium, and radioactive isotopes. The field of radiology was still in its infancy when she began her work, but Quimby's groundbreaking research ascertained the extent of radiation's ability to penetrate an object, thereby allowing physicians to use the smallest possible doses of the potentially dangerous substances. Quimby also cofounded the Radiological Research Laboratory at Columbia University, which studied the medical uses of radioactive isotopes (especially in cancer diagnosis and treatment).

Born on July 10, 1891, in Rockford, Illinois, Quimby was one of Arthur and Harriet Hinckley's three children. The family moved frequently during Quimby's childhood, and she completed high school in Boise, Idaho.

Quimby attended Whitman College in Walla Walla, Washington, majoring in physics and mathematics. After graduating in 1912, she taught high school in Nyssa, Oregon. In 1914, she won a fellowship to study physics at the University of California, where she met and married Shirley L. Quimby (a fellow physics student). Edith Quimby was awarded her master's degree in 1915 and then returned to teaching high school science for four more years.

Quimby's career changed course in 1919 when her husband was offered a faculty position at Columbia University. In New York, she accepted a position at the newly created New York City Memorial Hospital for Cancer and Allied Diseases. Working with chief physicist Dr. Gioacchino Failla, Quimby began to explore the medical uses of X rays and radium, particularly in treating tumors. Quimby and Failla would enjoy a 40-year scientific association.

Although physicians had begun to use radiation by this time to diagnose and treat certain diseases, their efforts were often haphazard at best since no standardized techniques had yet been developed. Quimby's work did much to alleviate this problem. In 1923, she instituted a film badge program so that X-ray film could be employed to gauge radiation exposures accurately. More importantly, she determined the specific radiation doses required to treat tumors. In 1932, she became the first to determine the distribution of radiation doses in tissue from various arrangements of radiation needles. The techniques she devised became the standard in the United States. During this period, she also

Edith Quimby's groundbreaking research ascertained the extent of radiation's ability to penetrate an object, thereby allowing physicians to use the smallest possible doses of the potentially dangerous substances. *(Center for the American History of Radiology, courtesy AIP Emilio Segrè Visual Archives)*

focused on measuring the penetration of radiation, quantifying the different doses required to produce the same biological effect, such as skin erythema (reddening). In the process, she formulated the concept of biological effectiveness of radiation, which is still employed by radiobiologists today.

After spending more than 20 years at Memorial Hospital, Quimby moved with Failla to the Columbia University College of Physicians and Surgeons in 1943, becoming an associate professor of radiology and later advancing to the rank of full professor. While at Columbia, she and Failla cofounded the Radiological Research Laboratory. At their new laboratory, the pair began working with the newly available artificial radioisotopes being produced by accelerators and reactors. They concentrated on the application of radioactive isotopes, such as radioactive sodium and iodine, to treat thyroid disease and diagnose brain tumors. These early clinical trials established Quimby as a pioneer of nuclear medicine. During her tenure at Columbia, Quimby was also involved with several other projects. In addition to working on the Manhattan Project, which created the atomic bomb, she joined the Atomic Energy Commission and advised the U.S. Veterans Administration on radiation therapy.

Quimby finished her career at Columbia University by teaching a new generation about radiation physics and the clinical use of radioisotopes. She retired in 1960, though she continued to research, write, and lecture. She coauthored a book, *Physical Foundations of Radiology*, and wrote a number of influential scientific papers. Acknowledged as a founder of radiobiology, Quimby received a slew of honors and distinctions. In 1940, she became the first woman to win the Janeway Medal of the American Radium Society. The Radiological Society of North America later awarded her a gold medal, citing her work "which placed every radiologist in her debt." She died at the age of 91 on October 11, 1982.

Richards, Ellen Swallow
(1842–1911)
American
Chemist, Home Economist

Ellen Swallow Richards is considered the mother of euthenics, or the study of environmental improvement as a means of human improvement, as well as of ecology and home economics. She taught housewives practical ways to use science to improve their families' health. She was also the first woman to graduate from the Massachusetts Institute of Technology (MIT), where she remained for the rest of her career, and the first woman in the United States to earn a bachelor of science degree in chemistry.

Ellen Henrietta Swallow was born on December 3, 1842, in Dunstable, Massachusetts. Her mother, Fanny Gould Taylor, was a teacher, as was her father, Peter Swallow. She helped her parents tend their farm and keep their grocery store, and in return they educated their only daughter at home, until her intellectual appetite outstripped their ability to feed it. At that point, the family moved to the nearby town, home to Westford Academy, which she attended. After graduating in 1863, she taught in Littleton, Massachusetts, where she worked various other jobs (nursing, tutoring, storekeeping, cooking, and housecleaning) to save money for furthering her education.

In September 1868, at the age of 26, Swallow enrolled at Vassar, a women's college, where she studied astronomy under Maria Mitchell and applied chemistry under Charles Farrar. She completed its four-year curriculum in two years to become a member of Vassar's first graduating class, earning her bachelor of arts degree in 1870. Seeking further education in chemistry, she became the first woman to enroll at the Massachusetts Institute of Technology in January 1871. MIT dubbed her studies as the "Swallow experiment," considering her a kind of guinea pig testing the educability of women. In 1873, she submitted a thesis entitled "Notes on Some Sulpharsenites and Sulphantimonites from Colorado" to become the first woman in the United States to earn a bachelor of science degree in chemistry.

That same year, Swallow isolated a new metal, vanadium, from iron ore, and composed a thesis on this research to earn a master of arts degree from Vassar College. She was granted an *Artium Omnium Magistra* by MIT, which retained her as an untitled and unsalaried instructor and laboratory assistant. She continued her studies, though MIT voted at that time to disallow women from receiving doctorates. However, her achievement did convince the institute to vote, on May 11, 1876, in favor of allowing women to officially enroll at MIT. The year before, she had married Robert Hallowell Richards, the head of MIT's department of mining engineering, on June 4, 1875.

Ellen Swallow Richards appealed to the Women's Education Association of Boston to fund a women's laboratory at MIT, which was established in 1876. Serving as an assistant to lab director John M. Ordway, she instructed a small group of women in chemical analysis, industrial chemistry, mineralogy, and biology. MIT closed the women's laboratory in 1883 (allowing all its students to matriculate into the regular curriculum), but the next year, it opened a new laboratory for the study of sanitary chemistry. As the lab's principal instructor, Richards taught air, water, and sewage analysis.

In 1881, Richards collaborated with Marion Talbot to form the Association of Collegiate Alumnae (now the American Association of University Women, or AAUW). She published her first book, *The Chemistry of Cooking and Cleaning: A Manual for House-keepers*, in 1882, following up with *Food Materials and Their Adulterations* in 1885 and *Home Sanitation* in 1887. That year, she commenced a two-year survey of Massachusetts's inland waters, analyzing some 40,000 water samples. Her study raised concern over the management of the environment, exposing more human damage than had been suspected. She fused these concerns with the more practical concerns of "home economics," or the management of the household, to found a discipline that she called ecology, though the title met with resistance (Melvill Dewey refused to include ecology in his classification system).

In Boston in 1890, Richards founded the New England Kitchen to teach women about proper food hygiene. Nine years later, she organized a conference on home economics at Lake Placid, New York; nine years after that, in 1908, she helped establish the American Home Economics Association, serving as president for its first two years. She even underwrote the publication of the association's *Journal of Home Economics*. She published almost a dozen books in the 20th century, including new editions of *Food Materials and Their Adulterations* in 1906, and *The Chemistry of Cooking and Cleaning* in 1910, as well as *Euthenics: Science of Controllable Environment* in 1910.

In 1910, Richards finally became a doctor when Smith College granted her an honorary doctorate. She died of heart disease in Jamaica Plain, Massachusetts, the next year, on March 30, 1911.

Ellen Swallow Richards is considered the mother of home economics. *(Courtesy The Massachusetts Institute of Technology Museum)*

Röntgen, Wilhelm Conrad
(1845–1923)
German
Physicist

Wilhelm Conrad Röntgen was the first scientist to recognize the phenomenon of X rays (which he so-named for their mysterious properties, such as the ability to penetrate solid objects such as human skin). Other scientists were certainly exposed to X rays, as Röntgen discovered them while conducting fairly commonplace research on the cathode rays emitted by a CROOKES tube, but none noticed their peculiar effects. For this discovery, which established the field of radiology (the application of X rays for medical diagnosis), Röntgen received the first Nobel Prize in physics in 1901.

Röntgen was born on March 27, 1845, in Lennep im Bergischen, Prussia (now part of Remscheid, Germany). He was the only son of Charlotte Constanze Frowein and Friedrich Conrad Röntgen, a cloth manufacturer and merchant. When Röntgen was three, the family moved to Apeldoorn, Holland, where he attended the Institute of Martinus Herman van Doorn. Röntgen entered the Utrecht Technical School in December 1862, but after about two years, he acted as the fall guy for a fellow student who caricatured an unpopular teacher, resulting in Röntgen's expulsion from school. Unfortunately, his sacrifice cost him dearly, as the University of Utrecht refused him admittance as a regular student, relenting in January 1865 by admitting him as an "irregular" student.

In November 1865, Röntgen transferred to the Swiss Federal Institute of Technology in Zurich, which admitted him as a "regular" student in mechanical engineering. He earned his diploma in August 1868, and, at the suggestion of his physics professor, August Kundt, he remained at the institute for graduate work in physics, not engineering. He wrote his doctoral dissertation on "Studies about Gases" to earn his

Wilhelm Conrad Röntgen was the first scientist to recognize the phenomenon of X rays. *(Photograph by Aufnahme u. Verlag v. A. Baumann Nacht, München, courtesy AIP Emilio Segrè Visual Archives, Landé Collection)*

Ph.D. in 1869. At about this time, he met Anna Bertha Ludwig, the daughter of a German revolutionary in exile in Zurich, and the couple married on January 19, 1872. They had no children together, but they adopted Ludwig's six-year-old niece, Josephine Bertha, in 1887.

After earning his doctorate, Röntgen maintained his relationship with Kundt by serving as his assistant, first in Zurich, then moving with his mentor to the University of Würzburg in Germany in 1871. After two years there, Röntgen again relocated to remain with Kundt, this time to France's newly established University of Strasbourg, which appointed him as a privatdozent. In

1875, he struck out on his own to serve as professor of physics and mathematics at the Agricultural Academy of Hohenheim, but within a year, he returned to the superior research facilities at Strasbourg, where he was promoted to the rank of associate professor of physics.

In 1879, Röntgen finally broke ranks with Kundt to accept a full professorship at the German University of Giessen in Hesse, where he remained for the next decade. In 1888, his last year there, he published a paper reporting what he considered his most important discovery: an experimental confirmation of JAMES CLERK MAXWELL's theory of electromagnetism. Henry Rowland claimed to have created a magnetic effect of electrostatic charges in motion in 1875, but none could repeat his results until Röntgen did so. In recognition, Hendrik Lorentz dubbed the phenomenon the "röntgen current."

In 1888, Röntgen returned to the University of Würzburg (which had previously refused him an academic appointment for lack of acceptable credentials) as a professor of physics and director of its Physical Institute. In 1894, the university appointed him as its rector, though he continued to conduct research on top of his administrative responsibilities.

In 1895, Röntgen was investigating the cathode rays emitted by a Crookes tube, a popular line of research at the time, although it was outside of his own specialty. On Friday, November 8, he darkened his laboratory and covered the tube in black cardboard; to his surprise, he noticed some barium platinocyanide crystals on a screen all the way across the room glowing when he turned on the apparatus. The chemical plate continued to luminesce even when he moved it into the next room, convincing Röntgen that he was observing not cathode rays, which travel only a few centimeters, but a wholly different, much stronger phenomenon.

Röntgen spent the next seven weeks confirming the extraordinary properties of this previously undiscovered effect. These X *strahlen*, or X rays (as he named them after their mysterious nature), traveled in straight lines as far as two meters, passed through glass and wood but not 20 metal, and could not be refracted, reflected or deviated by magnetism. They could also expose photographic plates, and Röntgen took photos of balance weights enclosed in a box, the chamber of a shotgun, and, most famously, the bones of his wife's hand. It was this last picture, taken on December 22, that captured the world's imagination when the popular press carried word of his discovery on January 6, 1896.

A week later, the kaiser summoned Röntgen to demonstrate X rays before the royal court, immediately earning the physicist the Prussian Order of the Crown, Second Class (though he refused to use the "von" title before his name). Röntgen himself followed up on his discovery with surprisingly little attention, writing only three papers on X rays, and declining to patent the discovery, preferring to allow free use of the innovation. However, the scientific community followed up on his discovery with intense activity, as ANTOINE-HENRI BECQUEREL discovered radioactivity later in 1896 while conducting X-ray studies.

The scientific community also followed up on Röntgen's discovery of X rays by honoring him with numerous awards, including the 1896 Rumford Medal of the Royal Society, the Royal Order of Merit of Bavaria and the Baumgaertner Prize of the Vienna Academy that same year, the 1897 Elliott-Cresson Medal of the Franklin Institute, and the 1900 Barnard Medal of Columbia University. Most significantly, the Royal Swedish Academy of Science granted Röntgen the first Nobel Prize in physics in 1901.

In 1900, the Bavarian government requested that Röntgen move from Würzburg to fill a similar position at the University of Munich as professor of physics and director of its Physical Institute, a position he retained until grief over his wife's 1919 death forced him to retire in 1920. The inflation in the wake of World War I bankrupted Röntgen, who died from intestinal cancer on February 10, 1923.

Rowland, F. Sherwood
(1927–)
American
Atmospheric Chemist

F. Sherwood Rowland shared the 1995 Nobel Prize in chemistry with his colleague Mario J. Molina, as well as Paul Crutzen, for their roles in realizing the erosion of the protective ozone layer by chlorofluorocarbons (CFCs), which the scientific community assumed until that point to be inert in the atmosphere. Their informing the world of this phenomenon elicited an extremely swift response from the international community, and within the quarter of a century since their initial discoveries, CFC production has ceased.

Frank Sherwood Rowland was born on June 28, 1927, in Delaware, Ohio, the middle son of three. The family had moved there the year before, when his father, Sidney A. Rowland, was appointed as the chairman of the mathematics department at Ohio Wesleyan University. His mother, Margaret Drake, taught Latin. Rowland's parents encouraged his scientific curiosity and study, and he developed his interest in atmospheric science by volunteering at the local weather station, measuring temperature highs and lows as well as the amount of precipitation. He attended an accelerated public school system that promoted him through its ranks to graduate from high school at the age of 16.

Ineligible for the draft, Rowland entered college at Ohio Wesleyan University in 1943, studying year-round for two years before he enlisted in the U.S. Navy to train radar operators. After World War II ended (he saw no action), he returned to Ohio Wesleyan, electing to finish his studies in two years instead of just one year. He graduated in 1948 with a bachelor of arts degree in chemistry, physics, and mathematics. He immediately matriculated at the University of Chicago to pursue graduate studies under some of the most prominent scientists of the day: the chemist Willard Libby, the physical chemists Harold Urey and Edward Teller, the inorganic chemist Henry Taube, and nuclear physicists Maria Goeppert Mayer and ENRICO FERMI.

Rowland earned his master's degree in radiochemistry in 1951 and then focused his dissertation on the chemical state of radioactive bromine produced in the cyclotron to earn his doctorate the next year. On June 7, 1952, he married a fellow graduate student, Joan E. Lundberg, and the couple embarked for Princeton University, where Rowland had secured an instructorship in chemistry. There, their first child, Ingrid Drake, was born in the summer of 1953; a son, Jeffrey Sherwood, joined the family in the summer of 1955, born on Long Island during one of Rowland's summers conducting research in the chemistry department of the Brookhaven National Laboratory.

Rowland remained at Princeton until 1956, when the University of Kansas hired him as an assistant professor. The university promoted him to a full professorship by 1964, when the University of California at Irvine appointed him as a full professor and the first chairman of its chemistry department. He chaired the department until 1970, when he stepped down to pursue a new direction in research. That year, he chanced upon that direction: after attending the International Atomic Energy Agency meeting on the environmental applications of radioactivity in Salzburg, Austria, he shared a train compartment with an Atomic Energy Commission (AEC) official, who introduced Rowland to the notion of studying the atmosphere at the chemical level, an approach the scientist immediately espoused.

The AEC representative invited Rowland to his 1972 Chemistry-Meteorology Workshop in Fort Lauderdale, Florida, the second in a series, which featured the speaker James Lovelock, who suggested the utility of following the atmospheric movement of an extremely stable molecule, chlorofluorocarbon (CFC), to track wind patterns. The idea fascinated Rowland, who wondered what happened when this molecule eventually destabilized, as it must through solar photochemistry. With financial support from the AEC, Rowland commenced a study of the atmospheric chemistry of CFCs, enlisting the services of his new postdoctoral researcher, Mario Molina.

Within three months, Rowland and Molina determined that CFCs drift into the stratospheric ozone layer, some eight to 30 miles above the earth, where ultraviolet radiation breaks down the bond, releasing an atom of chlorine. This free radical readily combines with ozone molecules, destroying their radiation-blocking properties. What is more, a single chlorine atom recombines with thousands of ozone molecules. That year, the United States alone produced 400,000 tons of CFCs, most of which was released into the atmosphere. Rowland and Molina announced these initial observations in a landmark paper that appeared in the June 28, 1974, edition of the prominent British scientific journal, *Nature*, creating an almost immediate sensation.

The National Academy of Sciences confirmed Rowland and Molina's findings in 1976; their testimony in legislative hearings convinced the Environmental Protection Agency (EPA) to ban production of aerosol CFCs in October 1978. They continued to study CFC-ozone depletion, as did a spawning generation of atmospheric chemists who realized the dire implications of such diminution. In late 1984, Joe Farman discovered what the CFC-ozone depletion theory inherently prophesied: a "hole" in the ozone above Antarctica, as revealed by satellite pictures.

Susan Solomon led two Antarctic expeditions in fall 1986 and summer 1987, gathering data that confirmed the disappearance of the ozone layer above the south pole. Rowland had collaborated with Solomon (as well as Rolando R. Garcia and Donald J. Wuebbles) on the paper announcing her hypothesis explaining the polar concentration of atmospheric destruction, "On Depletion of Antarctic Ozone," published in the June 19, 1986, edition of *Nature*.

In 1987, the industrial nations of the world adopted the Montreal Protocol to phase out CFC production before the end of the century, and 1992 amendments foreshortened this time frame. The University of California at Irvine recognized Rowland's accomplishments by naming him to endowed chairs: from 1985 through 1989, he served as the Daniel G. Aldrich Jr. Professor of Chemistry; from 1989 through 1994 as the Donald Bren Professor of Chemistry, and from 1994 on as the Donald Bren Research Professor of Chemistry. Also since 1994, he has served as the foreign secretary of the U.S. National Academy of Sciences. In 1995, he received global recognition for his contributions to the welfare of humankind with the awarding of the Nobel Prize in chemistry, which he shared with Molina and Paul Crutzen.

Sagan, Carl Edward
(1934–1996)
American
Astronomer, Exobiologist

Carl Sagan proposed a number of theories on scientific topics, ranging from the effects of nuclear war on the planet's climate to the potential for the discovery of life outside Earth. He was unique in his extensive knowledge of biology and genetics, which he applied to his primary field of astronomy. He maintained a lifelong, passionate interest in exobiology—the discipline devoted to investigating the possibility of extraterrestrial life. His greatest contribution, however, was his capacity to make scientific concepts available and intriguing to a popular audience. His television show *Nova*, for example, used a synthesis of special effects and accurate science to captivate a mass audience.

Sagan was born in New York City on November 9, 1934, to Samuel and Ruth Gruber Sagan. His father was a Russian emigrant who worked as a cutter in a clothing factory. As a child Sagan was drawn to the study of astronomy and devoured science fiction novels. He was married three times—to Lynn Alexander Margulis in 1957, Linda Salzman in 1968, and Ann Druyan. He had five children.

In 1955, Sagan received his B.S. from the University of Chicago, where he remained to pursue his doctorate. As a graduate student he became enchanted with exobiology. He was fortunate to receive support in this typically unrespectable pursuit from a number of respected scientists, including Hermann Joseph Muller, Gerard Peter Kuiper (one of the very few astronomers at that time who was also a planetologist), and Harold Clayton Urey, who had advised Stanley Lloyd Miller in his experiments into the origins of life. Sagan's dissertation, "Physical Studies of the Planets," earned him a Ph.D. in astronomy and astrophysics in 1960.

Sagan was the Miller Residential Fellow at the University of California at Berkeley from 1960 to 1962. He then accepted a faculty position at Harvard University, where he remained until 1968. During his time at Harvard he postulated two theories that were subsequently confirmed. He hypothesized that the color variations on Mars were caused by dust shifted in windstorms, and that Venus's surface was extremely hot because the atmosphere trapped the Sun's heat. In 1968, he became an assistant professor at the Center for Radiophysics and Space Research at Cornell University, where he stayed until his retirement. He was named the David Duncan Professor of Astronomy and

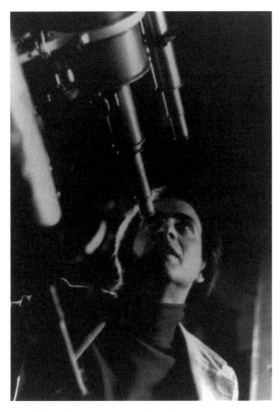

Carl Sagan is best known for his popular television show, *Nova*, which made scientific concepts accessible and intriguing to a broad audience. *(Cornell University, courtesy AIP Emilio Segrè Visual Archives)*

Space Science at Cornell in 1977, after he served as associate director of the center.

His later theoretical work focused on the possibility that Jupiter's moon Titan could sustain life, and also on the potential effects of nuclear war on the earth's atmosphere. In 1983, he coauthored a paper, "Nuclear Winter: Global Consequences of Multiple Nuclear Explosions," which galvanized both political leaders and the public. The paper asserted that even a "small" nuclear war would wreak havoc on the planet and ultimately cause a substantial temperature drop, a phenomenon termed *nuclear winter*.

A significant facet of Sagan's work occurred outside academia. He advised the National Aeronautics and Space Administration on many missions, including the *Mariner* Venus mission, the *Mariner 9* and *Viking* missions to Mars, and those involving the *Pioneer 10*, *Pioneer 11*, and *Voyager* spacecraft. He helped design a plaque affixed to some of these crafts that bore messages describing human life in case extraterrestrials encountered the spaceships. Sagan also was dedicated to spreading his love of science to a broad audience. He published several popular science books, such as *The Cosmic Connection* (1973), *The Dragons of Eden* (1977), and *Broca's Brain: Reflections on the Romance of Science* (1979). He frequently appeared on Johnny Carson's *Tonight Show*, serving as a sort of science adviser. In 1980 he launched the television show *Cosmos*, which introduced a huge audience to such formerly esoteric concepts as black holes.

Sagan's enthusiasm and receptiveness to new ideas were his true legacy. His books were widely read, and he moved exobiology closer to mainstream scientific thought. He was awarded the Pulitzer Prize for literature in 1978 and received numerous other awards, including the Rittenhouse Medal in 1980 and the UCLA Medal in 1991. Sagan died on December 20, 1996, at the Fred Hutchinson Cancer Research Center in Seattle, Washington, after a two-year battle with bone marrow cancer. He was 62.

Salk, Jonas Edward
(1914–1995)
American
Microbiologist

Jonas Salk's major accomplishment was his development of the first safe and effective vaccine against polio. With his collaborator, Thomas Francis Jr., Salk pioneered the use of a "killed virus" as a vaccine antigen. Salk's discovery made him a national hero. He also helped create a killed-virus flu vaccine and conducted epidemiological research. In 1960, he founded the Salk Institute for Biological Studies in La Jolla, California, a facility dedicated to cutting-edge research.

Born on October 28, 1914, in New York City, Salk was the eldest son of Orthodox Jewish emigrants from Poland. His father, Daniel Salk, was a garment worker with a talent for and a love of drawing. Dora Press, Salk's mother, took care to cultivate the considerable intellectual gifts the young Salk displayed.

Enrolled at the City College of New York, Salk received his bachelor's degree in science in 1933, when he was only 19. Although he had originally planned to pursue law, Salk had been captivated by biology during his undergraduate studies. In 1933, he entered New York University's School of Medicine, where he met his future collaborator, Dr. Thomas Francis Jr. He obtained his M.D. in 1939 and continued working for Francis while fulfilling a two-year internship at Mount Sinai Hospital in New York. He also married Donna Lindsay in 1939, and the couple had three sons. After his first marriage ended, Salk wed Françoise Gilot (Pablo Picasso's former mistress) in 1967.

Upon finishing his internship, Salk accepted a National Research Council fellowship and followed Francis to the University of Michigan, where the two developed a killed-virus vaccine for the flu in 1943. This achievement was significant. The killed virus was able to destroy live flu viruses at the same time that it stimulated the body to produce antibodies to ward off future infection. In 1946, Salk accepted a position as assistant professor of epidemiology at the University of Michigan, but he left for the University of Pittsburgh's Virus Research Laboratory the following year. He published papers on polio, some of which were read by Daniel Basil O'Connor, the director of the National Foundation for Infantile Paralysis, a well-funded organization devoted to eradicating polio. O'Connor contributed considerable money to Salk's fledgling attempts to devise a polio vaccine.

At the University of Pittsburgh, Salk joined researchers from other institutions to classify the more than 100 strains of polio virus. In 1951, he felt certain that all polio viruses could be arranged into three categories. He built on the work of Albert Bruce Sabin, John Franklin Enders, Thomas Huckle Weller, and Frederick Chapman Robbins, each of whom played a part in making cultivation of the polio virus easier in a lab. Salk grew samples of the three types in cultures of monkey kidneys, then exposed the virus in formaldehyde for 13 days, thereby killing it. He felt certain that the killed-virus technique would be safer than the live-virus method advocated by Sabin and others. In 1952 Salk began to administer his killed-virus vaccine to children—first to those who already had the disease, later to those who had never contracted it. In both cases he found that the vaccine definitively elevated antibody levels. These initial positive results were published in the *Journal of the American Medical Association* in 1953. The following year Salk spearheaded the implementation of a massive inoculation program, with 1 million children participating.

On April 12, 1955, Salk's vaccine was pronounced both potent and safe. A tragedy occurred that June, however, when 200 children contracted polio from the vaccine because one of the participating labs had used the wrong strain of the virus. This problem was quickly corrected and did not recur. In 1960, Salk founded his eponymous institute.

Salk's vaccine accounted for a 96 percent reduction in the number of cases of polio in the United States by the summer of 1961, and he was hailed as a hero. His chief scientific competitor, Sabin, created a live-virus vaccine, which gradually supplanted Salk's innovation because it could be administered orally. In addition to many other awards, Salk received the Presidential Medal of Freedom in 1977. He died of heart failure in 1995.

Sanger, Frederick
(1918–)
English
Biochemist

Frederick Sanger received the Nobel Prize in chemistry not once, but twice—first in 1958, for his determination of the structure of insulin, and later in 1980, for his development of a technique for rapidly sequencing deoxyribonucleic acid (DNA) by fragmentation. He was only the fourth double-Nobel laureate. He used his notoriety later in his career to advance the Human Genome Project by founding the Sanger Centre in 1993 to house research on the project.

Sanger was born on August 13, 1918, in Rendcombe, in Gloucestershire, England. His mother was Cicely Sanger, and his father, after whom he was named, was a physician. He received his early education at the Bryanston School, then matriculated in St. John's College of Cambridge University. He earned his bachelor of arts degree in natural sciences in 1939. The next year, he married Margaret Joan Howe, and together the couple eventually had three children—two sons and a daughter. Also in 1940, he commenced his pursuit of doctoral studies, conducting research on the metabolism of the amino acid lysine under Albert Neuberger. Cambridge granted him his Ph.D. in 1943.

At the instigation of Charles Chibnall, who took up Cambridge's chair of biochemistry in 1943, Sanger conducted postdoctoral research on insulin with free amino groups, in an attempt to identify and quantify the amino acid content of insulin. Chibnall had identified a significant discrepancy between the expected and the actual number of free amino groups, and Sanger developed a new chromatographic method for identifying groups of amino acids attached to the end of insulin's protein chains. He received support for this research from a Beit Memorial Fellowship for Medical Research (a grant he held from 1944 through 1951).

Sanger published his end-chain results in 1945, and a decade later, he published the complete sequence of insulin, a major scientific advancement allowing for the synthesis of insulin. This breakthrough promised to improve the health of diabetics, whose bodies could not produce enough insulin to sustain their needs. He received the 1958 Nobel Prize in chemistry "for his work on the structure of proteins, especially that of insulin," according to the Royal Swedish Academy of Sciences.

In 1951, the Medical Research Council hired Sanger, and in 1962, the MRC installed him at the helm of its division of protein chemistry in the newly constructed Laboratory of Molecular Biology. Inspired by his LMB colleagues—the likes of FRANCIS CRICK and John Smith—Sanger applied his analytical expertise to the nucleic acids, starting with ribonucleic acid (RNA) studies. In collaboration with G. G. Brownlee and B. G. Barrell, he developed a rapid method for fragmenting RNA—separating it into small pieces and then overlapping the pieces in sequence.

Sanger tried to apply this same method to DNA, with little success, until he developed the "didoxy" method, which "tagged" fragments of DNA with dideoxy triphosphate. This allowed Sanger to identify the composition of DNA fragments at different points in the strands, which he read on an autoradiograph connected to a gel containing the DNA. As it turned out, this became the most efficient means of analyzing the chemical structure of DNA.

In 1977, Sanger identified the nucleotide sequence in the DNA of a bacteriophage, phi-X 174, representing the first determination of an organism's entire nucleotide sequence. In 1980, Sanger won his second Nobel Prize in chemistry (which he shared with WALTER GILBERT and Paul Berg) "for their contributions concerning the determination of the base sequences in nucleic acids," according to the Royal Swedish Academy of Sciences.

Throughout his career, Sanger received numerous honors in recognition of the significance of his work. In 1951, he won the Corday-Morgan Medal and Prize of the Chemical Society. The Royal Society inducted him into its fellowship in 1954, and that same year, he became a fellow of King's College, Cambridge. He became a Companion of the British Empire in 1963. In 1981, he became a Companion of Honor, and in 1986, the Order of Merit was conferred upon him.

Sanger continued to advance his field late into his career. In 1993, he founded the Sanger Centre to support work on the Human Genome Project. Funded in part by the Wellcome Trust and the MRC, the Sanger Centre is housed at the Wellcome Trust Genome Campus in Cambridge.

Savery, Thomas
(1650?–1715)
English
Engineer

Thomas Savery invented the first working steam pump, although his version did not prove practical enough for widespread use. Despite the fact that Savery protected his invention with a broadly written patent, which effectively prevented others from profiting off any innovations and improvements to his design, THOMAS NEWCOMEN went ahead and made the improvements necessary to make the steam-driven engine for pumping water a workable reality. This, in turn, spurred the Industrial Revolution, as it allowed for the draining of mines and a drastic increase in the production of minerals needed as raw material to fuel new industrial activity.

Savery was born in Shilstone, near Modbury, in about 1650. His father was Richard Savery, and his grandfather was Christopher Savery. Savery joined the British military at the age of 16, serving as an engineer. He remained with the military for several decades, attaining the rank of captain of engineers in 1702.

On January 10, 1696, Savery patented a method of propelling boats by means of paddle wheels mounted on either side, a system that only worked under calm conditions. He tried to interest the British navy in his invention, to no avail. Later in 1696, he patented a device for cutting, grinding, and polishing mirror glass. He also invented a two-handed bellows for stoking foundry fires to temperatures that would melt metal or turn wood to charcoal.

Savery is best remembered for his introduction of a steam pump, which he patented on July 2, 1698. Savery wrote the patent in broad terms, covering the "raising of water . . . by the important force of fire," thus blocking subsequent innovations. The "Fire Engine," as he called it at first, consisted of a boiler connected by piping, fitted with a regulator valve, to a container that received the steam generated by an open fire under the boiler. The pressure of this steam forced air out of the receiving chamber through suction pipes that pulled water up and out of the system, through one-way valves. Cold water then doused the receiving chamber, condensing the steam to create a vacuum that sucked up a second round of water.

Savery demonstrated a model of his steam pump successfully before the Royal Society (which later inducted him into its fellowship, in 1706). With this stamp of approval, he embarked on the unusual course of advertising his invention, publishing a pamphlet called "The Miner's Friend," in which he explained the positive features and advantages of his steam pump. Printed in London in 1702, he distributed the fact sheet at mines, which were plagued at that time with flooding problems with no practical solution beyond using horses to haul hand-filled buckets up ropes. Savery noted that his system compared favorably against this existing system, performing the same work as 10 horses (Savery thus introduced the term *horsepower* into the language).

Unfortunately, the miner's friend was plagued with problems, mostly stemming from limitations of the contemporary technology. The tin solder joining the wooden pipes together could only withstand the pressure of about three atmospheres, or the equivalent of 40 pounds per square inch. The soldering blew out when pushed beyond its limits, or else the caution required to avoid such explosions severely restricted the usefulness of the pump. The miner's pump could lift water some 20 to 26 feet, according to Savery's calculations, thus requiring the installation of a series of pumps, staggered up the shaft of the mine. At a cost of 50 pounds per pump, this necessity of using multiple pumps made the technology prohibitively expensive. In addition, the pump also consumed a prodigious amount of fuel, adding to its expense. No evidence exists that even one of Savery's steam pumps was actually installed in any mine; their significance is more as a precursor to future developments, namely Thomas Newcomen's steam engine.

In 1705, Prince George of Denmark appointed Savery as "treasurer for sick and wounded seamen." While at this post, he patented an odometer for tracking how far ships have traveled. In 1714, he became surveyor of the Hampton Court waterworks, where he designed a waterwheel pump for its fountains. Savery died in early May 1715, as his will was dated on the 15th and approved by his widow, Martha, on the 19th.

Shockley, William
(1910–1989)
English/American
Physicist

William Shockley invented the transistor by altering semiconductor crystals to perform the function then filled by vacuum tubes. However, the transistor improved upon vacuum tubes, mainly by amplifying the electrical signal, thus allowing for reduction in the size necessary for a transistor to handle the same amount of energy. Transistors became ubiquitous in electronics applications, from computers to radios to satellites.

William Bradford Shockley was born on February 13, 1910, in London, England. His father, Massachusetts-native William Hillman Shockley, was a mining engineer, and his mother, Mary Bradford, was a deputy mineral surveyor in Nevada. The family moved back to the United States when Shockley was three years old. He attended the California Institute of Technology, earning his bachelor of science degree in 1932. J. C. Slater directed his doctoral studies at the Massachusetts Institute of Technology (MIT), where Shockley wrote his dissertation on the energy band structure of sodium chloride to earn his Ph.D. in 1936.

Bell Telephone Laboratories hired Shockley directly upon graduation from MIT. During World War II, he contributed to the war effort by serving as research director of the Antisubmarine Warfare Operations Research Group and remained a consultant to the office of the secretary of war thereafter. In 1946, he received the Medal of Merit for his wartime work. That year, Princeton University hosted him as a visiting lecturer, after which he returned to the Bell Laboratories.

In 1948, Shockley collaborated with his colleagues John Bardeen and Walter Brattain to invent the transistor. The trio wrapped a plastic knife in gold foil, then connected this to an electrically charged block of germanium. The resulting mechanism revolutionized electronics, as it performed the same function as a vacuum tube much more efficiently, in much smaller space, without wearing out nearly as often. Furthermore, the transistor allowed for the electrical flow to be turned on and off, as well as amplified.

The advent of the transistor created the microelectronics market. Computers run with vacuum tubes filled huge rooms and required constant air-conditioning to avoid overheating, whereas computers run with transistors consistently shrank until the introduction of the personal computer, both as a desktop unit and as a self-contained laptop unit. A half-century after the invention of the transistors, manufacturers managed to fit almost a billion transistors on one computer chip.

In 1954, Shockley's alma mater, Cal Tech, invited him to campus as a visiting professor. Over the next year, he served as deputy director and research director of the Weapons System Evaluation Group of the Defense Department. Then, in 1955, he resigned from his position at Bell as director of the transistor physics department to work for Beckman Instruments establishing a semiconductor laboratory and conducting research and development on a new transistor. He then founded his own company, the Shockley Transistor Company, to capitalize on his invention.

In the meanwhile, honors were showered on Shockley for his contributions to science and society: he received the 1952 Morris Leibmann Memorial Prize from the Institute of Radio Engineers; the 1953 Oliver E. Buckley Solid State Physics Prize from the American Physical Society; and the 1954 Cyrus B. Comstock Award from the National Academy of Sciences. Then in 1956, Shockley shared the Nobel Prize in physics with Bardeen and Brattain, splitting the $38,633 in prize money equally. In 1963, the American Society of Mechanical Engineers awarded him its Holley Medal.

That year, Stanford University named him its first Alexander M. Poniatoff Professor of Engineering Science. A decade later, he shifted his focus to eugenics, specifically investigating the discrepancy between races in intelligence quotient tests. He discovered that African Americans score an average of 10 to 20 points lower than Caucasians on IQ tests. Shockley's interpretation of this information embroiled him in controversy—he proposed government-enforced sterilization for those scoring low on IQ tests. Opponents to this stance vandalized his car with graffiti, burned him in effigy, and stormed his classes in protest. However, the scientific validity of his findings was irrefutable, despite the odiousness of his application of these results.

Shockley had married twice, first to Jean Bailey, who bore him three children, and second to Emmy Lanning. He held 90 patents by the time of his death on August 12, 1989. However, his most important patent was on the transistor.

Sikorsky, Igor
(1889–1972)
Ukrainian/American
Aeronautical Engineer, Transportation Industry

Igor Sikorsky promoted the growth of the aviation industry through several innovations as well as through his business acumen. He designed and built the first multiengine aircraft, which served in the Russian Imperial Army with distinction in World War I. After emigrating from Bolshevik Russia after the revolution, Sikorsky established several aviation companies in the United States that solidified the aviation industry with the introduction of amphibian and long-distance airplane models. Sikorsky made his most indelible mark by introducing the first successful helicopter, and then by designing and building a succession of helicopters to different specifications for diverse uses.

Igor Ivan Sikorsky was born on May 25, 1889, in Kiev, Ukraine, the last of five children born to Zinaida and Ivan Temrouk-Tcherkoss Sikorsky. His mother was a doctor, and his father was a psychology professor at St. Vladimir University. Seeing Leonardo da Vinci's sketches of a proto-helicopter, combined with his witnessing of Count Ferdinand von Zeppelin's dirigible while on a tour of Germany with his father, inspired his lifelong interest in aviation. In 1903, at the age of 14, Sikorsky entered the Petrograd Naval College, graduating three years later. He then spent two years studying aviation science at the Mechanical College of the Kiev Polytechnic Institute, but he abandoned the school's curriculum, with its theoretical focus, to concentrate on more practical engineering and design concerns.

After his second visit to Paris in 1909 (he met American aeronautical pioneer WILBUR WRIGHT on his first sojourn there in the summer of 1908), Sikorsky brought back a 25-horsepower Anzani engine, which ultimately proved too heavy to power his first two helicopter prototypes. He turned his attention to designing fixed-wing aircraft, building a succession of models (the S-1, a 15-horsepower biplane; the S-2, which succeeded in flying a short distance; and the S-3 and S-4) before arriving at his truly successful design, the 50-horsepower S-5 that could remain airborne as long as an hour, reach heights of 1,480 feet, and achieve record speeds of 70 miles per hour. The next model, the S-6A, won top honors at the 1912 Moscow Aviation Exhibition.

That year, the Russian-Baltic Railroad Car Works hired Sikorsky as head of its aviation division. Sikorsky then embarked on his most ambitious project yet: designing and building multiengine airplanes. He built first *The Grand*, a four-engined bomber, and then the even larger *Ilya Mourometz*, named after a 10th-century Russian war hero. Manned by a crew of five, powered by four 220-horsepower engines, defended by four machine guns, and armed with a bombload of up to 1,543 pounds, the bomber took its maiden flight on May 13, 1913. More than 70 versions of this airplane served in the Imperial Army in World War I, achieving a 60 percent on-target rating in more than 400 bombing missions by the end of the war.

Deemed an enemy of the state after the Bolshevik Revolution, Sikorsky fled first to France, then to the United States, landing in New York on March 30, 1919. Unable to secure a position in the aviation industry, suffering as it was from the postwar glut of airplanes built for the war, Sikorsky resorted to tutoring Russian immigrants in mathematics, astronomy, and even aviation. Within four years, he saved enough to establish the Sikorsky Aero Engineering Corporation, which he installed on Lieutenant Victor Utgoff's farm near Roosevelt Field on Long Island on March 5, 1923. The company engineered the S-29, a twin-engine, 14-passenger commercial aircraft. In 1924, he married his second wife, Elizabeth A. Semion, a Russian immigrant. Sikorsky had four sons and one daughter.

Within two years, Sikorsky founded the Sikorsky Manufacturing Corporation to build the S-29, and three years later, in 1928, he merged his engineering with his manufacturing company to establish the Sikorsky Aviation Corporation. Within a year, the United Aircraft Corporation bought out this enterprise, impressed by the success of the S-38 amphibian passenger plane that spurred the growth of air traffic between North and South America. Sikorsky retained much independence in running his subsidiary.

In the mid-1930s, Sikorsky revisited his dream of building a helicopter, backed now by a $300,000 budget. Sikorsky, who always piloted the first flights of aircraft he designed, manned the controls of the VS-300 on September 14, 1939, to navigate the first successful helicopter flight, rising a few feet in the air. Within two years, Sikorsky had improved his helicopters to remain airborne for up to 90 minutes, and by 1943, Sikorsky was producing the R-3, the world's first production helicopter. That year, the U.S. Army contracted Sikorsky to supply it with R-4 helicopters, though few served in World War II.

The Korean and Vietnam Wars saw the advent of advanced helicopter technology, as the Sikorsky Aeronautical Division designed and built successively improved models: the S-55 transport helicopter; the twin-engine S-56 with troop-carrying capacity of 50; the S-58 for transporting small numbers short distances; the S-61 antisubmarine helicopter; and the S-62 amphibious helicopter. A devoutly religious man, Sikorsky stressed helicopters' peaceful uses, such as for rescue missions, transporting air mail and passengers, dusting crops, fighting forest fires, and even hauling cargo (the S-64 "Skycrane" could carry up to 10 tons).

Sikorsky retired in 1957, but he remained active in the industry he helped spawn. This industry repaid him with numerous honors, including a 1948 Presidential Certificate of Merit, the 1951 Daniel Guggenheim medal, the 1967 Wright Trophy, and the 1967 National Medal of Science. Additionally, Sikorsky represented the helicopter industry in receiving the 1951 Collier Trophy. United Aircraft established the Igor I. Sikorsky International Trophy for helicopter design and endowed the Igor Sikorsky professorship in mechanical engineering at the University of Bridgeport in Connecticut. Sikorsky died of a heart attack at his home in Easton, Connecticut, on October 26, 1972.

Silbergeld, Ellen Kovner
(1945–)
American
Toxicologist

Ellen Kovner Silbergeld has combined a career in research science, focusing specifically on environmental toxicology, and public policy, advocating the implementation of laws based on hard scientific research. She commenced her scientific career studying lead neurotoxicity as well as the toxicology of food dyes.

Ellen Kovner was born on July 29, 1945, in Washington, D.C. Her mother, Mary Gion, was a journalist who had already given birth to Ellen's older brother. Her father, Joseph Kovner, was a liberal lawyer who was ousted from his governmental job by the House Un-American Activities Committee. He established a private practice in Concord, New Hampshire, from when Kovner was seven until she reached eighth grade, when the political climate had changed enough for the family to return to Washington, D.C.

Kovner studied modern history at Vassar College, where she earned her A.B. degree in 1967. She then studied quantitative history at the London School of Economics on a Fulbright Fellowship. In 1968, she returned to Washington, D.C., where she spurned history and economics in favor of working as a secretary and program director for the Committee on Geography at the National Academy of Sciences of the National Research Council until 1970. In 1969, she married Mark Silbergeld, and together they had two children—a daughter, Sophia, who was

Ellen K. Silbergeld's research as an environmental toxicologist has led to her work in public policy, advocating laws based on scientific facts. *(Courtesy of Ellen Silbergeld)*

born in 1981, and a son, Reuben Goodman, who was born in 1985.

Silbergeld's exposure to reports at the National Academy of Sciences peaked her interest in science, so she decided to pursue a doctorate in environmental engineering starting in 1968 at Johns Hopkins University, where she was the only woman in the program. She earned her Ph.D. in 1972, and she remained at Johns Hopkins under a postdoctoral fellowship conducting biochemical research on lead neurotoxicity.

In 1975, the National Institutes of Health (NIH) hired Silbergeld as a staff scientist, and by 1979, the NIH appointed her as the laboratory chief of her section, the Unit on Behavioral Neuropharmacology of the National Institute of Neurological Disorders and Stroke. She continued her research on lead while also investigating the toxicity of food dyes and conducting research on neurological disorders such as Huntington's and Parkinson's diseases.

At the risk of compromising her scientific credibility, Silbergeld assumed a public policy position in 1982 as the chief toxic scientist and director of the Toxic Chemicals Program for the Environmental Defense Fund, a position she retained until 1991. Although some oppose the mingling of science and politics, Silbergeld believes the two can be symbiotic, with politics cueing science on real-life problems to investigate and science providing hard facts for political decision making.

During the 1980s, Silbergeld held several concurrent positions at the National Institute of Child Health Development, at the University of Maryland School of Medicine, and at Johns Hopkins Medical Institutions. In the 1990s, she decided to return to a more academic focus, accepting more intensive appointments at the University of Maryland (as an affiliate professor of environmental law and a professor of pathology and epidemiology) and Johns Hopkins (as an adjunct professor of health policy and management). She retained her affiliation with the Environmental Defense Fund as a senior consultant toxicologist.

Silbergeld received a prestigious MacArthur Foundation Fellowship in 1993, validating the significance of her career. She has also received the 1987 Warner-Lambert Award for Distinguished Women in Science, the 1991 Abel Wolman Award, and the 1992 Barsky Award. In 1994, her early work was rewarded with a patent for her lead detection procedure.

Singer, Isaac Merrit
(1811–1875)
American
Inventor, Textile Industry

Although the Singer name is associated with the sewing machine so closely as to be almost synonymous, Isaac Merrit Singer did not invent the device. However, he did introduce numerous integral improvements to the design; more significantly, Singer marketed the sewing machine to the masses, thereby introducing it as a household item and revolutionizing the domestic workplace by increasing the speed of sewing exponentially. Sewing machines similarly revolutionized the textile industry, increasing production several-fold.

Singer was born on October 27, 1811, in Pittstown, New York, son of a millwright. He started working at the age of 12 as an unskilled laborer, then apprenticed to a machinist. He became a journeyman machinist, also finding work as an itinerant actor on his travels. He also applied his creativity to inventing new machinery. On May 16, 1839, while working in Lockport, Illinois, he patented a rock-drilling machine.

Over the next decade, Singer frittered away the money he made on his first invention, sparking another round of creativity. In 1849, while living in Pittsburgh, Pennsylvania, he patented a wood- and metal-carving machine that he had been designing over the previous five years. Bad luck struck when a boiler explosion destroyed his new machine, sending him back to square one.

By 1850, Singer had found employment with Orson C. Phelps in his Boston machine shop. When a customer brought in a Lerow & Blodgett sewing machine for repair, Singer immediately recognized numerous improvements he could implement in the design. Over the next 11 days, Singer incorporated these improvements into a prototype of a new sewing machine model and filed for a patent. The U.S. Patent Office issued patent number 8,294 to Singer on August 12, 1851.

In the meanwhile, Singer had already formed a partnership with Phelps and G. B. Zieber, establishing the I. M. Singer Company to market his sewing machine. By June 1851 (even before the issuance of the patent), Singer had bought out Phelps's interest, and Edward Clark had bought out Zieber's interest. The Singer and Clark partnership worked perfectly, as the latter excelled in marketing while the former worked on design improvements while also defending his patent in court.

The first practical sewing machine appeared in 1841, invented by the Frenchman Barthélemy Thimonnier to expedite the making of army uniforms. Four years later, Elias Howe imported the sewing machine concept to the United States, patenting a machine that performed a lockstitch method. Singer borrowed liberally from this design, though he incorporated significant improvements, including a method of continuous stitching. However, Howe recognized that Singer's design essentially replicated his own. A protracted lawsuit ensued, with Howe arguing successfully that Singer's improvements did not represent new ideas. Singer was thus forced to pay royalties to Howe, but by this time the Singer name had already established itself as the preeminent brand in the minds of consumers.

Fueled by Clark's effective advertising, which stressed the reliability and quality of Singer machines, the Singer Company captured the burgeoning market. In 1853, 810 Singer sewing machines sold; a decade later, that number had mushroomed to 21,000 machines. In order to induce new sales, as his machines were so well built that a second hand market was developing, Singer took in used machines and broke them down for parts, thus removing old machines from the market and encouraging sales of new machines.

Singer realized that commercial sales limited the potential market, so in 1856, he introduced a model aimed exclusively at home use, as well as introducing the first installment-plan payment system, which further encouraged sales. Also in 1856, the Great Sewing Machine Combination resulted in the pooling of patented technology by the major sewing machine manufacturers, allowing them to share design innovations. Singer donated his needle-bar cam patent to the collective.

When Singer retired in 1863, the I. M. Singer Company transformed into the Singer Manufacturing Company. Clark remained at the helm, replacing "on-off" production—whereby skilled laborers filed and fitted machine parts separately—with mass production. By 1869, more than 110,000 sewing machines had been produced in the United States. Singer spent his retirement at Torquay, in Devon, England, where he died on July 23, 1875. Singer remains the dominant brand name in sewing machines.

Skinner, B. F.
(1904–1990)
American
Psychologist

B. F. Skinner advanced the controversial psychological theories of behaviorism and operant conditioning, which propounded the control of behavior through consequences, particularly positive consequences. A failed novelist early in

B. F. Skinner's best-known book, *Walden Two*, promotes a community shaped by behaviorism, the psychological theory he developed and promoted. *(Courtesy B. F. Skinner Foundation)*

his career, he later returned to the fictional medium to present his psychological ideas in his best-known work, *Walden Two*, which described a community shaped by behaviorism as a kind of utopia. The book and its theoretical basis earned sharp criticism on the one hand, and on the other hand, extensive popularity. Although behaviorism has fallen from vogue, and never enjoyed the support of the entire psychology community, it influenced the course of psychological theory significantly.

Burrhus Frederic Skinner was born on March 20, 1904, in Susquehanna, Pennsylvania. His mother was Grace Madge Burrhus, and his father, William Arthur Skinner, was a lawyer who had apprenticed as a draftsman on the Erie Railroad. His younger brother died of a cerebral aneurysm at the age of 16. Skinner attended the same one-room schoolhouse through high school, where his teacher, Mary Graves, influenced him to major in English literature at Hamilton College. He received a letter of encouragement from Robert Frost in his senior year, influencing him to try his hand at writing a novel after graduating in 1926.

Unfortunately, "I had nothing important to say," Skinner reported of his "dark year," so he abandoned writing and moved to bohemian Greenwich Village in New York City, where he worked in a bookstore. There, he encountered John B. Watson's *Behaviorism*, which influenced him to pursue graduate studies in psychology.

Despite the fact that he had never taken any courses in the discipline, he gained admission to Harvard University, working in William J. Crozier's physiology laboratory.

At Harvard, he invented the Skinner box, which contained food-delivery levers to reward animals for certain behaviors. After Skinner earned his Ph.D. in 1931, Harvard's Society of Fellows retained him as an instructor. In 1936, he moved laterally to the University of Minnesota, which promoted him from this instructorship to an assistant professorship in 1937. The next year, he published *The Behavior of Organisms*, in which he discussed his theory of operant behavior, or the governance of behavior by means of positive and negative consequences.

In the fall of 1936, Skinner had married Yvonne Blue, a University of Chicago graduate. They had two daughters—Julie and Deborah; the latter became famous as the first infant raised in Skinner's "baby tender," little more than a climate-controlled crib. Media reports of Deborah's later suicide were greatly exaggerated—in fact, she became a successful artist and writer.

During World War II, Skinner served in the U.S. Office of Scientific Research and Development, where he developed a missile guidance system (which was never implemented) based on pigeon's vision. After the war, in 1945, Indiana University at Bloomington appointed him chairman of its psychology department. Three years later, Skinner returned to Harvard as a professor of psychology.

Also in 1948, Skinner published his most influential book, *Walden Two*—a work of fiction, depicting a utopian society based on operant principles, that became a mainstream best-seller. A decade later, Harvard named him Edgar Pierce Professor of Psychology. In the intervening time, he had published *Science and Human Behavior* in 1953, and *Schedules of Reinforcement* in 1957. The latter came under attack, most visibly by the linguist and social critic Noam Chomsky.

Chomsky later attacked Skinner's 1971 text, *Beyond Freedom and Dignity*, comparing its scientifically based cultural engineering to Nazism, while the philosopher Karl Popper condemned Skinner's "behaviorist dictatorship." Skinner defended his notion of cultural engineering by pointing out that, since all behavior is controlled (usually through random, uncoordinated efforts, such as imprisonment), it is better to acknowledge this control mechanism and apply scientific reason to arrive at the best behaviors.

Skinner had his supporters as well: the American Humanist Association named him the 1972 Humanist of the Year, and a 1975 poll identified him as the best-known American scientist by college students. Skinner retired in 1974, though he continued to write and lecture actively until he died of leukemia in Cambridge, Massachusetts, on August 18, 1990. He wished to be remembered as a "social inventor," as he applied an engineer's discipline to the questions of social organization.

Soddy, Frederick
(1877–1956)
English
Chemist

Frederick Soddy revolutionized science with his explanations of seemingly inexplicable phenomena: he proposed the theory of nuclear disintegration, which explained radioactive decay as elemental transmutation, and he also posited the notion of isotopes, accounting for the existence of identical chemical elements that nonetheless differed in physical properties. For this latter explanation especially, Soddy won the 1921 Nobel Prize in chemistry.

Soddy was born on September 2, 1877, in Eastbourne, Sussex, England. He was the youngest of seven children born to Hannah Green, who died 18 months after his birth, and Benjamin Soddy, a wealthy London corn merchant. Soddy studied science at Eastbourne College under R. E. Hughes, with whom he cowrote his first scientific paper (on the reaction between dry ammonia and dry carbon dioxide) in 1894, when he was a mere 17 years old. On Hughes's recommendation, he attended a postgraduate year at the University College of Wales at Aberystwyth, where he won an Open Science Postmaster Scholarship to Merton College of Oxford University. He matriculated there in 1895, graduating with first-class honors from the school of natural science in 1898.

Soddy remained at Oxford for two years of unremarkable postdoctoral research, after which he traveled to Canada to lobby in support of his candidacy for a post at the University of Toronto, which he failed to receive. On his return trip, however, he stopped over in Montreal, where McGill University offered him a position as a junior demonstrator in Ernest Rutherford's department. The two men collaboratively developed a theory to explain radioactive decay: they proposed an atomic explanation for radioactive disintegration, positing that the emission of an alpha or beta particle transforms a radioactive element into a new substance. This theory of nuclear disintegration, though revolutionary, achieved almost immediate acceptance in the scientific community.

In 1902, Soddy returned to London, working at University College with Sir William Ramsay, who had served as external examiner on his doctoral committee. In 1903, they confirmed spectrographically that radium always produces helium as the by-product of its radioactive decay, a phenomenon predicted by the joint work of Soddy and Rutherford. The next year, Soddy embarked on a tour of western Australia as an extension lecturer in physical chemistry and radioactivity for London University; upon his return later that year, he took up a lectureship in physical chemistry at Glasgow University. He produced some of his most significant results there over the next decade. He married Winifred Moller Beilby in 1908. No children resulted from their union.

In 1911, Soddy proposed the alpha-ray rule: elements decrease by two in atomic number upon emission of an alpha particle. When combined with A. S. Russell's beta-ray rule (elements increase by one in atomic number upon emission of a beta particle), these rules became known as the displacement law two years later. Also in 1913, in the February 28 issue of *Chemical News*, Soddy coined the term *isotope* (which means "same place") to describe elements that are chemically identical but differ in atomic weights. This proposal followed up on his 1910 paper in which he first proposed that radioactive decay can produce elements that are indistinguishable except for their distinct atomic

Frederick Soddy's work as a chemist, especially his discovery of isotopes, revolutionized modern science. (*Smith Collection, Rare Book & Manuscript Library, University of Pennsylvania*)

weights. This explanation clarified the confusion resulting from the existence of chemically identical elements, such as ionium and radium or thorium and radiothorium, which nevertheless exhibited different properties.

In 1914, Soddy finally ascended to a chair in chemistry, at the University of Aberdeen, where he continued to investigate radioactive elements. He demonstrated that the atomic weight of lead can vary significantly from the value that appears on the periodic chart, as lead derives from so many different radioactive reactions. World War I interrupted this line of research, as he contributed to the effort by devising means of converting coal gas into ethylene. After the war, Soddy returned to Oxford to inhabit the Dr. Lees Chair in Chemistry.

In 1921, the Royal Swedish Academy of Sciences recognized the significance of Soddy's explanation of isotopes by awarding him the Nobel Prize in chemistry. Soddy tried to translate his laureate status into cachet for advancing his political and social beliefs, but his causes—women's suffrage, Irish autonomy, and the exercise of restraint with nuclear energy—proved too advanced to find popular support. Ironically, human scientific achievement required commensurate advances in moral understanding, Soddy realized, but he had trouble convincing his scientific colleagues of the moral responsibility inherent in the discovery of radioactivity, a force more powerful than human reckoning.

Soddy's career came to an abrupt halt when his wife died unexpectedly of a coronary thrombosis in 1936; almost immediately, he took an early retirement. His scientific output at that point was practically nil, as he had not researched any original ideas throughout his time at Oxford, but he wrote prolifically on social causes, which he continued to do in retirement. His cautionary talk of the dangers of nuclear power, which at first sounded alarmist, now took prophetic airs as World War II bore out his warnings against the destructive applications of radioactivity. He continued to lobby scientists to take more responsibility for how their discoveries are applied. Soddy died on September 22, 1956, in Brighton.

Sperry, Elmer Ambrose
(1860–1930)
American
Engineer

Elmer Ambrose Sperry's pioneering inventions transformed the gyroscope from a scientific curiosity that interestingly stayed in sync with the earth's rotational plane into an indispensable tool for stabilizing the roll of ships and aircraft and for reading compass directions amid heavy metals and electrical charges interfering with the earth's magnetic pull. He founded eight companies to manufacture his inventions and secured 360 patents to protect them. He is considered one of the fathers of modern control theory.

Sperry was born on October 21, 1860, in Cortland, New York. His father, Stephen Sperry, was a farmer, and his mother, Mary Burst, died soon after he was born. He went to local schools before attending the State Normal and Training School in Cortland, where he became interested in electrical engineering. He graduated in January 1880 and then sat in on lectures at Cornell University to continue his education.

During this time period, Sperry invented a generator to power arc lighting and established the Sperry Electric Company (which later folded into General Electric). However, the developing field of electricity distribution changed its standards so rapidly that he abandoned this line of business in 1887. On June 28 of that year, he married Zula Goodman. The next year, he established the Sperry Electric Mining Machine Company to manufacture his newest inventions—a continuous chain undercutter and an electric mine locomotive. In 1890, the Sperry Electric Railroad Company began producing streetcars, complete with his patented electric brake and control system, in Cleveland, Ohio, where he moved his family in 1893. The next year, he began designing and manufacturing electric motorcars.

In 1896, Sperry concentrated his attention on the gyroscope, a set of wheels spinning on two or three axes that maintained its plane of rotation in line with the earth's. Before Sperry, scientists had found little practical use for this interesting effect. Over the next decade, Sperry applied the gyroscope to the motion of ships: when the gyroscope detected sway, it tripped an electric motor to counteract the motion as soon as it started.

Sperry's active gyrostabilization system superseded the German naval engineer Ernst Otto Shuck's passive system, developed in 1904. In 1908, Sperry submitted a huge patent application for the gyrostabilizer and received patent number 1,242,065. The year before, he had moved his family to Brooklyn. He continued to investigate other applications for the gyroscope.

In 1903, German engineer Hermann Anschutz-Kaempfe had applied the gyroscopic principle to compasses, which had fared well enough in wooden ships but experienced magnetic interference from electrical systems and the iron hulls of modern ships. In 1908, Anschutz-Kaempfe introduced the gyrocompass on the German navy's flagship, *Deutschland*, inspiring Sperry to develop similar technology for the U.S. Navy. In 1910, he founded the Sperry Gyroscope Company, installing a gyroscope in the Dominion Line *Princess Anne* in 1911, and another in the USS *Utah* in November of that same year. Duly impressed, the U.S. Navy installed a gyrostabilizer prototype in its USS *Worden* in 1912, complete with two 4,000-pound gyro wheels that diminished its 30-degree roll to a mere six degrees.

Sperry expanded gyroscopic applications to other modes of transportation, installing air-compressed gyrostabilizers in aircraft to control both roll and pitch (the former component was tested on a Curtiss aircraft on Thanksgiving Day 1912. The next winter, Sperry and his son designed and built a four-gyro system that became the standard for guidance systems in aircraft, submarines, and missiles. The U.S. Navy continued to finance Sperry, who then developed the Curtiss pilotless aircraft, a missile capable of delivering a thousand pounds of explosives at targets up to a hundred miles away.

Sperry contributed to the effort in World War I by inventing gyroscopic applications that allowed for navigating aircraft "blindly," in conditions when the pilot could not see the ground or the horizon. His gyroscopic autopilot steering system became known as "Metal Mike," the mechanical pilot. After the war, he invented high-intensity searchlights.

In 1926, Sperry retired as president of the Sperry Gyroscope Company to chair its board. His hunger for innovation remained insatiable, though, as he invented a device for identifying defective train rails that the American Railway Association called "one of the most important safety moves in years." He served as president of the American Society of Mechanical Engineers from 1929 through 1930. On March 11, 1930, his wife died, and on June 16 of the same year, Sperry died in Brooklyn. In honor of his contributions to navigation, the U.S. Navy named the USS *Sperry* after him.

Sperry, Roger
(1913–1994)
American
Psychobiologist

Although people now commonly identify themselves as "left brain" types (analytical, logical) or "right brain" types (intuitive, visually oriented, musical), this understanding did not exist before Roger Sperry identified split-brain functions in his research. Earlier in his career, he had identified neurospecificity, or the fact that nerves are not interchangeable but rather perform specific functions. One of Sperry's early students, Robert W. Doty, memorialized him in a "American Psychological Society Observer" newsletter obituary by stating that his discovery "was on an intellectual par with the Copernican and Darwinian revelations that helped define man's place in nature."

Roger Wolcott Sperry was born on August 20, 1913, in Hartford, Connecticut, the elder of two sons born to Francis Bushnell Sperry, a banker, and Florence Kraemer. He attended public grade school in suburban Elmwood, where he collected and raised large American moths and later trapped and raised wild animals. At the age of 11, Sperry lost his father, and his mother, who had attended business school, supported the family as a high school principal's assistant. Sperry graduated from West Hartford's William Hall High School in 1931.

Sperry won the four-year Amos C. Miller scholarship at Oberlin College in Ohio, where he majored in English but also studied psychology under speech physiologist R. H. Stetson. After earning his bachelor of arts degree in English in 1935, he remained at Oberlin as Stetson's graduate assistant while studying for his master of arts degree in psychology, which he received in 1937. Oberlin continued to host him for a third year of graduate study in zoology to prepare himself for doctoral study at the University of Chicago under Paul A. Weiss, who had recently overturned the theory of neural specificity, which held that nerves could perform only their appointed functions, by surgically crossing nerve ends without disrupting neural function.

In an effort to confirm Weiss's results by switching rats' right and left hind-leg nerves, Sperry ended up disproving them, as the rats' right nerves (now installed in the left legs) continued to respond to right-leg stimuli. Sperry wrote up these findings in his doctoral dissertation, entitled "Functional results of crossing nerves and transposing muscles in the fore and hind limbs of the rat," to earn his Ph.D. in zoology in 1941. Sperry confirmed his own results in a series of "dramatically brilliant experiments" (according to Doty) by severing the optical nerves to newts' eyeballs, which he rotated. When the nerves healed, they failed to adjust to the change, and the newts approached the world as if they were upside down. Much to his doctoral adviser's dismay, Sperry conclusively proved the specificity of neural connections.

Sperry similarly shot holes in the research of his next mentor, Harvard University's Karl S. Lashley, who believed in the critical role of electrical waves in neocortical processes. As a National Research Council postdoctoral fellow, Sperry conducted experiments inserting mica plates and tantalum pins into the cortex and found that these electrical elements created negligible effects. After his one-year fellowship ended, he moved to the Yerkes Laboratories of Primate Biology in Orange Park, Florida, where he maintained his affiliation with Harvard as a biology research fellow. During these war years, from 1942 through 1946, he simultaneously served the U.S. Office of Scientific Research and Development conducting research on the surgical repair of nerve injuries.

In 1946, Sperry returned to the University of Chicago as an assistant professor of anatomy; the university promoted him to an associate professorship in 1952, the same year he served as section chief of neurological diseases and blindness for the National Institutes of Health. In the meanwhile, he had married Norma Gay Deupree on December 28, 1949; together, the couple had two children—Glenn Michael (Tad) in 1953, and Janeth Hope in 1963.

In 1954, the California Institute of Technology hired Sperry as its Hixon Professor of Psychobiology, a position he held for the remainder of his career. Over several decades at Cal Tech, Sperry performed his pioneering split-brain research, which proceeded from the invention of a surgical procedure severing the bundle of nerves in the corpus callosum, also known as the great cerebral commissure, or the connection between the two hemispheres of the brain. Neurosurgeons performed this so-called callostomy to treat epileptic patients from their seizures; Sperry capitalized on the method to study each hemisphere of the brain separately.

Contrary to the conventional scientific wisdom that the commissure performed no practical function, Sperry discovered that the hemispheres of the brain not only communicate to integrate their knowledge but they also serve significantly different functions. Reporting on a cat callostomy study in a 1964 edition of *Scientific American*, Sperry wrote that "it was as though each hemisphere were a separate mental domain operating with complete disregard—indeed, with a complete lack of awareness—of what went on in the other. The split-brain animal behaved in the test situation as if it had two entirely separate brains."

It was previously believed that the right hemisphere, which controls the left side of the body, was, if anything, inferior to the left hemisphere, which controls the right side of the body. Sperry's research revealed quite the opposite, attributing to the left brain such vital functions as intuitive processing, musical understanding, visual recognition, and spatial orientation, while the right brain controls analytical reasoning and logical thought. This realization carried profound implications for education, revealing that monolithic teaching methods may not suffice to teach skills that themselves vary physiologically.

Sperry eventually received due recognition for the significance of his discovery, winning the 1979 Albert Lasker Medical Research Award, as well as the 1981 Nobel Prize in physiology or medicine, which he shared with fellow neuroscience researchers Torsten N. Wiesel and David H. Hubel. He also received the 1989 National Medal of Science.

Sperry retired to emeritus status in 1984. Over the last three decades of his life, he suffered from neurological loss of motor control, though he managed to minimize the adverse effects of this condition on his life. Ironically, he was deprived of the very physiological system that he focused his studies on, the neural system. In the end, however, he succumbed not to neural failure but to a cardiac arrest, dying on April 17, 1994, in Pasadena.

Spode, Josiah
(1733–1797)
English
Potter

Josiah Spode and his son introduced some of the most important innovations in the earthenware industry in the 18th century. Josiah Spode I (so-named to distinguish him from his son, despite the fact that he was not the first Josiah Spode) introduced blue transfer-printing, a process that replaced hand-painting with a mechanized process that greatly increased production capabilities. He also facilitated the building of a canal adjacent to his factory, thereby increasing efficiency in product shipping. Josiah Spode II established retail outlets for selling Spode earthenware, a method that increased demand and sales. Finally, at the turn of the century, Spode the younger introduced bone china, which became the standard in English earthenware.

Spode was born on March 23, 1733, at Lane Delph near Fenton, in what is now Stoke-on-Trent in Staffordshire, England. As the only son (he had three older sisters), he was named after his father, who had become destitute by the time he died when Spode was only six years old. At the age of 16, Spode apprenticed to the preeminent potter in the region, Thomas Whieldon. By 1752, he had reached journeyman status, and in 1754, he left Whieldon to work under the potters R. Banks and John Turner. On September 8 of that year, he married Ellen Findley, who eventually ran her own haberdashery business while raising their eight children.

In 1761, Spode struck out on his own to establish a pottery manufacturing business at Shelton. Within three years, he moved his business to Stoke, where he took over the works of his former employers, Banks and Turner. In 1772, he commenced a partnership with Thomas Mountford by setting up another works back in Shelton. In 1776, he moved his Stoke factory to a new location on a tract of land he bought from Jeremiah Smith, the sheriff of Staffordshire. Spode's early success rested on the quality of his creamware, or cream-colored earthenware.

By 1778, Spode came to realize that he could expand business by marketing his products directly to the consumers who could afford his work, so he sent his first son, Josiah Spode II, to London, the seat of English wealth. There, Spode set up a showroom and shop at number 29 Fore Street. The shop moved down Fore Street in 1784 and again in 1788, each time to a larger location; Spode the younger also rented warehouse space for storing inventory awaiting sale.

In 1784, Spode the elder introduced a major innovation in earthenware. Up until that point, blue designs had to be painted onto the dishes by hand. Spode improved upon the transfer printing process first used by Ralph Baddeley, whereby blue designs were transferred onto pearlware (white-glazed earthenware) by a gelatin pad, or bat, which created a stippled effect. This process essentially mechanized the decoration process, increasing efficiency and consistency in the manufacturing process, not to mention increasing production significantly. These efficiencies lowered the price of Spode earthenware, thus making it accessible to a wider range of people in different economic circumstances. Henry Daniel, Spode's lead decorator, introduced multiple other decoration innovations, including new gilding and enameling techniques.

Spode introduced other innovations to the earthenware manufacturing process, including the contribution of lands adjacent to his Stoke factory for the building of a canal, which revolutionized the shipping of products from the factory and raw materials to the factory. Not only was water transportation quicker and more direct than ground transportation but it also was gentler, without the bumps inherent in dirt roads—water proved to be much smoother.

In 1796, Josiah Spode II again moved the retail outlet, this time to a ritzier location in a converted theater on Portugal Street. The next year, Josiah Spode I died unexpectedly, prompting his son to return to Stoke to take over the business. It was Spode the younger who introduced perhaps the most significant innovation to the industry: in 1799, he brought out bone china, which became the standard in English porcelain-ware in the 19th century. Josiah Spode III, who lost an arm to a newly installed steam engine in the family pottery works, inherited the company from his father, but he sold the company to William Taylor Copeland and Thomas Garrett when his son, Josiah Spode IV, was 10 years old.

Starley, James
(1830–1881)
English
Inventor, Transportation Industry

James Starley is often called the "father of the bicycle industry," not because he invented the bicycle but because many historians consider his *Ariel* "highwheeler" the first design to feature the elements that came to define the bicycle—such as center-pivot steering and spoked wheels. He subsequently invented an open differential system for his *Salvo* quadricycle that was adopted in the gear system of automobiles. Starley's nephew, John Kemp Starley, designed the *Rover* bicycle, which popularized equal-sized wheels and rear-wheel chain drives, the final elements that survived into the modern bicycle design.

Starley was born in 1830 in Sussex and grew up on his parents' farm. Discontent with country life, he ran away to Coventry as a teenager, convinced that his mechanical ingenuity would be in high demand in this more industrialized region. However, he did not manage to escape his agricultural roots immediately, as he landed a job with John Penn as a gardener. In his early twenties, he married Jane Todd, and to support her, he took on extra work fixing clocks and other mechanical instruments.

Aware of his employee's reputation for working with mechanical gadgetry, Penn asked Starley to fix his wife's sewing machine, which had broken soon after its purchase. Starley not only fixed it but he also identified numerous weaknesses in the design. Penn, as it happened, knew Josiah Turner, the London manufacturer of the sewing machine, to whom he introduced Starley. Turner hired Starley in 1855, and by about 1861, the two formed a partnership to open a sewing-machine company back in Coventry.

Starley introduced many improvements in sewing-machine design, including the model that he called *The European* but was commonly known as the *Lady Godiva* after the renowned 11th-century radical who rode naked through the streets of Coventry in protest over the town's high taxes. The *Lady Godiva* was one of the first sewing machines to incorporate a free arm. Earlier, in 1868, Starley had patented his sewing-machine design that exploited the use of cams in conjunction with a Wheeler & Wilson rotary shuttle. This model operated quietly due to its belt-driven connection between the flywheel and main shaft.

The year before, in 1867, Josiah Turner's nephew, Rowley Turner, returned from a trip to France with a "boneshaker," the velocipede invented by the MICHAUX brothers. Starley, who worked as foreman of the Coventry Sewing Machine Company at the time, recognized several design improvements that might reduce the instability that earned the prototypical bicycle its nickname. The Coventry works contracted to produce 400 boneshakers for the French market, but the advent of the Franco-Prussian War prevented the exporting of the bicycles, which Starley and Turner therefore introduced to the English market.

Starley's design changes earned the bicycle a new nickname, the "penny-farthing," after the difference between the size of the large front wheel and the small back wheel, resembling the size difference between a penny and a farthing coin. The official name of Starley's model (which he designed in collaboration with William Hillman) was the *Ariel*, after the Shakespearean sprite. First marketed in 1871, it was also known as the "highwheeler," after the tall front tire. This differential in wheel size allowed for more efficient pedaling, as one revolution of the pedals could power two revolutions of the wheels.

Starley incorporated numerous other improvements in the *Ariel*, including a mounting step (which he had invented in 1869), center-pivot steering, foot rests on the fork blades, rubber-covered pedals, and rounded solid rubber tires. Starley also improved upon the wheels themselves by introducing iron-wire tension spokes, tightened by screwing threads on the ends of the spokes. In 1874, Starley introduced a further improvement to the spokes, incorporating eyed and threaded nipples for securing them to the wheel rim.

In 1876, Starley introduced the Coventry Lever Tricycle, which sold for 15 pounds. It incorporated spokes arranged tangentially (instead of in a simple radial formation) to strengthen the integrity of the wheels, a system he patented and which is still used in bicycles. The next year, he improved upon this model with the Coventry Rotary Tricycle, as well as the *Salvo* quadricycle, which incorporated a differential gear system that allowed the wheels to rotate at different speeds. This differential system, which Starley patented in 1877, presaged the differential system later employed in cars.

Starley died in 1881, but his influence continued. He had employed his nephew, John Kemp

Starley, who in 1877 started his own bicycle business with William Sutton. Three years after his uncle's death, John Kemp Starley designed the *Rover* safety bicycle, based on Harry J. Lawson's 1879 "safety" bicycle design. Although Starley copied many of Lawson's features, including wheels of equal size (which prevented "headers," or riders plunging forward headfirst, hence the "safety" moniker), a rear-wheel chain drive, and the distinctive diamond-shaped frame, it was the *Rover* that became popular, acting as the precursor to modern bicycles.

⊠ Stephenson, George
(1781–1848)
English
Engineer

George Stephenson was most famous for his role in building the *Rocket*, the locomotive that won a time-trial competition that is generally regarded as the birth of the modern railroad era. He served as the chief engineer on many of the railway lines spawned in the early years of this era. He also earned a degree of fame as inventor of the miner's safety lamp, a distinction he holds with Sir HUMPHRY DAVY, who concurrently but independently produced a safety lamp of a different design.

Stephenson was born on June 9, 1781, in the village of Wylam, near Newcastle upon Tyne, England. He was the second son of six children born to Mabel and Robert Stephenson. His father was a colliery fireman who operated a NEWCOMEN atmospheric steam engine pumping water from the coal mine. Stephenson learned firsthand all of the facets of transporting coal from the mines, as he started working at the age of eight, driving cows off the wagon-way that carried coal past High Street House, the dirt-floored tenement house his family shared with three other families.

At 14, Stephenson joined his father at the Dewley colliery as an assistant fireman, becoming a fireman himself the next year. At 17, he became a "plug man," moving ore through the mine chutes. The next year, he became a "post," feeding ore into the smelting furnace. In 1802, he married Frances Henderson, a servant at a nearby farm who gave birth to their only son, Robert, in 1803. Stephenson struggled to support the family, mending boots, watches, and clocks on the side while attending night school to learn to read and write.

In 1804, Stephenson moved his family to the Killingworth colliery. Two years later, his wife died of consumption. In 1808, Stephenson became the colliery's engine man. To teach himself the workings of the engine, he disassembled it every Saturday and reassembled it before it went back on-line at the end of the weekend. This knowledge paid off in 1812, when he was appointed engine wright at a salary of 100 pounds a year.

The year before, John Blenkinsop had invented a steam locomotive with toothed wheels that traveled on a racked track. Stephenson modified this design to build the *Blutcher* in 1813, which used flanged instead of toothed wheels to travel on a smooth track by adhesion. In 1814, despite popular skepticism, the *Blutcher* pulled 30 tons of coal in eight wagons uphill at four miles per hour.

Besides building one of the first locomotives, Stephenson also invented a miner's safety lamp, which he tested in the "blowers," or the plumes of gas, of the Killingworth mines on October 21, 1815. He survived this trial, proving the efficacy of his design, which sealed a flame behind glass, supplying air by narrow tubes too small to ignite gas outside the lamp. Sir Humphry Davy had designed a similar miner's lamp at about the same time, causing an argument over priority. It is now generally accepted that they produced their designs concurrently but independently.

In 1817, Stephenson designed another locomotive for the Kilmarnock & Troon railway, and two years later, he designed the "steam blast," which exhausted up the chimney to produce a draught that stoked the fire more efficiently. He introduced this innovation on the eight-mile railroad from Hetton to the River Wear at Sunderland in Country Durham.

On April 19, 1821, an Act of Parliament ordered Edward Pease to build a horse railway from West Durham in Darlington to Stockton. When he heard of this scheme, Stephenson demonstrated his locomotive, the *Blutcher*, to Pease, who was sufficiently impressed to appoint Stephenson as the chief engineer of the Stockton & Darlington Company and to persuade Parliament to amend the act to read "make and erect locomotive or moveable engines." And instead of running the train on rails he made, Stephenson chose those manufactured by John Birkinshaw by rolling wrought iron rails in 15-foot lengths. At this time, Stephenson also introduced the four-foot, eight-inch gauge, which became the standard for railways.

In 1823, Stephenson joined forces with Pease, Michael Longridge, and his son to found Robert Stephenson & Company, which established a locomotive works at Newcastle. Two years later, the *Locomotion* rolled out of production as the first railway locomotive, and on September 27, 1825, the Stockton & Darlington line opened as the world's first public line with Stephenson as the engineer of the *Locomotion*, carrying 36 wagons of coal and flour over the nine-mile route in some two hours, reaching speeds of 15 miles per hour. With this success, Stephenson was appointed engineer of the Bolton & Leigh railway the next year and chief engineer of the Liverpool & Manchester railway in 1826 as well.

The most memorable event in Stephenson's life happened in October 1829, at the Rainhill Trails. In order to decide which locomotive design to use on the Liverpool & Manchester line, officers of the company held a locomotive competition with prize money of 500 pounds. Locomotives weighing less than six tons had to carry three times their own weight at a speed of at least 10 miles per hour 20 times up and down a two-and-a-half mile track (the equivalent of the distance between Liverpool and Manchester). Stephenson's *Rocket*, with its multi-tubular boiler, beat out Hackworth's *Sans Pareil* and Braithwaite and Ericsson's *Novelty* to win the 500-pound prize. The Stephenson Company earned another 550 pounds when the railway purchased the *Rocket*, which traveled 12 miles in 53 minutes, reaching a top speed of 29 miles per hour.

On September 15, 1830, the Liverpool & Manchester line opened amid fanfare that was marred by the death of William Huskisson, president of the Board of Trade (and ardent Stephenson supporter in Parliament), who was killed when he stepped in front of the locomotive. Despite this mishap, the opening of the line was generally hailed as a triumph, especially in consideration of Stephenson's solution to the Chat Moss bog dilemma, as he managed to run the line through the swampy area by maximum weight distribution. Over the remainder of his career, he served as chief engineer of the Manchester & Leeds, Birmingham & Derby, Normanton & York, and Sheffield & Rotherham lines.

In 1838, Stephenson moved into the Tapton House, near Chesterfield, where he lived in semiretirement as a "gentleman" farmer. Although he allowed himself this degree of luxury, he refused both a knighthood and fellowship in the Royal Society. His second wife, Elizabeth Hindley, died in 1845, and he married a third time just before he died on August 12, 1848, at Tapton House.

Sutherland, Ivan Edward
(1938–)
American
Electronics Engineer, Computer Scientist

Ivan Sutherland introduced interactive graphics into computing, thus replacing complex mathematical symbols with pictorial symbolism understandable to mainstream computer users. His doctoral dissertation introduced "Sketchpad" technology, which used a electronic pen to enter a user's drawings into computer programs, which could then manipulate the graphics. He later extended this technology into "virtual reality," which allowed users to interact with computerized landscapes and scenarios. "A display connected to a digital computer gives us a chance to gain familiarity with concepts not realizable in the physical world," wrote Sutherland. "It is a looking glass into a mathematical wonderland." Sutherland's invention of interactive computer graphics revolutionized society, simulating actual experience to entertain and to inform.

Ivan Edward Sutherland was born on May 16, 1938, in Hastings, Nebraska. His mother was a teacher, and his father held a doctorate in civil engineering. His father's blueprints and mechanical drawings peaked Sutherland's interest, inspiring his resolution to become an engineer. He first encountered a computer when Edmund Berkeley lent his family SIMON, a relay-based mechanical computer with 12 bits of memory that could add up to 15. Sutherland programmed SIMON to divide by feeding it a division algorithm on an eight-foot paper tape—one of the first big computer programs written by a high school student. His favorite subject in high school was geometry, as he considered himself a visual thinker: "if I can picture possible solutions, I have a much better chance of finding the right one," he later said.

Sutherland attended the Carnegie Institute of Technology (now Carnegie-Mellon University) on full scholarship, graduating with a bachelor of science degree in electrical engineering in 1959. He then held a summer internship with International Business Machines before pursuing graduate work at the California Institute of Technology (Cal Tech) to earn his master of science degree in electrical engineering in 1960. He transferred to the Massachusetts Institute of Technology (MIT) for doctoral work, in large part to work on the Lincoln Laboratory's TX-2, one of the most advanced computers of its day due to generous funding from the U.S. Air Force.

Sutherland fused his visualization skills with the TX-2's capabilities to develop a novel idea in computing: interactive graphics, or what is now known as Graphical User Interface. Sutherland invented a light-sensitive pen for "drawing" on the screen; in fact, Sutherland's computer program tracked the movement of the pen across a grid, translating its place on vertical and horizontal axes into its position on the screen. Sutherland's software could extrapolate the drawn shape into a three-dimensional representation, which could then be manipulated in any direction, transforming it into any desired shape.

Sutherland presented his dissertation, "Sketchpad: A Man-Machine Graphical Communication System," to a committee comprised of Marvin Minsky, Steven Coons, and Claude Shannon, the inventor of information theory, to earn his Ph.D. in electrical engineering in 1963. He also presented his invention at a meeting between MIT faculty and the Department of Defense's Advanced Research Projects Agency (ARPA, later DARPA), which was so impressed that it appointed him director of its Information Processing Techniques Office in 1964. First Lieutenant Sutherland, who was merely fulfilling a Reserve Officer Training Corps obligation from Carnegie, found himself with a secretary, a $15 million budget, and a vague order to "go sponsor computer research."

Bell Helicopter Company received one of Sutherland's inventions—a head-mounted display system connected to infrared cameras that moved in conjunction with pilots' head movements, thus acting as their "eyes" at night and in thick fog. Sutherland advanced this idea after moving to Harvard University in 1966 as an associate professor of electrical engineering—in collaboration with undergraduate Robert Sproull (son of a DARPA associate), the work transformed this "remote reality" into "virtual reality" by replacing the cameras with computer-generated scenes with which the viewer could interact.

In 1968, Dave Evans invited Sutherland to join the computer science department he had founded at the University of Utah as an associate professor. The two also founded Evans & Sutherland Computer Corporation, where they continued to develop Sutherland's computer visualization technology into flight simulation equipment for pilot training. By 1972, the university had promoted Sutherland to a full professorship in recognition of his helping to transform Utah into a mecca for computer graphics research.

In 1976, Cal Tech recruited Sutherland to direct its new computer science department as Fletcher Jones professor of computer science. Within four years, he had established a strong enough department to depart in order to found Sutherland, Sproull & Associates with Robert Sproull in Pittsburgh. Two years later, Sutherland's brother, Bert, joined the firm, which subsequently moved to Palo Alto. The partners secured numerous patents that made the company attractive enough for Sun Microsystems to buy them out intact to establish Sun Laboratories, a research and development facility. Sutherland serves as a Sun fellow and vice president, continuing to conduct pure research that gets imported from the lab to the corporate offices when deemed commercially viable—as was the case with the Java programming language.

Sutherland, who is married with two children, has received much recognition for his pioneering innovations. In 1972, the National Academy of Engineering granted him its first ZWORYKIN Award, and in 1988, the Association for Computing Machinery gave him its A. M. Turing Award. When the Smithsonian Institution and *Computerworld* bestowed Sutherland and Evans with the Price Waterhouse Information Technology Leadership Award for Lifetime Achievement in 1996, they noted that without "visualization, there would not have been mass computer usage; without modeling, engineers would still have to build prototypes to find out if their designs worked—*and if they were safe.*"

Swan, Sir Joseph Wilson
(1828–1914)
English
Electrical Engineer, Chemist, Energy and Power Researcher

Sir Joseph Wilson Swan invented the incandescent electric lamp concurrently with THOMAS EDISON, though the latter is more often acknowledged for this feat due to his greater commitment to developing electricity delivery systems to take advantage of electric light bulbs. As with Edison, Swan produced numerous inventions across a wide technological range: he invented the dry-plate photographic processing, the electric miner's safety lamp, electrolytic cells (or batteries), and the artificial fiber that was later developed as rayon.

Swan was born on October 31, 1828, in Sunderland, England. He attended a dame school administered by three elderly women, and then a larger school run by the Scottish minister Dr. Wood. At the age of 14, he apprenticed to a pair of Sunderland druggists who died before he completed his apprenticeship, so he joined the Newcastle chemical firm run by his brother-in-law, John Mawson.

Eager to expand his knowledge, Swan attended lectures at the Athenaeum in Sunderland, including one in 1845 when W. E. Staite demonstrated electric-arc and incandescent lighting, which captured Swan's scientific imagination. By 1848, he started experimenting by baking strips of cardboard and paper (sometimes in a tar syrup) at high temperatures and charring them into coiled carbon filaments. He then arduously attached the delicate filament to wire and encased the setup in a glass bottle, which he evacuated of air and sealed with a cork. Unfortunately, in the absence of a complete vacuum, the filament burned quickly, blackening the glass.

In 1865, the German chemist Hermann Sprengel invented the mercury vacuum pump, capable of sustaining a more complete vacuum than was previously attainable. Charles Stearn, who had become quite adept at this process, helped Swan employ the Sprengel pump. They discovered that continued heating of the filament during the extraction of air, evacuating the exhausted gases while creating the vacuum, thus improved the light produced.

Swan continued to refine his incandescent electric lamp until it was finally ready for exhibition at the December 18, 1878, meeting of the Newcastle Chemical Society. The next year, he demonstrated it before an audience of 700 at the Newcastle Literary and Philosophical Society. However, Swan neglected to patent his invention. On October 21, 1879, Thomas Edison successfully illuminated his light bulb, based on a similar principle to Swan's, for the first time. By the next month, he had secured a British patent but did not commence production.

On January 2, 1880, Swan received British patent number 18, covering his incandescent electric lamp. Production began in a factory at Benwell in Newcastle and then moved to London. By 1881, the House of Commons was illuminated by Swan's bulbs, and by 1882, so was the British Museum. Swan lit his own house with his incandescent electric lights, the first private home so illuminated, and Lord KELVIN soon followed suit. In 1882, Edison filed suit against Swan for patent infringement, but the two settled out of court, and indeed they joined forces to form the Edison and Swan Electric Light Company. Swan benefited from this arrangement, as Edison had developed infrastructure for delivering electricity to homes and businesses, a necessary component for capitalizing on electric light technology.

Besides inventing the electric light, Swan contributed to society with several other important inventions. While developing his carbon filament, he also experimented with a carbon (or autotype) process for developing photographic film, which he patented in 1864. In 1871, he improved upon the messy wet plate photo developing process by introducing a dry plate process, a much neater system. In 1879, he patented bromide printing paper, which remains the standard for photographic printing.

Swan also invented an electric miner's safety lamp that, although it was too expensive to produce in his time, anticipated the modern miner's safety lamp. In the process of developing this invention, he also invented improved plates for the Planté cell that prevented the spilling of acid, which he patented on May 24, 1881. That year, the French government made him a Chevalier de la Légion d'Honneur.

In 1883, Swan patented a synthetic material he had invented while developing the carbon filament; his wife and daughter crocheted swaths of this "artificial silk," which he exhibited in 1885. Originally developed for carbonization and use as filaments, the cotton was nitrated to produce nitrocellulose, a volatile substance that Swan extruded in acetic acid through a grid of small holes to form thin threads that were then treated with a coagulating fluid to transform them into inert cellulose. When these filaments proved unsuccessful in his incandescent bulbs, he abandoned the fiber and sold the patent rights to Courtaulds, which later developed it into rayon, one of the first synthetic fibers.

Swan received numerous honors at the end of his career. The Royal Society inducted him into its fellowship in 1894. He served as president of the Institution of Electrical Engineers in 1898 and of the Faraday Society in 1904. That year, he was knighted, and he also received the 1904 Hughes Medal from the Royal Society. A decade later, Swan died at home in Warlingham, Surrey, on May 27, 1914. His light bulb continued to illuminate the world, and it served as the precursor to the modern light bulb, which was based on the exact same technology, merely replacing the vacuum with inert nitrogen and the carbon filament with tungsten.

Swedenborg, Emanuel
(1688–1772)
Swedish
Metallurgist, Physiologist, Theologist

Emanuel Swedenborg commenced his career as a scientist with wide-ranging interests, from metallurgy to physiology to invention. Many of his scientific contributions have made a lasting impact on society—for example, a tank he invented for experimenting on ships is still in use. However, his continuing relevance rests more on his spiritual writings, which form the basis of the Swedenborgian Church, a distinct Christian denomination.

Swedenborg was born Emanuel Swedberg on January 29, 1688, in Stockholm, Sweden, the second son of Jesper Swedberg. At the age of 11, Swedenborg entered the University of Uppsala, where his father served as a professor of theology before becoming the Lutheran bishop of Skara, necessitating a move. Thereafter, Swedenborg lived with his sister and brother-in-law, Erik Benzelius, the university librarian who influenced him to shift his major from humanities to science. He presented his dissertation on June 1, 1709, after which he toured Europe, a common postgraduate practice of the time.

First, Swedenborg visited England, where he stayed for the next three years, then continued to Holland, Belgium, Germany, and France. He returned with a portfolio of designs for various scientific instruments and machines, including a submarine, a prototypical steam engine, an air gun with a 70-bullet magazine, an airplane, a calculator, and a mercury air pump, among other things. He worked with CHRISTOPHER POLHEM, who greatly appreciated the assistance on his mechanical experiments. Swedenborg fell in love with and became engaged to the eminent scientist's youngest daughter, though he called off the wedding upon realizing that his love was unrequited and remained single the rest of his life.

In 1716, the Royal College of Mines appointed Swedenborg its Extraordinary Assessor, a nonsalaried position (his family had sufficient wealth from mining investments to support him and his scientific activities). That year, he established Sweden's first scientific journal, *Daedalus Hyperboreus*, publishing numerous accounts of Polhem's mechanical inventions. Also in 1716, he entered a British contest for accurately determining longitude at sea, though his solution, using the moon as a reference point, earned the scorn of the judging committee. The next year, he turned down an offer of a mathematics professorship at his alma mater in favor of his assessor's position, which he retained for the remainder of his career. In 1719, Queen Ulrica Eleanora elevated his family to the nobility, resulting in the name change from Swedberg to Swedenborg, as well as his installment in the Swedish House of Nobles, the governing body he actively contributed to for the rest of his life.

Swedenborg devoted himself to his mining duties wholeheartedly, touring mines and smelting facilities throughout Europe as metallurgical research for two volumes, one on copper (*De Cupro*) and one on iron (*De Ferro*—which the French Académie des Sciences "considered to be the best about this subject"). A companion volume, *Principia Rerum Naturalium*, published in 1734, acted as a kind of introduction (as well as an answer) to NEWTON's own *Principia* by establishing the philosophical foundations of physics and chemistry. In fact, Swedenborg's *Principia* anticipated the atomic theory and the wave theory of light, among other scientific theories, and is considered a crowning achievement in science for that time period.

Over the next decade, Swedenborg increasingly fused matters of science with matters of spirit in an attempt to locate the physiological seat of the soul, starting with an exhaustive, multivolume examination of the animal kingdom (Regnum Animale). As the decade ended, a spiritual crisis hovered on the horizon; from June 1743, he began recording his dreams, which were taking on a prophetic air. Finally, on Easter weekend of 1744, while staying at a London inn, a vision of Jesus Christ visited him, charging him with the duty of interpreting the Bible, and then the gates of heaven and hell opened before him.

Swedenborg spent the rest of his life fulfilling this duty by applying scientific rigor to his spiritual writings. He published detailed exegeses of the books of Genesis and Exodus, then he skipped to the book of Revelations when it became clear that he would not complete the Bible in his lifetime at his pace. He addressed the widest possible audience by touching on everyday issues; for example, one of his most popular and controversial works, published in 1768 at the age of 80, concerned conjugal love, as he simultaneously lauded the ideal of fidelity while also accepting sexual desire as a driving force in life.

Swedenborg died on March 29, 1772, in London. There, in 1787, some 16 years after his death, the Swedenborgian Church was established to practice his particular version of Christianity. He did not intend for his writings to form the basis of a new denomination, also called the Church of the New Jerusalem, but his followers found that his beliefs spoke very strongly to them. Today, the General Convention of Swedenborgian Churches promotes his religious vision worldwide.

Swinburne, James
(1858–1958)
Scottish
Electrical Engineer, Plastics Industry

Sir James Swinburne is considered the "Father of British Plastics" for his invention of this synthetic material that ended reliance on raw materials derived from nature. Swinburne cannot be accorded primary credit for inventing plastic, as LEO HENDRIK BAEKELAND filed a patent for the same chemical formulation just a day before Swinburne. However, Baekeland recognized Swinburne's common claim to the invention by coordinating their marketing efforts and eventually collaborating professionally.

Swinburne was born on February 28, 1858, in Inverness, Scotland, the third of six sons of a naval captain. He grew up speaking Gaelic on the isolated island of Eileen Shona in Loch Mordart. He attended Clifton College and then commenced his career in engineering as an apprentice at a locomotive works in Manchester, England. He then moved into electrical work at a Tyneside engineering firm.

In 1881, Swinburne went to work for JOSEPH WILSON SWAN, inventor of the incandescent electric lamp (concurrently with THOMAS EDISON), at his Newcastle lamp factory. So impressed was Swan with Swinburne's work that he charged the young engineer with the duty of setting up new lamp plants in Paris, and then in Boston the next year. Upon his return to England, Swinburne went to work for Hammond's lamp factory.

In 1885, Rookes Crompton hired Swinburne as his technical assistant and later as manager of his dynamo factory in Chelmsford. There, Swinburne invented a watt-hour meter and also reputedly introduced the terms *motor* and *stator* into the electrical engineering vernacular. In 1889, Swinburne invented the "hedgehog" transformer, a device for translating medium-voltage alternating current (AC) into high-voltage power for transmission over long distances. His claim that the hedgehog transformer maintained the highest all-day efficiency met opposition in 1892, when John Ambrose Fleming counterclaimed that the hedgehog was the worst type of transformer. A heated dispute ensued, with political issues coloring the debate as much as technical issues. Baron KELVIN supported Fleming's calibrations of AC measurement, effectively squelching Swinburne's claims.

In 1894, Swinburne established his independence by setting up his own laboratory and freelancing as a consulting engineer while also offering technical evidence to the High Court. After the turn of the 20th century, he conducted research on the thermodynamics from an engineering perspective, and in 1904, he published his first book, *Entropy*.

In 1902, a patent agent had introduced Swinburne to a synthetic resin derived by combining phenol with formaldehyde. Swinburne spent the next half-decade experimenting with the compound, trying to develop commercial applications for the malleable substance. He finally applied for a patent on this invention in 1907 only to discover that Leo Baekeland had filed a patent for the same invention the day before!

Swinburne and Baekeland reached a tacit agreement to divide rights to the innovation, with the former establishing and serving as chairman of the Damard Laquer Company to produce a liquid lacquer in Birmingham for distribution in Britain, while the latter produced a solid resin in the United States. In 1926, Baekeland bought out Damard and established Bakelite Limited, installing Swinburne as its first chairman and H. V. Potter of Damard's research department as managing director. Thereafter, plastics entered the British market in earnest.

In 1934, Swinburne succeeded his cousin as the ninth baronet, adding the title "Sir" to his name. He resigned as Bakelite chairman in 1948, though he retained a seat on the company's board of directors until his 1951 retirement. Throughout his career, he had served his field in numerous capacities: from 1898 through 1899, he served as honorary editor of *Science Abstracts;* from 1902 through 1903, he presided over the Institution of Electrical Engineers (IEE); and from 1909 through 1911, he served as president of the Faraday Society. In 1906, the Royal Society inducted him into its fellowship.

In February 1958, the scientific community recognized Swinburne's 100th birthday by honoring his contributions. The IEE council sent a letter of congratulations to its member of 73 years, the longest-standing corporate membership on record. The Royal Society paid tribute to him, and the *New Scientist* published a long profile on him. He died a month after this birthday, on March 30, 1958, in Bournemouth. By this time, plastics had transformed modern society by allowing for the synthesis of durable materials, films, and liquids.

Telford, Thomas
(1757–1834)
Scottish
Civil Engineer

Thomas Telford, nicknamed the "Colossus of Roads," worked prolifically in building roads, bridges, and canals in Scotland, England, and Wales. His work advanced transportation and communications, connecting previously remote regions to one another, and thereby promoting traffic in commerce. His most famous works, the Pont Cysylltau Aqueduct and the Menai Suspension Bridge, continue to impress observers as breathtaking engineering feats—Sir Walter Scott called the former "the most impressive work of art I have ever seen." Despite his prodigious output, Telford maintained modest financial circumstances throughout his life, often working without compensation.

Telford was born on August 9, 1757, in the village of Bentpath, in Dumfriesshire, Scotland. (His father, the shepherd John Telford, died that November.) Telford attended the Westerkirk parish school while working as a shepherd. At the age of 14, he became a stonemason, and his work brought him to Edinburgh and then, in 1782, to London, where he distinguished himself as a conscientious and skilled laborer working under William Chambers on additions to Somerset House. Two years later, he moved to Portsmouth, where he worked on a house for the Commissioner of Portsmouth's dockyard.

In 1786, the county of Shropshire (bordering England and Wales) appointed Telford as its surveyor of public works, a post that carried no salary at first. His first major project, bridging the River Severn at Montford in 1790 using convict labor, earned him a reputation as an extraordinary engineer. He completed two more bridges over the Severn, at Buildwas and Dewdley. Although Telford worked on innumerable other projects throughout his career, he retained this post in Shropshire for the rest of his life.

In 1793, the Ellesmere Canal Company hired Telford as its engineer and architect, charging him with the duty of constructing aqueducts over the Ceirog and Dee valleys in Wales. Telford rose to this challenge by innovating a new construction method, employing troughs of cast-iron plates, which he fixed in masonry. In 1805, he completed one of his more famous works, the Pont Cysylltau (or "connecting bridge") Aqueduct over the Dee Valley on Shropshire Union Canal. Supported by 18 piers, it is both the highest (at 121 feet) and longest (at 1,007 feet) aqueduct in Britain. Telford's excellent work earned him a concurrent appointment in 1795 as engineer for the Shrewsbury Canal Company.

In 1801, Telford was hired to survey the rural roads of Scotland, and two years later, he was appointed the civil engineer for an ambitious governmental project to improve the roads connecting the Scottish Highlands, many of which dated back to Roman times. Over the next two decades, Telford built almost a thousand miles of road and 120 bridges. His method consisted of digging to the foundation of the old road, grading it level, then laying down a base of cobblestone-sized rock, covered by seven inches of broken stone, overlaid with a top layer of three inches of fine stone, resulting in smooth and durable roads. One testament to the success of his method was the speed with which mail coaches could traverse his roads, averaging eight miles per hour and achieving top speeds of 12 miles per hour. Telford was responsible for laying new roads between Carlisle and Glasgow, as well as between London and Holyhead.

In 1818, Telford helped found the Institute of Civil Engineers and served as its first president. In this latter period of his career, he achieved his crowning glory in the construction of the Menai Bridge. Built between 1819 and 1826, it spanned 580 feet over the Menai Straits to join the island of Anglesey to the Welsh mainland. As with the nearby bridge over the River Conwy, also completed in 1826, he dipped the wrought-iron links in oil to prevent rusting. As it was one of the first suspension bridges, Telford prayed on his knees the day the suspension links took on weight; the bridge continues to carry its weight.

Telford also completed many projects outside of Britain, such as the Gotha Canal from Lake Malaren in Stockholm to the central-Swedish lakes Vaner and Vettir, for which King Gustav of Sweden knighted him. Telford died in Westminster on September 2, 1834, and was accorded the honor of burial in Westminster Abbey.

⊠ **Tesla, Nikola**
(1856–1943)
Croatian/American
Physicist, Electrical Engineer

Nikola Tesla defied the contemporary belief in direct-current (DC) electricity generation, whose proponents included THOMAS EDISON, by inventing the induction motor, which generated alternating current (AC). AC ultimately prevailed over DC as the standard for electricity generation. In 1912, both Tesla and Edison were considered for the Nobel Prize in physics, a fact that infuriated Tesla, though neither ended up receiving the honor.

Tesla was born at midnight between July 9 and 10, 1856, in the village of Smitjan, in Lika, Croatia (part of the former Austro-Hungarian Empire). His mother was Djuka Mandić, and his father, Milutin Tesla, was a priest of the local Serbian Orthodox Church. When Tesla was seven, the family moved to Gospić, where he continued grammar school. He matriculated in the Real-Gymnasium and then the Higher Real-Gymnasium in Karlovac. Although his father intended him to enter the priesthood, he allowed Tesla to attend the Joanneum Technical University in Graz, Austria, to study engineering.

In 1878, one of Tesla's professors demonstrated a Gramme direct-current electric dynamo; when Tesla pointed out the inefficiency of its sparking brushes and suggested the design of a motor without a commutator, the professor laughed. The next year, Tesla transferred to the University of Prague to study philosophy, though he left before completing his degree due to the death of his father. In 1881, he moved to Budapest to work as an engineer in a new telephone company, where he invented a telephonic repeater.

The idea Tesla had proposed to his professor continued to ferment in his mind, reaching fruition instantaneously. While walking in a park with Antony Szigety, the words to Goethe's *Faust* welled up in Tesla's mind spontaneously, and as he recited them, "the idea [for his alternating current induction motor] came like a lightning flash. In an instant I saw it all, and drew with a stick on the sand the diagrams which were illustrated in my fundamental patents of May, 1888, and which Szigety understood perfectly."

In the meanwhile, Tesla moved to Paris in 1882 to work at the Continental Edison Company, which dispatched him to Strasbourg the next year to make repairs on its electric plant. There, he constructed the first prototype of his AC induction motor, which gave him "the supreme satisfaction of seeing for the first time rotation effected by alternating currents without commutator," said Tesla. In 1884, Tesla immigrated to the United States, penniless but for the value of his letter of introduction to Thomas Edison, who hired Tesla despite his espousal of a competing standard of electrical production. Tesla spent the next year at the Edison Machine Works redesigning DC dynamos.

In 1885, Tesla left Edison to establish his own company producing industrial arc lamps, but he was railroaded out of the business, leaving him teetering on the verge of poverty. After two years working odd jobs, he had saved enough to establish his own laboratory, where he finalized his design and construction of the alternating-current motor. In 1888, he filed a set of complicated patents covering all aspects of alternating-current electricity generation, receiving them that same year. The next year, Tesla gained his American citizenship.

In 1891, Tesla explained his AC system at a lecture before the American Institute of Electrical Engineers; when word of this talk made its way to GEORGE WESTINGHOUSE, he immediately obtained Tesla's patents for commercial exploitation. Westinghouse exhibited Tesla's system by lighting the entire 1893 Chicago World Columbian Exposition with AC power. Within months, Westinghouse had secured a contract to provide AC electricity generated at Niagara Falls to power the nearby city of Buffalo.

This development represented the death knell for Edison's DC, as it clearly established AC as the standard in electrical generation.

Tesla pushed the notion of AC power generation to its logical extreme, promoting the use of very high frequencies. At the turn of the century, he used the so-called Tesla coil at his Colorado Springs laboratory to produce a 135-foot, million-volt electric spark, the greatest bolt of artificial lightning ever produced. At the same time, he experimented with wireless communication. In 1900, he wrote: "I have no doubt that [wireless] will prove very efficient in enlightening the masses, particularly in still uncivilized countries and less accessible regions, and that it will add materially to general safety, comfort and convenience, and maintenance of peaceful relations."

The scientific credibility of Tesla's wireless experiments was called into question when he announced his reception of signals from alien beings and his invention of a "death-ray" in 1934. He spent his final years in seclusion, moving from one hotel to the next in New York City, where he died on January 7, 1943.

Thomson, Elihu
(1853–1937)
English/American
Electrical Engineer

Elihu Thomson helped found the electrical manufacturing industry in the United States. He invented the recording wattmeter, a device for tracking (and billing) electrical output, and almost 700 other innovations (some not in the field of electricity) that he patented throughout his career. He also founded the company that merged with THOMAS EDISON's company to become General Electric, one of the largest producers of electricity and electrical products. Although Thomson favored the laboratory over the boardroom, he did much to advance the electrical industry as an economic sector.

Thomson was born on March 29, 1853, in Manchester, England, the second of 10 children. His parents, skilled artisans who originally hailed from Scotland, moved the family to Philadelphia, Pennsylvania, when Thomson was merely five years old. He attended the Central High School, where he served as science professor Edwin J. Houston's assistant. After Thomson's graduation, he continued to collaborate with Houston, joining the school's faculty as a fellow science professor (a title he retained for the rest of his life).

Thomson and Houston conducted joint research, inventing arc lights as well as disproving Thomas Edison's claim to have discovered wireless transmission. Most significantly, the pair founded the American Electric Company in New Britain, Connecticut, in 1880. Thomson served as the company's chief electrical engineer, while owning 30 percent of the company's stock. Two years later, a group of investors bought out the company and moved it to Lynn, Massachusetts, where it was renamed the Thomson-Houston Electric Company in 1883.

In 1888, Thomson invented the walking beam meter, which used heat to evaporate alcohol in bottles attached to a balance beam, thereby measuring lamp-hours. Although this apparatus ultimately failed as a practical device, it acted as a direct precursor to the Thomson recording wattmeter. Edison had invented a motorized meter in the later 1870s, but he abandoned this project in favor of a chemical meter. After the failure of the walking beam chemical meter, Thomson took up where Edison had left off, collaborating with Thomas Duncan to improve upon the design to measure both alternating and direct current. This wattmeter proved sturdier and more accurate than any of its counterparts. This innovation of a means of measuring electrical output transformed the electrical industry, as it allowed for the tracking and billing of actual electrical consumption. Thomson followed up on this innovation with the invention of the high-frequency generator in 1890.

In 1892, the Thomson-Houston Electrical Company merged with the Edison General Electric Company to form the General Electric Company. This development consolidated the interests of the foremost pioneers in the electrical industry. Thomson remained with the company as a consultant, but throughout his corporate career, he maintained an aloof distance from business dealings, preferring to focus his interest on the scientific front. However, Thomson was also an astute businessman who managed to amass a significant fortune.

In 1893, Thomson exhibited a high-frequency coil that could produce arcs more than six feet long. Frederick Finch Strong later developed therapeutic applications for these currents. In addition to these innovations, Thomson invented the three-coil generator and the process for incandescent electric welding. He did not limit his research to the electrical field, though, as he contributed to radiology by improving X-ray tubes and innovating stereoscopic X-ray pictures. He also recognized the relationship between the intake of helium and oxygen and caisson disease, or the bends.

During World War I, Thomson promoted the establishment of an engineering school hosted jointly by Harvard University and the Massachusetts Institute of Technology, but the plan was foisted when courts ruled against the use of funds donated by the industrialist Gordon McKay for this purpose. MIT was sufficiently impressed by Thomson's work on this initiative to offer him the school's presidency in 1919. Thomson refused the offer, but he did serve as acting president from 1921 through 1923.

Thomson was married twice—first to Louise Peck, and then to Clarissa Hovey. His second wife increased his participation in the social milieu, though he retained an air of detachment from the world beyond the laboratory. Throughout his career, he secured 696 U.S. patents, making him one of the most prolific inventors ever. Thomson died on March 13, 1937, at his home in Swampscott, Massachusetts.

Trevithick, Richard
(1771–1833)
English
Engineer

Richard Trevithick harnessed high-pressure steam to construct the first railway locomotives. Up until the turn of the 18th century, JAMES WATT, who controlled the steam engine patent, insisted that high-pressure steam engines were prohibitively dangerous (an opinion that benefited his monopoly on low-pressure steam engine technology). After Watt's patent expired in 1800, Trevithick built high-pressure steam engines that proved small enough and light enough to propel themselves. Although he is not nearly as renowned as Watt, he is considered by many to be the "father of railroads," as he was the first person to build a working locomotive engine.

Trevithick was born on April 13, 1771, at Illogan, in Cornwall, England. He was the first son and the fifth of eight children born to Ann Teague, a mine captain's daughter, and Richard Trevithick, an engineer and captain at the Dolcoath mine. He attended the Camborne School, but he showed more aptitude for athletics than for academics—at six feet two inches in height, he was known as the "Cornish giant," one of the best wrestlers in Cornwall. After school, he joined his father at the Wheal Treasury mine, where he followed in his father's footsteps in engineering.

Trevithick became fascinated with the notion of making a steam locomotive, and by 1796, he had built a miniature steam locomotive model. To fuel the engine, he dropped a red-hot iron in a tube beneath the boiler, causing steam to rise. The next year, Trevithick married Jane Harvey, the daughter of a foundry owner, and together the couple eventually had six children. In 1798, they moved to Camborne Churchtown, where Trevithick demonstrated his locomotive model to Lady Basset.

In 1800, James Watt's patent on steam engines expired, opening the door for experimentation with high-pressure steam engines, technology that Watt had opposed—purportedly due to its risk. In order to generate more power, low-pressure steam engines were simply built larger, a suitable technique for immobile applications, such as clearing mines of floodwater. However, Trevithick hit upon the idea of using steam engines to move themselves, thus requiring a smaller, lighter engine than could be built using low-pressure technology. The advent of high-pressure technology allowed for the construction of engines small and light enough to propel themselves.

On Christmas Eve 1801, Trevithick drove seven friends up Camborne Hill in the *Puffing Devil*, a locomotive powered by high-pressure steam, which required no condenser or vacuum pressure, as did low-pressure steam systems. A boiler filled a single horizontal cylinder with the steam necessary to move a piston linked by connecting rod to a crankshaft that turned a large flywheel, which powered the wheels. However, the compactness of the system prevented it from keeping up steam for very long—the *Puffing Devil* ran out of steam while his friends were celebrating Trevithick's success in a nearby inn.

A little over two years later, in February 1804, Trevithick produced the first railway locomotive, or high-pressure steam engine to run on rails, rising to the challenge that he could not haul 10 tons of coal from the Penydarren Ironworks near Merthyr Tydfil, in south Wales, to the Merthyr-Cardiff Canal. With a single vertical cylinder, an eight-foot flywheel, and a steam jet up the chimney to create an exhaust draught that improved efficiency, the locomotive hauled 10 tons (including 70 people) in five wagons over the nine miles of rails in four hours and five minutes, reaching speeds of almost five miles per hour. However, the weight of the locomotive, at seven tons, broke the brittle cast-iron rails three times, after which it was retired from use as a locomotive, though it continued in other functions long afterward.

Trevithick ran his next locomotive in Northumberland, along the existing five-mile wooden wagon-way from Wylam Colliery to the River Tyne. At five tons, the locomotive still proved too heavy for the iron-enforced rails. This locomotive trial was not a complete failure, however, as it was witnessed by GEORGE STEPHENSON, who grew up alongside these rails and went on to incorporate many of Trevithick's innovations into his own locomotive designs.

In July and August 1808, Trevithick operated his latest locomotive, *Catch Me Who Can*, along a circular track he had erected at Euston Square in London. He charged one shilling for the novelty ride, which reached speeds of 12 miles per hour. Unfortunately, this locomotive also broke its tracks—it was not until the advent of wrought iron rails that tracks could withstand the weight of locomotives. However, Trevithick had already abandoned the development of locomotives in favor of other diverse applications for his steam-engine technology: river-dredging (he received sixpence for every ton lifted from the bottom of the Thames River), raising sunken ships, and rock-boring, among other applications.

In 1810, Trevithick came down with typhus, and he returned from London to Cornwall, where he installed his family at Penponds on the property of his mother, who had just died. He fell into bankruptcy the next year but persevered with plans to apply his steam-engine technology to mining in South America. In 1814, he sent nine of his engines to Peru, and two years later, he followed to oversee their transportation up treacherous mule trails and installation at the silver mines above 14,000 feet at Cerro de Pasco. His success earned him enough money to buy his own silver mine, but he lost all his bullion to revolutionary forces that invaded in 1826. Robert Stephenson, son of George Stephenson, paid for Trevithick's passage back to England, where his wife had become the landlady at the White Hart Inn at Hayle.

In February 1828, George Stephenson petitioned the House of Commons to establish a government pension for the "father of railroads," but the parliamentary body refused this request. Trevithick sunk into extreme poverty, though legend of his penniless death and burial in an unmarked pauper's grave are exaggerations. In fact, when he died on April 22, 1833, at the Bull Inn at Dartford, in Kent, England, the pawning of his gold watch paid for a funeral as well as a simple marked grave.

Tull, Jethro
(1674–1741)
English
Agriculturist

Considered one of the founding fathers of modern agriculture, Jethro Tull was responsible for developing a variety of agricultural machinery. Tull's inventions helped spark the agricultural revolution, considerably changing farming methods by allowing farmers to work more efficiently and more productively in the field. His methods were not immediately accepted by his fellow farmers but were later adopted into general use.

Baptized on March 30, 1674, Tull was born at Basildon, in Berkshire, England. Tull's father owned a farm in Oxfordshire, which Tull later inherited. He attended St. John's College, Oxford, then entered Gray's Inn in London to pursue a career in law. Though he was called to the bar in 1699, he chose to return to the country to run his father's farm. Tull's career change was due in part to poor health, from which he had suffered for much of his life.

Tull's first contribution to modern agriculture was the seed drill, which he developed in

1701. The drill mechanized and simplified the process of sowing seeds, which traditionally were broadcast by hand. The drill was capable of performing three functions, which had previously been handled separately: drilling, sowing, and covering the seeds with soil. The seed drill, which was designed to be pulled by a horse, was made up of two hoppers, funnel-shaped vessels into which seed was placed before being dispensed. The openings of the hoppers were controlled by a spring-loaded mechanism. As the seed drill was pulled forward, a wooden gear turned, thereby allowing the seed to drop out at regular intervals. A plow and harrow, a sharp instrument used to break up the ground, were also part of the machine, employed to cut the groove, or drill, in the soil for the seed and to turn the soil to cover the seed. Tull's mechanical seed drill sowed in rows, a revolutionary practice at the time, and was capable of sowing three rows of seed at one time.

After creating the seed drill, Tull moved on to develop machinery to facilitate plowing. Recognizing the existence of different types of soil and the need for equipment specialized for particular types of soil, he first created a two-wheeled plow designed for lighter soils such as those that could be found in southern England and in Midland areas. The plow was capable of cutting to a depth of about seven inches. Tull also developed a swing plow more appropriate for heavy clay soils. These soils were typically found in the eastern regions of England. Another breakthrough instrument created by Tull was a horse-drawn hoe. At the time farmers were often forced to resort to using a breast plow to cut through top layers of grass and weeds before the primary plowing could take place. Tull was able to develop a hoe using four coulters, blades that created vertical cuts in the soil, that were arranged in such a manner as to pull up the grass and weeds and permit them to dry on the surface.

In 1709, Tull moved to Berkshire, where he purchased his own farm. In 1733, after more than three decades working in agriculture, Tull published a book on farming, *The New Horse Hoeing Husbandry: Or an Essay on the Principles of Tillage and Vegetation*. The French philosopher Voltaire practiced Tull's farming methods on his estate, but in general, many farmers were skeptical of Tull's principles and were slow to adopt them. Despite his initial struggle for acceptance of his agricultural techniques, Tull greatly influenced farming for centuries to follow, and his inventions helped to increase farm productivity. Tull died at Prosperous Farm in Berkshire in 1741.

V

Venter, J. Craig
(1946?–)
American
Biochemist, Genetics Researcher

J. Craig Venter initiated what some might call the race to map the human genome. In reality, only after the Human Genome Project had rebuffed Venter's offer to use his whole-genome shotgun sequencing approach to help in mapping the human genome did Venter strike out on his own, finding corporate backing to do so. As it turned out, Venter's shotgun method proved cheaper and more accurate, not to mention faster, than the methods employed by the Human Genome Project. Venter is credited with speeding up the discovery of the whole human genome by some five years, thereby quickening the pace whereby scientists can apply this information to their research, potentially finding cures for diseases or creating helpful drugs much sooner than would otherwise be possible.

After barely graduating from high school, Venter went to Southern California to surf, a move that practically ensured that he would be drafted into the Vietnam War. From 1965 through 1968, he served in the U.S. Navy Medical Corps, where he first encountered how tenuous a hold humans have on life. He also encountered the inanity of bureaucratic authority and rebelled against it, resulting in his imprisonment in the brig twice for insubordination when he refused to follow orders.

Returning to the United States with a renewed sense of purpose, he attended the University of California at San Diego, earning his bachelor's degree in biochemistry in 1972, and his Ph.D. in physiology and pharmacology in 1975. He then served as a professor of pharmacology at the State University of New York, Buffalo, and subsequently as a professor of biochemistry there.

In 1984, he joined the National Institutes of Health (NIH) as a section chief of the Institute of Neurological Diseases and Stroke. There, he innovated a new way to isolate gene fragments—the expressed sequence tag (EST)/complementary DNA (cDNA) approach, which allowed him to identify 25 or more genes a day, an unheard-of rate. However, the scientific community balked when the NIH tried to patent the discovered gene fragments, and the ensuing controversy created a political fracas that prevented Venter from receiving funding for his research. "If I had a choice, I would definitely have stayed at NIH" to continue genomic research, Venter later stated, but his ambitions had eclipsed the limitations of his title, not to mention the politics of government-sponsored research.

So in 1992, Venter and his colleague (also his wife), Claire Fraser, founded The Institute of Genomic Research (TIGR), with Venter serving as president and director of this nonprofit center for genetic research based in Gaithersburg, Maryland. There, he innovated yet another new technique for sequencing genes, the so-called shotgun approach, which called for the shredding of complete DNA sequences into fragments that could be identified and then reassembled back into the proper sequence. Using this method, TIGR fully sequenced the first free-living organism, *Haemophilius influenzae*, by late May 1994.

After his offers to join forces were refused by the publicly funded National Human Genome Research Institute (host of the Human Genome Project), Venter founded Celera Genomics in 1998 with the mission of sequencing the entire human genome by 2001, several years before the target date of 2005 set by the Human Genome Project (which was already behind schedule). Celera (named after the Greek for "speed") armed itself with new ABI PRISM® 3700 DNA Sequencers, and on September 8, 1999, Hamilton O. Smith, Mark Adams, Gene Myers, Granger Sutton, and Venter—essentially the same team that invented the shotgun approach at TIGR—commenced sequencing the human genome with this radical technique.

Francis Collins, director of the National Human Genome Research Institute, charged that Celera would arrive at a *Mad Magazine* version of the genome by its shotgun method. However, when Celera published its sequencing of the *Drosophilia* genome in *Science* magazine on March 24, 2000, it became clear that the opposite was true—Celera's work was accurate and complete. In fact, whereas the Human Genome Project was mapping only a single human genome, Celera was mapping the genome of five different individuals—three females and two males, who identified themselves as Hispanic, Asian, Caucasian, and African American.

"We did this . . . not in an exclusionary way, but out of respect for the diversity that is America, and to help illustrate that the concept of race has no genetic or scientific basis," stated Venter in his remarks delivered at the White House on June 26, 2000, when Celera and the Human Genome Project jointly announced the complete mapping of the human genome. Although Celera and the Human Genome Project received equal billing, it was well known that Venter had generated a more complete version of the genome than Collins.

Although Celera was often accused of limiting access to its human genome sequence, Venter pointed out in his April 6, 2000, testimony before the U.S. House of Representatives Committee on Science that Celera intended to make the entire human genome sequence freely available on-line at its World Wide Web site, where noncommercial scientists could use the information for basic research. Celera's economic model generated profits by offering bioinformatics tools and software that would allow researchers to manipulate the information in Celera's databases. Celera would offer sliding scale fees, charging corporate entities much more than academic institutions for subscriptions. "Our goal is to have all major commercial life science companies and academic biomedical research institutions as subscribers in the future," said Venter in his testimony, much as businesses, law firms, and universities alike subscribe to LexisNexis.

Celera did not rest on its laurels but immediately started sequencing the mouse genome after finishing the human genome. Interestingly, human genes differ from those of a mouse only slightly—only a few hundred genes separate the two. In 2001, Ventner shared with Francis Collins the Biotechnology Heritage Award from the Chemical Heritage Foundation and the Biotechnology Industry Organization.

Volta, Count Alessandro
(1745–1827)
Italian
Physicist

Alessandro Volta was best known for inventing the electric battery, which provided not only the first reliable source of electricity but also the first source of readily available continuous current. For his pioneering work he was made a count by Napoleon and granted numerous honors and awards. Volta's discoveries greatly advanced the understanding of electricity and paved the way for further contributions in the field, such as William Nicholson's studies of electrolysis and Sir HUMPHRY DAVY's work in electrochemistry.

Volta was born on February 18, 1745, in Como, Italy, to Filippo Volta and Maddalena de' Conti Inzaghi, who raised him in an aristocratic and religious environment. Many members of his family were involved in the church, and most of the male members became priests, including his three paternal uncles and three brothers. Two of his sisters became nuns. After Volta's father died when Volta was about seven, one of his uncles directed his education.

Though Volta's family persuaded him to study law, a field in which many family members on his mother's side were involved, at the age of 18 he was already determined to investigate electricity. In 1757, he began study at the local Jesuit college. He later attended the Seminario Benzi.

In 1774, Volta took a job teaching physics at the Royal School of Como. A year later he invented the electrophorus, an instrument used to produce static electricity, which provided the most effective way to store electric charge at that time. Also in 1778, Volta encountered and isolated methane gas. In 1779, he accepted the chair of physics at the University of Pavia. It was in Pavia that Volta conducted his breakthrough research on electricity. In 1780, his friend Luigi Galvani found that contact of a frog's leg muscle with two metals, copper and iron, generated an electric current. Galvani believed that the animal's tissues produced the electricity. Volta was interested in discovering the source of the electricity and began conducting experiments in 1794, using only metal. He discovered that he could produce electric current without animal tissue; instead, he placed different metals in contact with one another, and this arrangement generated electricity. Volta introduced his battery, the voltaic pile, in 1800. It was made up of alternating disks of silver and zinc separated by brine-soaked cardboard. The voltaic pile successfully generated electric current and resolved the controversy regarding the source of electricity; some believed it was produced by animal tissue, and others thought electric current was generated by metals. Volta exhibited his voltaic pile in 1801 in Paris and provided Napoleon with a demonstration of the battery. Napoleon was sufficiently impressed to make Volta a count and senator of Lombardy. Napoleon also awarded Volta the medal of the Legion of Honor.

After presenting his voltaic pile, Volta did little to improve upon the battery. In 1815, he began working at the University of Padua. He received many honors for his work, including the Royal Society of London's Copley Medal, which he received three years after becoming a member in 1791. He also was a member of the Paris Academy and was a correspondent of the Berlin Academy of Sciences. He died in Como, Italy, on March 5, 1827. The volt, the unit of electric potential, was named in Volta's honor.

Von Neumann, John
(1903–1957)
Hungarian/American
Mathematician

John Von Neumann was considered a genius because of his ability to understand complex mathematical and scientific theories, abstracting them down to simpler expressions comprehensible

"Von Neumann architecture," which serves as the foundation for most applications of computer design, is named for its creator, John Von Neumann. *(AIP Emilio Segrè Visual Archives)*

to lesser minds. His career bridged the gap between pure mathematics and applied mathematics, which subsequently developed into separate disciplines. His formulation of Von Neumann architecture set the foundation for most applications of computer design. What became known as Von Neumann algebras were derived from quantum mechanics. He was also instrumental in the development of the first atomic bomb, and he devoted the latter part of his career to implementing sane policies regarding nuclear power.

Von Neumann was born on December 28, 1903, in Budapest, Hungary. He was the eldest of three sons born to Margaret Von Neumann and Max Von Neumann, a successful banker. In 1929, Von Neumann married Mariette Kovesi of Budapest, and together the couple had one daughter, Marina, in 1935. They divorced in 1937, and in 1938 Von Neumann married Klara Dan, also of Budapest.

In 1921, Von Neumann commenced study of mathematics at the University of Budapest and at the University of Berlin, and in 1925 he received his degree in chemical engineering from the Swiss Federal Institute of Technology in Zurich. In 1926, he earned his Ph.D. in mathematics from the University of Budapest, where he wrote his dissertation on set theory. That year he received a Rockefeller grant for postdoctoral work at the University of Göttingen under David Hilbert, with whom he developed axiomatizations that satisfied both Erwin Schrödinger's wave theory and Werner Karl Heisenberg's particle theory in quantum mechanics.

Von Neumann inaugurated his professional career as a privatdozent at the University of Berlin from 1927 to 1929 and at the University of Hamburg from 1929 to 1930. He then spent the rest of his career at Princeton University, first as a visiting professor, then as a professor of mathematics at Princeton's Institute for Advanced Study from 1933 until his early death in 1957. In 1932 he published his influential text *The Mathematical Foundations of Quantum Mechanics,* which represented the first distillation of quantum theories applied to mathematics. While at Princeton, Von Neumann became a naturalized citizen of the United States.

In 1937, Von Neumann began a lasting relationship with the U.S. government, which was anticipating a war, as a consultant to the Ballistics Research Laboratory of the Army Ordnance Department. In 1941, he became a consultant on the theory of detonation of explosives for the National Defense Research Council, and on mine warfare for the Navy Bureau of Ordnance. He joined the Los Alamos Scientific Laboratory in 1943 to work on the Manhattan Project; there he contributed the idea of implosion as the detonation technique for the atomic bomb. Von Neumann continued work on nuclear policy as a member of the Atomic Energy Commission from 1954.

In 1944, Von Neumann took a brief respite from his wartime activity to publish *The Theory of Games and Economic Behavior* on the "minimax theorem" that he had formulated in 1928, which stated that the rational choices of game players would make it pointless to play if the outcome is determined not by their efforts but by the rules of the game alone. His coauthor Oskar Morgenstern applied this idea to economics, though it also applied to politics and other human endeavors. That same year he continued his wartime work in conjunction with the Moore School of Engineering at the University of Pennsylvania, transforming the electronic numerical integrator and computer (ENIAC) into the electronic discrete variable automatic computer (EDVAC) to improve calculations involving wartime technologies. Von Neumann's 1945 report on EDVAC contained the first statement of what became known as Von Neumann architecture, or computer memory storage. He also pioneered the technology for the central processing unit (CPU) and random access memory (RAM).

Von Neumann received the 1956 Enrico Fermi Science Award and the 1956 Medal of Freedom from President Eisenhower. In the year before, however, he had been diagnosed with bone cancer, and he died on February 8, 1957, in Washington, D.C.

W

 **Watson-Watt, Robert Alexander**
(1892–1973)
Scottish
Radio Engineer

Robert Alexander Watson-Watt invented radar, a system that uses the deflection of radio waves to locate and track objects. He originally developed the idea on the eve of World War II for locating enemy aircraft, though the technology has since expanded to encompass diverse applications, from police radar that use Doppler shift to catch speeding vehicles to meteorologists who forecast weather patterns to radar telescopes that scan the sky, tracking spacecraft or gathering data on distant planets.

Watson-Watt was born on April 13, 1892, at Brechin, in Aberdeenshire, Scotland. His mother was Mary Small Matthew, and his father, Patrick Watson Watt, was a master carpenter who passed both his mother's and father's surnames (respectively) on to his son. Watson-Watt was educated at the Damacre Road School and then the local Brechin high school. He attended University College, Dundee (then affiliated with the University of St. Andrews), on scholarship, graduating in 1912 with a bachelor of science degree in electrical engineering. The university retained him briefly as an assistant professor of physics.

In 1914, with the advent of World War I, Watson-Watt volunteered and was assigned to the Meteorological Office. He theorized a method for locating the signature short bursts of atmospheric energy emitted by thunderstorms through the triangulation of radio-wave signals sent by cathode-ray oscilloscopes, and he patented this "radiolocator" in 1919, though it was some years before the development of cathode-ray technology that could perform the tasks he envisioned. Also in 1919, after he had ended his wartime duties, he earned a bachelor of science degree in physics from the University of London.

Watson-Watt spent the postwar years working in a field observation station at Ditton Park, Slough. When the Department of Scientific and Industrial Research integrated this unit with a nearby division of the National Physical Laboratory in 1927, it appointed Watson-Watt as director of this newly formed Radio Research Station. He continued to conduct research and, in fact, he contributed to the scientific lexicon by coining the term *ionosphere* for the reflective layer of the upper atmosphere discovered by Sir Edward Appleton in 1924 using the frequency-shift method. Watson-Watt extended these studies by using a pulse method.

In 1935, H. E. Wimperis, a research director for the Air Ministry, inquired of Watson-Watt whether concentrated radio waves could disable an aircraft. Watson-Watt reported the unfeasibility of this technique, based on calculations performed by his assistant, A. F. Wilkins. Far from leading them to a dead end, though, this suggestion directed Watson-Watt to a related application of radio waves—for locating aircraft (and other objects).

Just as the ionosphere reflects radio waves (which allows for long-distance broadcasting), so too do solid objects. Since radio waves travel at a constant rate (the speed of light), the elapsed time between transmission and receipt of reflected radio waves determines the distance, and the direction the waves bounce back from completes the picture to pinpoint the exact location of the object. Watson-Watt outlined this idea in a secret memorandum to the Air Ministry dated February 12, 1935, which he later identified as the birth and invention of "radio detection and ranging," or radar, its common acronym.

As the threat of war hovered increasingly, the Air Ministry funded Watson-Watt's secret research to develop radar technology. The Battle of Britain in 1940 put the newly established radar stations to the test, which they passed by locating incoming German planes as readily by night as by day, in fog as in clear skies. Consensus among historians attributes radar as the deciding factor in the outcome of the battle. By 1943, microwave radar became operational, further enhancing the Allies' ability to locate enemy aircraft.

In honor of Watson-Watt's achievement, the Royal Society inducted him into its fellowship in 1941, and the next year, he was knighted (at which point he hyphenated his last name). In addition, he received the U.S. Medal for Merit, the Hughes medal of the Royal Society, and the Elliott Cresson medal of the Franklin Institute. After the war, in 1946, he formed a private consulting firm, Sir Robert Watson-Watt and Partners.

In 1952, Watson-Watt emigrated from England and spent his remaining years as a sometime consultant in Canada and in the United States. Also in 1952, he divorced Margaret Robertson, whom he had married in 1916. He subsequently married Jean Smith, who died in 1964, and in 1966, he married the former head of the Women's Royal Air Force, Air Chief Commandant Dame Katherine Jane Trefusis-Forbes, who died in 1971. Watson-Watt died in Inverness, Scotland, on December 5, 1973.

Watt, James
(1736–1819)
Scottish
Mechanical Engineer

The steam engine, though not invented by him, is closely associated with James Watt, as he developed an innovation that transformed a machine that wasted almost as much energy as it generated into an efficient and powerful mechanism. Before Watt introduced the separate condenser, the steam engine heated water to boiling in its main cylinder during each and every piston stroke, thus expending much of its energy production simply on raising the water temperature. Watt recognized "that in order to make the best use of steam, it was necessary first, that the cylinder should be maintained always as hot as the steam which entered it; and, secondly, that when the steam was condensed, the water of which it was composed, and the injection itself, should be cooled down to 100 degrees, or lower where it was possible." The watt, or the expenditure of one JOULE per second, was named after him, and he standardized calculations for horsepower (one unit of horsepower equals about 746 watts).

Watt was born on January 19, 1736, in Greenock, Scotland, the sixth of eight children but one of only two to survive childhood. His mother was Agnes Muirhead (also spelled Muireheid), and his father, James Watt, was a shipwright, merchant, and supplier of nautical instruments, as well as chief magistrate and treasurer of Greenock at various times. Due to his fragile health (he suffered from severe migraines), Watt attended Greenock Grammar School only sporadically, receiving a more important education in his father's workshop, where he learned woodworking, metalworking, smithing, and instrument-making.

At the age of 17, Watt went to live with relatives in Glasgow, where local guilds prevented him from apprenticing in instrument-making, so he worked for an optician doing odd jobs. In 1755, he moved to London, where he ran into more resistance from guilds until John Morgan of Finch Lane, Cornhill, agreed to take him on (at the price of 20 guineas) for a one-year apprenticeship in mathematical instrument-making. The next year, he returned to Greenock, then to Glasgow, where in 1757 he secured the position of "mathematical instrument maker to Glasgow University." There, he met professors John Robison, who later introduced him to the steam engine, and Joseph Black, who later lent him the money to develop his steam engine.

Watt also established his own instrument-making business in partnership with John Craig, at first employing one journeyman, though by 1764, the company had grown to 16 employees. In July 1764, Watt married his cousin, Margaret Miller. Also in that year, he first worked on a defective NEWCOMEN steam engine, and in the process of fixing it, he recognized its design flaw: "the waste of fuel which was necessary to bring the whole cylinder, piston, and adjacent parts from the coldness of water to the heat of steam, no fewer than from 15 to 20 times in a minute." Later, he related his moment of inspiration, when the idea for the separate condenser struck him: "I had gone to take a walk on a fine Sabbath afternoon [. . .] when the idea came into my mind that, as steam was an elastic body, it would rush into a vacuum, and, if a communication were made between the cylinder and an exhausted vessel, it would rush into it, and might be there condensed without cooling the cylinder. I then saw that I must get rid of the condensed steam and injection water if I used a jet, as in Newcomen's engine [. . .] I had not walked farther than the Golf house, when the whole thing was arranged in my mind."

Unable to work on the Sabbath, Watt proceeded to build a separate condenser on Monday morning. Though hastily constructed, the prototype proved his theory, improving efficiency while increasing strength (it lifted an 18-pound weight!). He rented a room in a deserted pottery works near Broomielaw, where he employed Folm Gardiner as a mechanic, and built a small

James Watt is famous for transforming the steam engine into an energy-efficient machine. The "watt," a term used to measure electricity, is named for him. (*AIP Emilio Segrè Visual Archives, E. Scott Barr Collection*)

engine by August 1765 and a larger engine by October 1765.

In 1767, Watt filed a patent for "A New Invented Method of Lessening the Consumption of Steam and Fuel in Fire Engines." By this time, he had depleted all his funds, forcing him to work as a civil engineer surveying canals. In 1768, he formed a partnership with John Roebuck, who sought to use the new steam engine to drain his coal mines at Bo'ness in West Lothian. However, in 1772, Roebuck went bankrupt, opening the door for MATTHEW BOULTON, who ran the Soho Engineering Works near Birmingham, to buy out Roebuck's stake in the patent. The next year, Watt's first wife died, and he subsequently married Ann MacGregor.

Parliament finally granted the separate condenser patent in 1775, fueling the Industrial Revolution in the process—particularly mining, waterworks, canals, and ironworks, all of which relied on the increased power and efficiency of the Boulton & Watt steam engine. While Boulton promoted the engine, Watt (who had quit his engineering job by this time) worked to improve its design: in 1782, he innovated the double-acting system, making use of the piston's forward and backward strokes, and in 1788, he invented the centrifugal (or "flyball") governor. The former employed a sun-and-planet gear (one orbiting the other) to create rotary motion, a more efficient mechanism; the latter automatically regulated engine speed by a servomechanism, the basic concept of all subsequent automation.

The Royal Society inducted Watt into its fellowship in 1785. In 1800, when his patent rights expired, he and Boulton passed the business on to their sons, James Watt (the only child that survived Watt) and Matthew Robinson Boulton. Watt set up a workshop in the garret of his Heathfield home, where he experimented with copying text by chemical means. Watt died on August 19 (or 25), 1819, at Handsworth Heath, in Birmingham. The fact that he is often remembered as the inventor of the steam engine, although unfair to THOMAS SAVERY and Thomas Newcomen, demonstrates the ultimate importance of his innovation of the separate condenser, which transformed the steam engine into a machine efficient enough to generate an Industrial Revolution.

⊠ Wells, Horace
(1815–1848)
American
Dentist

Although the peculiar effects of "laughing gas" were known since the turn of the 19th century, Horace Wells was the first person to apply nitrous oxide as a general anesthetic in surgery when he induced unconsciousness through inhalation of the gas. Unfortunately, his first public demonstration of the procedure went awry, casting doubt on his discovery. While traveling in Europe demonstrating the effects of anesthetics, Wells became addicted to these and other intoxicating chemicals, which quickly eroded his already-fragile mental state. It was only after he committed suicide that the medical community recognized the efficacy of nitrous oxide and exonerated his memory.

Wells was born on January 21, 1815, in Hartford, Vermont. He attended private schools in Massachusetts and New Hampshire until the age of 19, when he moved to Boston to apprentice under some of the best dentists in the city (as there were no dentistry schools at the time). In 1836, he established a dentistry practice on Main Street in Hartford, Connecticut, and soon became recognized as one of the best dentists in the city, promoting his practice with a slogan that continued in general use into the 20th century: "the clean tooth does not decay."

In 1838, Wells published *An Essay on Teeth*, a diatribe against quackery in dentistry. As early as 1840, he evinced an interest in identifying an anesthetic to aid in surgical procedures in a conversation with fellow Hartford resident Linus P.

Brochett. The existing anesthetics included alcohol, opium, and even hypnotism, none of which proved satisfactory, forcing many doctors and dentists to hurry through unanesthetized operations to minimize the patient's pain. Ironically, Sir HUMPHRY DAVY had discovered the strange effects of nitrous oxide in 1800, but he failed to apply it as an anesthetic.

On December 10, 1844, Wells brought his wife, Elizabeth, to a presentation by Garner Colton demonstrating the effects of nitrous oxide. Although the show was meant to entertain, Wells came up with a brilliant idea when he saw one of the volunteers who had inhaled nitrous oxide bang his leg against a chair without registering any pain: Wells witnessed the gas's anesthetic effect. He approached Colter afterward, and later that night the pair returned to Wells's office, where Colter administered "laughing gas" (as it was called) while John Riggs, who had studied dentistry under Wells, removed his mentor's third molar.

"I did not feel so much as the prick of a pin. A new era in tooth-pulling has come," stated Wells after Colter and Riggs finished the procedure. Over the next month-and-a-half, Wells successfully employed nitrous oxide as an anesthetic on 15 of his patients. Confident of his technique, he arranged a demonstration in Boston during Dr. John C. Warren's class on "The Use of Nitrous Oxide for the Prevention of Pain" on January 31, 1845. In an observation room at the Massachusetts General Hospital filled with Harvard Medical School students, Wells prepared to perform the first public exhibition of anesthetized surgery.

Later, in the December 9, 1846, edition of the *Daily Hartford Courant*, Wells described the fateful event: "A large number of students, with several physicians, met to see the operation performed—one of their number to be the patient. Unfortunately for the experiment, the gas bag was by mistake withdrawn much too soon, and he was but partially under its influence when the tooth was extracted. He testified that he experienced some pain, but not as much as usually attends the operation." Other reports claim that the patient cried out, though not in pain.

Nevertheless, the observers judged the technique a failure, and Wells returned to Hartford discouraged, selling his home and dental practice. His sense of defeat heightened when one of the observers, William T. G. Morton—who had been his partner in 1842 and 1843, and corecipient of an 1844 award for the design of a dental instrument case—announced his "discovery" of ether as an anesthetic, which he had already used successfully in 160 tooth extractions. Wells deserved credit for originating the idea of inhalation anesthesia, and after receiving no recognition in the United States, he traveled to Paris to establish his priority of introducing the first general anesthetic. He further supported his cause by publishing his account in the 1847 text, *A History of the Discovery and Application of Nitrous Oxide Gas, Ether and Other Vapours to Surgical Operations*.

When he returned to the United States, where the medical community continued to dismiss his claims, he moved his family to New York City while supporting them as a traveling sales representative, selling canaries and shower baths in Connecticut. However, after countless demonstrations, Wells had become addicted to the intoxicating effects of ether and chloroform, another intoxicating chemical, and he slowly lost his mind. In a chemical-induced stupor, Wells splashed sulfuric acid on some prostitutes, resulting in his imprisonment. In jail, he managed to procure some chloroform, and under its effects, he deliberately sliced his groin artery with a razor and bled to death on January 23, 1848, two days past his 33rd birthday. He was buried in the Cedar Hill Cemetery in Hartford.

When the medical community realized nitrous oxide's efficacy as an anesthetic, it resurrected Well's reputation, with one statue erected in Paris, and another, sculpted by T. H. Bartlett in 1874, in Hartford. However, the darker side of his reputation may have lived on as well, as some claim that his descent into insanity served as a model for Robert Louis Stevenson's 1876 novel, *Dr. Jekyll and Mr. Hyde*.

⊠ Westinghouse, George
(1846–1914)
American
Engineer, Energy and Power Industries

George Westinghouse's name is most commonly associated with the electric company he founded, but he first gained renown for inventing the continuous braking system for the railway, which used pneumatic pressure to halt trains. After he invented a means to stop trains automatically even after they broke apart, his system became the standard in the industry. However, his work promoting alternating current for electricity distribution proved even more significant, as AC prevailed over DC, or direct current, the delivery system advocated by THOMAS EDISON, as the industry standard.

Westinghouse was born on October 6, 1846, in Central Bridge, New York, the eighth of 10 children. At the age of 15, he ran away to join the Union Army in the Civil War, only to be brought back home by his parents. After a year and a half, they allowed him to reenlist, and he served a short while in the army before joining the navy in 1864. By 1865, he had reached the rank of acting third-assistant engineer, when he quit to pursue studies at Union College in Schenectady, New York. He only lasted there three months before deciding to join his father's agricultural implement manufacturing firm.

In October 1865, Westinghouse filed a patent (the first of his 400) on a railroad steam locomotive that ultimately proved unsuccessful. That year, he also invented a rerailing device for railway rolling stock, and between 1868 and 1869, he invented cast-steel frogs for railway points to replace the less durable cast-iron frogs.

At about this same time, Westinghouse happened upon a magazine article describing compressed-air pneumatic tools in use on the construction of the Mont Cenis Tunnel, which inspired him to apply this same concept to railroad brakes.

Until this time, railroads relied on manpower for braking: when the driver applied the brakes, operators in each car applied the brakes on cue. Westinghouse modernized this system by developing the continuous brake system, using compressed-air brakes that operated simultaneously in all cars. This system created uniform braking along the entire line, which allowed trains to travel much faster while still braking safely.

When Westinghouse pitched his continuous braking system to Cornelius Vanderbilt, the railroad tycoon scoffed at the notion of air alone stopping an entire train. Undeterred, Westinghouse prepared a prototype by 1869, which he demonstrated by driving a train toward a horse-drawn dray, stopping just in the nick of time. That year, he founded the Westinghouse Air Brake Company. In 1872, Westinghouse invented the triple valve, which would activate the brakes automatically if a car was detached—this final addition completed the system, which soon became standard in the industry.

In 1893, the Railroad Safety Appliance Act made air brakes compulsory, essentially giving control of the market to Westinghouse brakes (although in Britain, which lacked such a law, railroad companies continued to use inferior but cheaper vacuum brakes). In the meanwhile, Westinghouse had invested interest in other railway innovations: in 1880, he began buying patents for railway signals and points, and in 1884, he consolidated the various signaling methods into one system, run by pneumatics.

Just as Westinghouse identified the optimal brake system for trains, he similarly identified and promoted the optimal system for distributing electricity. Following the lead of Europe, he collaborated with NIKOLA TESLA to develop high-voltage single-phase alternating current

(AC), defying Edison's belief in direct current. In 1886, he established Westinghouse Electric Company, which installed this single-phase AC system, using French transformers and generators, in Pittsburgh.

Westinghouse's crowning achievement in electricity came in 1893, when he won the Niagara Falls contract for supplying power to Buffalo via alternating current. Edison, who considered this a fatal blow to the direct current cause, considered himself "Westinghoused," a term that he also applied to the use of the electric chair. Westinghouse similarly applied his systemic approach to gas distribution, transporting gas at high pressure, then decreasing pressure for delivery by incrementally increasing the diameter of the gas mains.

Westinghouse received much recognition for the significance of his contributions to society during his lifetime: he was inducted into the Légion d'Honneur and received the Order of the Crown of Italy and the Order of Leopold. Westinghouse died on March 12, 1914, in New York City. His systems continued to see use through the 20th century, earning him a place in the Hall of Fame for Great Americans in 1955.

⊗ Wheatstone, Sir Charles
(1802–1875)
English
Physicist

The Wheatstone bridge, an apparatus for accurately measuring electrical resistance, is named after Sir Charles Wheatstone, but ironically he did not invent the instrument. He did, however, bring the device popular attention and thus deserves credit for the widespread use of the tool. In another ironic twist, Wheatstone developed a cipher that bears another man's name, as LYON PLAYFAIR popularized the Playfair cipher that Wheatstone invented. Wheatstone conducted wide-ranging scientific investigations,

Sir Charles Wheatstone's inventive work contributed to the development of telegraphy and spectroscopy, among other fields. He also patented the first concertina, a small version of the accordion. *(Photographische Gesellschaft in Berlin, Lawrence del., courtesy AIP Emilio Segrè Visual Archives)*

inventing important new devices and making significant discoveries.

Wheatstone was born on February 6, 1802, in Gloucester, England. His family of musical instrument makers moved to London when he was four years old. In school, he distinguished himself early on in physics and mathematics and later won a French prize, though his reticence to deliver a speech in French prevented him from receiving the award. In 1816, he apprenticed at the musical instrument manufacturing plant of the uncle whose name he bore. When his uncle died in 1823, Wheatstone and his brother, William, inherited the business.

Wheatstone's factory mainly produced flutes, sparking his interest in the vibration of sound waves through air. In 1827, he invented the kaleidophone, an apparatus that projected sound onto an illuminated sounding surface, thus visually displaying the vibration of sound by the patterns of light produced by the persistence of vision. Two years later, he patented the concertina, a small version of the accordion.

In 1834, Wheatstone became a professor of experimental physics at King's College in London, a part-time position he retained for the rest of his life. That same year, he measured the velocity of electricity through wire by using a rotating mirror at the end of a quarter-mile of coiled copper wire. He estimated the velocity to be more than 250,000 miles per second, or one-and-a-third times the speed of light. Armand Fizeau used this same method in 1849 to correct Wheatstone's estimate, which was off due to the coiling of the wire.

At the 1835 British Association meeting in Dublin, Wheatstone presented a paper entitled "Prismatic analysis of electric light," in which he described his discovery that spark discharges from different metal electrodes generate different spectral lines and colors. "We have a mode of discriminating metallic bodies more readily than by chemical examination and which may hereafter be employed for useful purposes," he wrote, correctly anticipating the invention of spectroscopy by ROBERT WILHELM BUNSEN and Gustav Kirchoff in 1860. The Royal Society inducted him into its fellowship in 1836.

In 1837, Wheatstone collaborated with WILLIAM FOTHERGILL COOKE to produce the first telegraph patent, on their five-needle telegraph, which indicated letters when electrical impulses induced any two needles to point to the same letter. The pair developed the two-needle telegraph in 1845 and single-needle telegraph the year after that. Wheatstone continued to work to improve telegraphy throughout his life.

In 1843, Wheatstone published experimental verification of Ohm's Law, thereby introducing British science to the concept that was already well known and accepted in Germany. In further investigation on electrical resistance, he invented the rheostat to measure variable resistance. That same year, the *Transactions of the Royal Society* published his description of a mechanism for accurately determining electrical resistance. He readily acknowledged Samuel H. Christie's invention of this device a decade earlier, nonetheless it became known as the Wheatstone bridge, attributing credit to the popularizer over the inventor.

Wheatstone continued to innovate: in 1848, he invented the polar clock, which determines time by the angle of sunlight polarization, and in 1854, he invented what became known as the Playfair cipher, named after his neighbor, Lyon Playfair, who promoted the strategic use of this simple yet foolproof system. He was the first scientist to follow up on Leonardo da Vinci's observation that depth perception results from the reception of slightly different images by each eye, prompting him to invent the stereoscope in 1860. This device displayed two slightly different images that converged to create a three-dimensional illusion.

Wheatstone was knighted in 1868. While on a business trip to Paris, he fell ill and died on October 19, 1875. However, his name lives on in the instrument named after him (despite the fact that he did not invent the Wheatstone bridge), and his legacy lives on in telegraphy and spectroscopy.

Whitehead, Robert
(1823–1905)
English
Engineer

Robert Whitehead invented the self-propelled torpedo, thereby transforming the face of naval battle by reducing even the largest and best-protected warships to vulnerability. Before Whitehead's innovation, torpedoes amounted to immobile underwater mines or boats filled with explosives launched toward targets. Whitehead combined these two ideas, sending a self-propelled projectile at its target under cover of the water, thereby introducing a stealth factor into the naval consciousness.

Whitehead was born on January 3, 1823, at Little Bolton, in Lancashire, England, one of eight children. His father owned a cotton-bleaching business. He attended the local grammar school until the age of 14, when he left for Manchester to join his uncle, William Smith, at the engineering firm he managed, Richard Ormond and Son. While serving an apprenticeship there, he studied drafting in evening classes at Manchester's Mechanics Institute.

In 1844, Whitehead joined the Marseilles engineering firm of Philip Taylor and Sons, where his uncle had moved as a manager. In 1847, he moved to Milan (then in the Austrian Empire), where he founded his own engineering firm. The Lombardy marshes were drained with pumps of his design, and he improved the design of silkweaving looms. Throughout this period, Whitehead secured patents with the Austrian government that were rendered void by the Milanese revolution of 1848.

Whitehead thus moved to Trieste, where he worked for the Austrian Lloyd Company. In 1850, he left this position to manage the Studholt plant until 1856, when he moved to the nearby port of Fiume (then part of Austria), where he constructed ship engines for the Austrian navy at the Stabilimento Tecnico Fiumano. In 1864, he teamed up with Austrian naval captain Giovanni de Lupins to design a "coastal fire ship," or "Der Kustenbrander," a self-propelled boat loaded with explosives that could be directed at enemy ships, like unmanned kamikaze ships. The Austrian navy did not consider this weapon promising enough to support further development.

Instead of abandoning this idea, Whitehead pursued it further in secret experiments conducted in collaboration with his son, John, and another mechanic. At this time, star torpedoes (or charges mounted on the end of a long pole) were being employed in the American Civil War. Other torpedoes amounted to underwater mines. Whitehead pushed all these approaches to the logical conclusion: a self-propelled explosive that traveled under the cover of the water. In 1866, he introduced this new type of torpedo, a four-foot-long device weighing some 330 pounds (including 18 to 20 pounds of dynamite mounted on the tip), powered by a compressed-air motor turning a simple propeller some 700 yards at seven knots.

The only problem with the Whitehead torpedo was its inability to maintain a uniform depth. Whitehead solved this problem within two years by developing a simple "balance chamber," the key to torpedo technology, and thus kept in strictest confidences for many years. Whitehead licensed manufacturing rights to the Austrian government in 1870, and two years later he partnered with his son-in-law, George Hoyos (and later his own son, John), to buy out the Fiume factory where he worked and transform it into a torpedo manufacturing plant. He thus capitalized financially on his invention, licensing rights to England (which ordered 254 torpedoes) in 1871, France (which ordered 218) in 1872, and Germany (which ordered 203) and Italy (70) in 1873. He built torpedoes for numerous other countries, as well: Russia (250), Denmark (83), Greece (70), Portugal (50), Argentina (40), and Belgium (40).

Whitehead continued to improve upon the design of his torpedo, inventing a servometer for steering in 1876. By 1889, he had refined torpedo design such that they could reach speeds of 29 knots and travel for 1,000 yards. The next year, Whitehead built a factory at Portland Harbor in England; he eventually moved back to England and settled on an extensive estate in Worth, Sussex.

Whitehead's work received recognition even before he gained renown for his torpedo. In 1866, the Austrian Empire granted him a diamond ring for his design and construction of the ironclad warship *Ferdinand Max*, which rammed the Italian *Re d'Italia* in the Battle of Lissa. In May 1868, he received the Austrian Order of Francis Joseph in honor of his engineering exhibits at the Paris Exhibition of 1867. In 1884, he was named to the French Légion d'Honneur.

The Japanese navy made the first large-scale demonstration of the wholesale destructive power of torpedoes when it destroyed the Russian naval fleet outside Port Arthur on February 9, 1904. Whitehead died the next year, on November 14, 1905, near Shrivenham, Berkshire, and was buried at Worth. Whitehead torpedoes continued in use as tactical weapons throughout the 20th century: in the Falklands conflict of 1982, it was a Whitehead Mark VIII that sank the Argentine cruiser *Belgrano*.

Wilkinson, John
(1728–1808)
English
Steel and Iron Industrialist

John Wilkinson gained the nickname "Iron Mad" Wilkinson for his obsession with iron. He backed the building of an all-iron bridge, an iron-hulled barge for hauling iron, and iron coffins for his own burial. Despite these eccentricities, his understanding of iron helped fuel the Industrial Revolution, as he invented a hole-boring machine that produced incredibly accurate cannons as well as cylinders for steam engines, perhaps the most important machines of the period.

Wilkinson was born in 1728 at Clifton, in Cumberland, England—purportedly in a cart on its way to market at the Workington Fair. His father, Isaac Wilkinson, was a farmer who also worked at a local iron furnace. A Presbyterian, Isaac refused to educate his sons at the local Anglican academy, sending them instead to Dr. Caleb Rotherham's Dissenters' Academy at Kendal. When Wilkinson was 10 years old, his father patented a box-iron for laundering the frilled shirts fashionable at the time (some speculate that the idea for the patent belonged to John Wilkinson), and he established his own works for manufacturing these irons.

Although Wilkinson worked in his father's factory while growing up, he never officially apprenticed as a foundryman, the common practice of the day. Instead, he started working at the Bradley iron works in Bilston, Wolverhampton, in 1748. There, Wilkinson built the region's first blast furnace. After much experimentation, he managed to replace wood charcoal with coke derived from mineral coal, which burned much hotter and hence fueled the furnaces much more effectively. On June 12, 1755, he married Ann Maudesley, who died a year later while giving birth to their daughter, Mary. Wilkinson inherited wealth substantial enough to finance his own entrance into the iron industry.

Also in 1756, Wilkinson joined his father's iron works at Bersham, in Denbigh, Wales. Within about five years, he had ousted his father from the business and simultaneously founded his own iron works in Brosley, near Ironbridge, one of the first to use a Boulton & Watt steam engine. Wilkinson prospered by filling government contracts for producing cannons, mortars, and shell casings.

Wilkinson produced cannons by the traditional method, casting them in a single piece, leaving a hole in the middle. This technique could not prevent imperfections that affected the trajectory and safety of the cannons. In 1774, Wilkinson invented and patented a boring machine that drilled holes with far greater accuracy than the old casting method. While this innovation transformed the manufacture of cannons, it truly revolutionized the production of cylinders for steam engines. On the strength of his reputation for the accuracy of his hole-boring technology, Wilkinson entered into an agreement with MATTHEW BOULTON and JAMES WATT to become their exclusive supplier of cylinders for their steam engines.

Controversy surrounded Wilkinson's career. In 1779, he used his influence as a major shareholder in the Iron Bridge project to convince the other shareholders to agree to build the bridge completely (instead of partially) of iron, and iron smelted at his works, at that. He also became embroiled in controversy over supplying the French with water pipes, raising suspicion that he was really supplying this enemy with iron for conversion into ordnance. Finally, William Wilkinson exposed his brother's side business of manufacturing pirated Boulton & Watt steam engines, in violation of James Watt's patent. This episode led to Wilkinson's falling out with Boulton and Watt, who not only sued him but also established their own cylinder-manufacturing plant in Soho.

None of these controversies quelled Wilkinson's passion for iron. In July 1787, he reconciled with his father to build the first iron-hulled barge, much to his detractors' disbelief. Two years later, he patented a method for making cannonballs with spiral grooves for straighter and further projection, but production proved commercially unviable.

As Wilkinson neared the end of his life, he became more and more bizarre. He began collecting iron coffins, and he financed the installation of iron windows and even an iron pulpit in the Methodist chapel in Bradley. At the age of 72, he took up a mistress (he had married for a second time in 1763) named Ann Lewis, who bore him two daughters and a son, a development that complicated his inheritance. Wilkinson died on July 14, 1808, at Bradley, and his body was transported to Castle Head in Cumbria in a wooden casket that proved too large for the iron coffin he had prepared for his burial. In the end, Wilkinson was disinterred and reburied in iron three times.

Wozniak, Stephen
(1950–)
American
Computer Scientist

Stephen Wozniak and STEVEN JOBS transformed society with their introduction of Apple computers to the masses. Their smaller size and user-friendly interface shifted computer use from a specialized function performed only by institutional professionals and electronics enthusiasts into an everyday activity of mainstream users. In many ways, Wozniak and Jobs were responsible for ushering in the information age of the later 20th century. "It was like a revolution that I'd never seen," Wozniak said of the computer revolution. "You read about technological revolutions, the Industrial Revolution, and here was one of those sort of things happening and I was a part of it."

Wozniak was born in 1950 in Sunnyvale, California, a small suburb in the Santa Clara Valley (what is now known as Silicon Valley, after the material used in computer chips). He was the oldest of three children (he has a sister named Leslie and a brother named Mark) born to Margaret Wozniak, president of the local Republican women's club, and Jerry Wozniak, an engineer at Lockheed who worked in missile control systems. He attended school in the Cupertino district, pursuing his interest in electronics in his free time: in the fifth grade, he built a voltmeter from a kit, and the next year, he bought a 100-watt ham radio and obtained an operator's license. He even wired speakers between neighboring houses: "We didn't know why," he said, "but we had house-to-house intercoms."

At the age of 13, Wozniak won third prize at the Bay Area Science Fair for a transistorized calculator that he transformed from a one-bit into a 10-bit circuit board. He graduated from Homestead High School and then flunked out of the University of Colorado within a year. He returned to Cupertino to study at De Anza College briefly before abandoning formal education altogether, in favor of working as a programmer at a small computer company. In 1971, he returned to college, studying engineering at the University of California at Berkeley but again dropped out.

That summer, Wozniak worked at Hewlett-Packard (HP), where he met 16-year-old Steven Jobs, a fellow electronics enthusiast. Together with John Draper, "Woz" (as Wozniak was called) had constructed a "blue box," or a telephone attachment that erased long-distance charges. "Woz's first call was to the pope," Draper remembered. "He wanted to make a confession." Jobs agreed to market the product, but its quasi-legal status prevented it from succeeding on a large scale.

In 1975, Wozniak and Jobs again crossed paths as members of the Homebrew Computer Club, a group of amateur computer enthusiasts. That year, the January issue of *Popular Electronics* featured a cover story on the Altair 8800, the first kit for building a computer small enough for home use, thanks to the Intel 8080 microprocessor. Wozniak rose to the challenge of building a more powerful, cheaper computer, which he assembled in Jobs's garage; it featured an eight-bit circuit board, a 6502 microprocessor, four K of random access memory (RAM), and BASIC programming written by Wozniak himself. He demonstrated this prototype to his supervisors at HP, who showed little interest, so Jobs began marketing it through Homebrew contacts. Their big break came when Paul Terrill, owner of the newly established Byte Shop, ordered 25 units.

To raise the $1,300 in capital necessary to finance production, Jobs sold his Volkswagen microbus and Wozniak his HP scientific calculator, and together they founded Apple Computer, Inc. (named after the Beatles' record label or Jobs's favorite fruit, according to differing accounts). The Apple I, which grew out of Wozniak's prototype, sold for $666 in 1976. The next year, Wozniak left HP to work full time for Apple as vice president of research and development, improving their computer design with financial backing from Mike Markkula. That year, Wozniak and Jobs introduced the Apple II, featuring 16 K of RAM; Apple sold more than 600 units at $1,195 (without a monitor), earning some $775,000.

The Apple I and II transformed computers from huge monstrosities affordable only by large businesses, or specialized devices workable only by electronics wizards, into small, user-friendly machines that allowed mainstream users to make much more efficient use of their time. Apple thus hit on a formula that brought it great success. By 1980, when Apple went public, its stock surged to a value of $117 million, with Wozniak's own personal stock valued at $88 million.

By 1981, Apple was worth $335 million. However, tragedy struck that year when the single-engine plane Wozniak was piloting crashed on takeoff. He sustained injuries and suffered from amnesia, requiring two years to recuperate. During this time, however, he was able to finish his college education, earning bachelor's degrees in computer science and electrical engineering from the University of California at Berkeley in 1982. The next year, he returned to Apple as an engineer in the Apple II division, but just two years later, in 1985, he resigned from Apple to found his own computer company, Cloud 9 Inc.

In January 1985, Wozniak and Jobs shared the first National Technology Medal, presented by President Ronald Reagan. Since then, Wozniak has continued to contribute to the computer field, most notably by cofounding with Lotus developer Mitch Kapor the Electronic Frontier Foundation, an organization dedicated to preserving First Amendment rights in computer applications, especially on the Internet and the World Wide Web. He also enjoyed family life with his third wife and six children in their Los Gatos, California, home.

Wright, Wilbur and Orville
(1867–1912; 1871–1948)
American
Aeronautical Engineers

The Wright brothers were the first to make manned, powered flights in an airplane. In order to achieve this feat, they conducted wind-tunnel experiments on wing lift and drag, revealing major errors in the existing data, which they replaced with their own tables. They also conducted four years worth of glider trials to test their design innovations, such as the warped wing, the elevator (navigational equipment that is adjustable on the horizontal), and the tail rudder, a triaxial system they patented that remains the standard in the industry. After their first flight, they promoted this new technology, ushering in the age of flying.

Wilbur Wright was born on April 16, 1867, in Millville, Indiana, and Orville Wright was born on August 19, 1871, in Dayton, Ohio. They had two older brothers, Reuchlin and Lorin, and one younger sister, Katharine. Their mother, Susan Catherine Koerner, suffered from tuberculosis, and their father, Milton Wright, was a teacher and a pastor who became a bishop of the United Brethren Church. He bought his boys a toy helicopter in 1878 that sparked their interest in flying. Both boys loved learning, though they were largely self-taught, as Wilbur only finished high school, and Orville did not even pursue his education that far. Orville built printing presses, and in the late 1880s, the brothers collaboratively edited, published, and printed local newspapers (the *West Side News* and *The Evening News*) out of their home.

In 1892, the Wright brothers initiated their careers in mechanized self-propulsion by opening a bicycle shop. Within four years, the Wright Cycle Company was manufacturing the Wright Special, an $18 bicycle. On August 10 of that year, 1896, the German aviator Otto Lilienthal

Orville Wright, circa 1905, began his career in mechanized self-propulsion by opening a bicycle shop with his brother Wilbur. The pair would later be the first to make manned, powered flights in an airplane. *(Wright Brothers Collection, Special Collections & Archives, Wright State University)*

died after a glider accident, an event that rekindled the Wrights' interest in flying. They studied Lilienthal's wing-lift tables, scoured the literature on aviation, and even contacted Samuel Pierpoint Langley, an aviator who directed the Smithsonian Institution, and Octave Chanute, a French-American aviator.

Realizing the necessity of maneuvering the wings (after observing birds in flight), the Wrights designed and built a biplane kite with maneuverable wings of a five-foot span, which they tested in August 1899. This innovation allowed them to create lateral balance, or aileron control, by adjusting the angles at the wingtips, a key to manned flight. Next, they asked the U.S. Weather Bureau to suggest an ideal location for flight testing; Kill Devil Hills at Kitty Hawk, North Carolina, combined steady winds with sandy dunes, the bureau replied.

After conducting trials there in the fall of 1900 (using a glider they had constructed with a 17-foot wingspan) and the summer of 1901 (using a glider with different wing camber and a hand-operated "elevator" on the plane's tail), the Wright brothers concluded that Lilienthal's lift tables were erroneous. After Wilbur reported this opinion to the Western Society of Engineers at its Chicago meeting that year, the brothers experimented with an 18-inch wind tunnel and 48 different wing cambers to compile their own correct tables of lift and drag.

Wilbur Wright, circa 1905, and his brother Orville were the first to make manned, powered flights in an airplane. *(Wright Brothers Collection, Special Collections & Archives, Wright State University)*

When the Wrights returned to Kitty Hawk in the fall of 1902, their glider had a 32-foot wingspan and a double vertical fin behind the wings, which Wilbur redesigned as a single tail rudder operated by the wing-control wires. They also designed and built propellers (which were actually wings operating on the vertical axis, they realized) and a 12-horsepower engine, which they mounted on "Flyer," their 750-pound aircraft with a 40-foot wingspan, for their 1903 trials.

At 10:35 on the morning of December 17, Orville Wright took off, rising about 10 feet in the air, and landing about 100 feet away, thus navigating the first manned, powered flight. On the last flight of that day, Wilbur Wright stayed airborne for just under a minute, traveling about a half-mile at a speed of some 10 miles per hour. After this flight, the wind smashed "Flyer." The brothers announced their achievement on January 5, 1904, to an indifferent world.

Over the next several years, the Wright brothers improved upon their airplane design, building the Flyer II (which Wilbur flew three-quarters of a mile in about a minute-and-a-half in 1904) and the Flyer III (which flew just over 24 miles in just over 38 minutes). In 1906, they patented their triaxial control system—consisting of an elevator, rudder, and wing-warping—which remains the standard for airplanes. By then, airplane manufacturers had sprouted up in Europe as well as in the United States, many of them infringing on the Wright brothers' patent, which they spent years defending. During this protracted litigation, Wilbur contracted typhoid fever and died on May 30, 1912.

Before his brother's death, Orville and his brother Wilbur shared a medal granted by the French Academy of Sciences in 1909. After Wilbur's death, Orville received numerous other honors, including the 1910 Langley medal from the Smithsonian Institution, the 1914 Elliot Cresson medal from the Franklin Institute, and a

Orville and Wilbur Wright are shown here circa 1909 in Pau, France. In 1924, many years after Wilbur's death, Orville received an officer's cross from the French Legion of Honor. *(Wright Brothers Collection, Special Collections & Archives, Wright State University)*

cross of an officer from the French Legion of Honor in 1924. In 1930, he received the first Daniel Guggenheim medal.

Orville sold the Wright Company in October 1915; he then founded the Wright Aeronautical Corporation and subsequently became director of the Wright Aeronautical Laboratory. In this position, he made his last major contribution to airplane technology: split flaps. Orville died of a heart attack on January 30, 1948. Both brothers died in their hometown of Dayton, Ohio, and neither ever married.

Young, James
(1811–1883)
Scottish
Chemical Engineer

James Young, popularly known as "Paraffin Young" for his discovery of a process for converting coal shale into paraffin, is considered by many to be the father of the modern petrochemical industry. After patenting his "retortion" process of distilling oil from coal (which remains in use in the oil industry), he established one of the first petroleum refineries and proceeded to amass a fortune.

Young was born on May 11, 1811, in the Drygate section of Glasgow, Scotland. His father, John Young, was a carpenter and coffin-maker who took on Young as an apprentice. Young concurrently attended night courses in chemistry at Anderson's University (now the University of Strathclyde), where he studied under Thomas Graham. In 1832, Graham appointed Young as his assistant; Young distinguished himself not only in this capacity but also as an expert instrument-maker. In 1837, when Graham moved to University College, London, he took his young assistant with him.

University College installed Young as a junior lecturer, but this salary proved insufficient to support his family after he married in 1838. That year, he left academia for industry, landing a job as manager of James Muspratt's chemical works in St. Helens, near Liverpool. In 1844, he moved to the Manchester firm of Tennant, Clow and Company, where he discovered a more efficient means of producing stannate of soda and chlorate potash. Having thus proven himself, he was appointed to a consultancy, providing him with increasing scientific independence.

At this time, increasing demand for natural oils was depleting supplies (mainly from whale oil) and jacking up the price, thus inspiring scientists to investigate ways to synthesize oils. Young turned his attention to this problem after another former student of Graham's, LYON PLAYFAIR, consulted Young about the oil he had found oozing out of the coal deposits in the Riddings pit, a coal mine near Alfreton, in Derbyshire, owned by James Oakes. Instead of waiting for the earth to squeeze the oil from the coal, Young reasoned, the oil could be refined by a scientific method.

After two years of experimentation, Young discovered a refining process for extracting oil from shale and bituminous coal and converting it into paraffin oil and wax for use in industrial applications, as well as for lighting and heating. This process, called "retortion," involved heating chunks of shale or coal to extract oil from them by "dry distillation." Young patented this process in 1850, and he then collaborated with Oakes to establish the world's first oil refinery, in Derbyshire.

This deposit did not yield the amounts of petroleum Young had envisioned, so he continued to search for better methods, materials, and locations for oil refining. He discovered cannel coal to be the best raw material, and Hugh Bartholome of the Glasgow City and Suburban Gas Company, another former student of Graham's, gave Young a barrel of newly mined cannel coal from Boghead in Torbanehill, near Bathgate. This turned out to be the best oil-bearing shale.

Young joined forces with Edward William Binney to establish the Paraffin Light and Mineral Oil Company, a shale-mining and oil-refining operation opened at Bathgate in February 1851. Protective of his patent, Young maintained strict secrecy around the refinery, though entrepreneurs still managed to copy his process that was producing 50,000 gallons of oil per year.

Both geography and the law sided with Young, as the Bathgate shale deposits yielded the highest output, and the courts defended Young's patent until its expiration in 1864, when as many as 97 oil companies sprouted.

Young established himself as the first oil tycoon, though he squandered his opportunity to become the world's wealthiest oil tycoon by insisting on cash as royalty payments for the licensing of his process in the United States. Although he collected the equivalent of 50,000 British pounds, he could have positioned himself at the head of the oil industry in the United States if he had accepted some of the numerous offers of partnerships instead.

After Young retired in the 1870s, he donated 10,000 guineas to his alma mater, Anderson Institute, to establish a chair of technical chemistry, thereby insuring the continuation of his legacy. Young died in 1883, but his influence continued on—by the 1890s, Scotland was mining two million tons of oil shale per year. Young's "retortion" method continues to be used in the petroleum industry.

Zeiss, Carl Friedrich
(1816–1888)
German
Optician

Carl Zeiss gained a reputation for making the most exacting microscopes available, spurring the advancement of science through advances in instrumentation. He also had a knack for hiring promising young scientists, including Ernst Abbe, who applied mathematics and physics to microscopy, resulting in the development of new theoretical models on which to base the design. Zeiss's influence continued long after his death, as his company remained on the forefront of technological innovation.

Carl Friedrich Zeiss was born on September 11, 1816, in Weimar, Germany. After attending grammar school, he apprenticed to Dr. Friedrich Körner, a mechanic who built simple microscopes. Zeiss augmented his education by attending lectures in mathematics, experimental physics, anthropology, mineralogy, and optics at Jena University. He then spent several years traveling as a journeyman, working with instrument-makers in Stuttgart, Darmstadt, Vienna, and Berlin, after which he settled in Jena, in the Thuringia district of Germany.

There, on May 10, 1846, Zeiss applied to Weimar state authorities to open a mechanics workshop at seven Neugasse Street, on the banks of the Saale River. He purportedly had to borrow a hundred thalers to start his business, but he eagerly opened shop on November 17, two days before receiving official permission from the state. During his first year in business, he built and sold eyeglasses, magnifiers, and balances, as well as constructing and repairing scientific apparatus for the university, succeeding well enough to hire an apprentice and move to a bigger location, at 32 Wagnergasse Street, in September 1847.

There, Zeiss commenced building single-lens microscopes perfectly suited for dissecting, much like the simple microscopes he had built for his former mentor, Dr. Körner, who died that year. In the first production year, Zeiss sold 23 units, encouraging him to plan his next step: constructing compound microscopes, consisting of two optical elements—an eyepiece and an objective. By 1857, the "Stand I" model entered the market; four years later, this compound microscope won a gold medal at the 1861 Thuringian Industrial Exhibition as "among the most excellent instruments made in Germany."

In 1863, the Grand Duke appointed Zeiss as the official microscope manufacturer for the court. By the next year, Zeiss had again outgrown his shop, prompting a move to a larger location at 10 Johannisplatz, where some 200 employees worked. Although scientists throughout the continent recognized Zeiss microscopes as the best available, Zeiss himself recognized the limitations imposed by the lack of scientific bases for microscope design. In 1866, as the Zeiss works produced its one thousandth microscope, Zeiss hired a research director in the person of Ernst Abbe, a 26-year-old physics and mathematics lecturer at Jena University.

Zeiss charged Abbe with the task of applying scientific reasoning to the design of microscopes. Within six years, Abbe had determined that increased illumination would create better image resolution, leading to the introduction of a new illumination apparatus. In 1872, Abbe formulated his wave theory of microscopic imaging, which relied on the mathematical formulation of what became known as the Abbe sine condition. In 1875, Zeiss invited Abbe to become a partner in the business.

In 1881, Zeiss added another partner to the firm—his son, Roderich. That year, Abbe recruited the glass chemist Otto Schott to solve the last remaining major impediment to optimal optical quality: the glass used for lenses. By 1886, Zeiss introduced the next generation of microscopes, apochromates, capable of eliminating color distortion by means of compensating eyepieces and a variety of objectives, from dry to water immersion models. That year, the company's 250 workers produced its 10,000th microscope.

Zeiss died in Jena on December 3, 1888. However, his company and his legacy lived on. Nobel laureates in physiology or medicine (including Robert Koch in 1905, and Erwin Neher and Bert Sakman in 1991), chemistry (Richard Zsigmondy in 1925 and Manfred Eigen in 1967), and physics (Frits Zernike in 1953) used Zeiss microscopes in their research. Zeiss's influence also extended to the social realm, as he implemented progressive employment practices (including worker profit sharing as well as prototypical health and retirement plans) that continued after his death, thanks to the establishment of the Carl Zeiss Foundation by Abbe and Zeiss's son, Roderich.

Zeppelin, Ferdinand
(1838–1917)
German
Aeronautical Engineer

Count Ferdinand Zeppelin freed balloon flight from the whims of the wind with his invention of the dirigible, a framed balloon that could be directed in flight. "Zeppelins," as they were named after their inventor, proved effective for passenger flight before and after World War I, but they were eventually displaced by airplane flight, a safer, more efficient alternative. However, the majestic beauty of zeppelins made an indelible mark on popular consciousness, which retains the image in the name of the hugely popular rock and roll band, Led Zeppelin, and in the appearance of the Goodyear "blimp" at sporting events.

Ferdinand Adolf August Heinrich Zeppelin was born on July 8, 1838, at Constance, Baden, in what is now Germany. Hailing from an aristocratic family, he received the traditional military education in Stuttgart and then entered the military as an infantry officer. In 1863, he traveled to the United States to serve in the Union Army of the Potomac in the Civil War. He also fought for the independent state of Württemberg in alliance with Austria in the Seven Weeks War of 1866, losing to Prussia. He then joined Prussia to fight against France in 1870.

Also in 1870, Zeppelin was invited back to the United States to participate in an expedition exploring the Mississippi River. The journey ended upriver at Fort Snelling in St. Paul, Minnesota, where he took his first ascent in a balloon, a catalytic experience that shaped his later life after he retired from the army in 1891, having reached the rank of brigadier general. Thereafter, he dedicated his efforts to the invention of a dirigible balloon, or one that could be directed instead of floating on the whim of the winds.

By 1893, Zeppelin had devised designs with engineering assistance from Theodore Kober, though when he presented his plan to German authorities, they rejected it outright. He continued undeterred, financing the work with his pension and inherited wealth. In 1898, he patented his design of an elliptical, cigar-shaped balloon and founded a company to produce them. For the next two years, he directed construction in a floating hangar on Lake Constance. In a key innovation, Zeppelin framed the balloon with aluminum, a strong but lightweight metal that had been prohibitively expensive and scarce until the 1880s discovery of new extraction and processing methods. Zeppelin covered the frame with a light fabric, and within the framework inserted individual cells of gas-filled airbags made of rubberized fabric.

On July 2, 1900, *Luftschiff Zeppelin No. 1* (*LZ1*) was launched on its maiden flight. Earlier attempts at manned, directed flight had met with limited success: in 1852, Henri Giffard piloted a balloon propelled by a coke-fired steam engine some 17 miles, from the Hippodrome in Paris to Trappe, but his craft could not fly in anything but the mildest weather conditions. In 1876, GOTTLIEB WILHELM DAIMLER and NIKOLAUS AUGUST OTTO invented the four-stroke internal-combustion engine, which replaced the steam engine in almost all applications, including the first "zeppelin," as the dirigibles were then known.

Journalist Hugo Eckener reported on the second flight of *LZ1* in October of 1900. Duly impressed by this invention, he joined Zeppelin's company, which spent the first decade of the 20th century refining zeppelin design and construction. By 1906, Zeppelin had solved most of the problems, and he tested his zeppelin in a 12-hour flight to Lucerne, Switzerland, in 1908. The next year, DELAG (as his company's name was abbreviated) introduced passenger flight, with Eckener serving as director of flight operations.

In mid-June 1910, *LZ7*, or the zeppelin *Deutschland*, expanded passenger service to Frankfurt, Baden-Baden, and Düsseldorf. By 1914, some 35,000 passengers had traveled in zeppelins with no fatalities. Despite this success, Zeppelin still had to appeal for public support, as he had depleted his considerable personal fortune in developing the zeppelin. Kaiser Wilhelm II offered him patronage, and the general public contributed to the National Zeppelin Fund, generating some six million marks, which went to found the Zeppelin Institution.

The German military tried to utilize the zeppelin as a strategic weapon in World War I, but in 1914 alone, 13 zeppelins were destroyed in action, as their large size and use of flammable gases made them vulnerable targets. Nevertheless, Germany produced about a hundred zeppelins throughout the war, which made successful air raids on France and Britain (reaching London) from 1915 through 1916. However, it was clear to Zeppelin that his invention would not win the war for Germany. He died in Charlottenburg, near Berlin, on March 8, 1917, before the end of the war.

Eckener inherited many of Zeppelin's responsibilities, and he offered a zeppelin (subsequently named *Los Angeles*) as war reparation. After the war, in 1918, he introduced regular long-distance passenger service, which proved a better role for zeppelins than military applications. *Graf Zeppelin* (named after the dirigible's inventor, as "graf" is German for Count), made its maiden flight in 1928, and by the next year it had circumnavigated the entire globe. The most successful craft of its kind, *Graf Zeppelin* made 144 transatlantic flights and maintained a perfect safety record in more than a million miles in flight. However, 12 other zeppelins crashed or were lost at sea. The year 1936 saw the launching of the largest dirigible ever, the 803-foot-long *Hindenburg*, capable of carrying 70 passengers. The zeppelin era ended in tragedy when the *Hindenburg* exploded in flight above Lakehurst, New Jersey, in May 1937, killing 35 people.

Zworykin, Vladimir
(1889–1982)
Russian/American
Physicist, Electrical Engineer

Vladimir Zworykin is considered one of the "fathers of television," as his invention of the iconoscope and the kinescope led to the development and commercial distribution of electronic television. He based his system on a theory proposed by ALAN ARCHIBALD CAMPBELL-SWINTON in 1908, though at that time the technology did not exist to realize this theory. Zworykin also directed the laboratory responsible for the invention of the electron microscope, an instrument that allowed for the examination of samples at the atomic level. He also invented the electric eye (used in security systems and with electric door openers) as well as the infrared electron image tube, or night-vision technology.

Vladimir Kosma Zworykin was born on July 30, 1889, in Murom, Russia, near Moscow. His mother was Elaine Zworykin, and his father, Kosma Zworykin, owned and ran a riverboat operation on the Oka River. Zworykin went to the St. Petersburg Institute of Technology, where he studied electrical engineering under Boris Rosing, an early television pioneer who passed on to his student the belief in the superiority of electronic methods of transmitting streams of charged particles with cathode-ray tubes over mechanical transmission methods.

After Zworykin graduated from the institute in 1912, he traveled to Paris to conduct X-ray research under Paul Langevin at the Collège de France. Upon the outbreak of World War I, he returned home to work for the Russian Marconi Company in St. Petersburg, and then in Moscow. When Russia entered the war, he served in the Russian Army Signal Corps; also during the war, he married Tatiana Vasilieff, with whom he had two children. After the war, Zworykin and his family escaped the chaos of the Bolshevik Revolution and traveled the world until they settled in the United States, where he secured work as a bookkeeper for the financial agent of the Russian embassy.

In 1920, the Westinghouse Corporation hired Zworykin to work in its research laboratories developing radio tubes and photoelectric cells. He also followed up on his earlier work in television, developing the iconoscope, which used an electron beam to detect electrical impulses emitted from photoelectric cells mounted on a mica plate containing an image focused there by a lens. The electron beam scanned top-to-bottom in parallel lines, with the number of lines corresponding to the resolution. He applied for a patent on December 29, 1923, but the U.S. patent office did not grant the iconoscope patent until some 15 years later, in December 1938.

In 1924, Zworykin gained his U.S. citizenship. That same year, he filed a patent for the kinescope, which received and displayed the iconoscope's transmission. The kinescope consisted of a hard-vacuum cathode-ray tube that focused a steady electron beam on a signal-modulation grid; this device came to be known as the picture tube. He demonstrated his new electronic television system to his superiors at Westinghouse. "I was terribly excited and proud," he remembered. "After a few days I was informed, very politely, that my demonstration had been extremely interesting, but that it might be better if I were to spend my time on something 'a little more useful.'"

Zworykin persevered, patenting a color television system in 1925 (that was granted in 1928) and earning his Ph.D. from the University of Pittsburgh in 1926 with a dissertation on the improvement of photoelectric cells. In 1929, the

Radio Corporation of America (RCA) hired him as associate director of electronic research, in large part to develop television technology for commercial distribution. When asked how much to budget for this project, Zworykin estimated "about $100,000." According to RCA chairman David Sarnoff, "RCA spent $50 million before we ever got a penny back from TV."

After laying the theoretical foundation for electronic television, Zworykin left the development to other RCA engineers. In the meanwhile, he oversaw his research group member

Vladimir Zworykin (left) is considered one of the fathers of television for his invention of the iconoscope and the kinescope. *(AIP Emilio Segrè Visual Archives, gift of Manfred von Ardenne [1977])*

James Hillier's invention of the electron microscope, which took advantage of the extremely short wavelength of electron beams to achieve extremely high magnification of a millionfold, enabling scientists to make observations at the atomic level. The scanning electron microscope used an electron beam to scan the specimen and displayed the image by the same theory as electronic television. The first industrial electron microscope appeared in 1941.

Zworykin spent World War II on the Scientific Advisory Board of the U.S. Air Force, developing radar, aircraft fire-control, and TV guided missiles. After the war, in 1946, RCA promoted him to director of electronic research, and the next year, it elevated him yet again to a vice presidency and a technical consultancy.

Upon his 1954 retirement to emeritus status at RCA, he became director of the Rockefeller Institute for Medical Research (now Rockefeller University), a position he retained until 1962.

Zworykin collected numerous honors and awards for his pioneering work: the 1934 Morris Liebmann Memorial Prize from the Institute of Radio Engineers; the 1952 Edison Medal from the American Institute of Electrical Engineers; the 1965 Faraday Medal from the Institute of Electrical Engineers; and the 1967 U.S. Medal of Science. Throughout his career, he secured more than 120 patents. When asked in 1981 for his opinion of television programming, he replied that it is "awful." Zworykin died the next year, on July 29, 1982, just one day short on his 93rd birthday.

Entries by Field

AGRICULTURE
Bakewell, Robert
Lawes, Sir John Bennet
Tull, Jethro

ANTHROPOLOGY
Mead, Margaret

ASTRONAUTICS
Lucid, Shannon W.

ASTRONOMY
Galilei, Galileo
Huygens, Christiaan
Marcy, Geoffrey
Sagan, Carl Edward

ASTROPHYSICS
Alfvén, Hannes

ATMOSPHERIC CHEMISTRY
Rowland, F. Sherwood

BACTERIOLOGY
Avery, Oswald Theodore
Domagk, Gerhard
Fleming, Sir Alexander
Lancefield, Rebecca Craighill

BIOCHEMISTRY
Chain, Sir Ernst Boris
Hoobler, Icie Gertrude Macy
Sanger, Frederick
Venter, J. Craig

BIOLOGY
Daily, Gretchen
Medawar, Peter Brian

BIOPHYSICS
Quimby, Edith H.

BOTANY
De Vries, Hugo

CARTOGRAPHY
Mercator, Gerardus
Ortelius, Abraham

CHEMISTRY
Baekeland, Leo Hendrik
Bunsen, Robert Wilhelm
    Eberhard von
Carothers, Wallace Hume
Crookes, Sir William
Dalton, John
Daniell, John Frederic
Davy, Sir Humphry
Elion, Gertrude Belle ("Trudy")
Franklin, Rosalind Elsie
Gilbert, Walter
Haber, Fritz
Hodgkin, Dorothy Crowfoot
Kipping, Frederic Stanley
Lavoisier, Antoine-Laurent
Leblanc, Nicolas
Lebon, Philippe
Niepce, Joseph
Nobel, Alfred Bernhard
Pasteur, Louis
Pennington, Mary Engle
Perkin, William Henry
Playfair, Lyon
Richards, Ellen Swallow
Soddy, Frederick
Swan, Sir Joseph Wilson

CINEMATOGRAPHY
Lumière, Auguste and Louis

COMMUNICATIONS INDUSTRY
Armstrong, Edwin
Baird, John Logie
Berliner, Emile
Campbell-Swinton, Alan
    Archibald
Cooke, Sir William Fothergill

Farnsworth, Philo (television)
Lear, William
Nipkow, Paul Gottlieb
Popov, Alexander Stepanovich

Computer Science and Engineering
Aiken, Howard Hathaway
Allen, Paul
Babbage, Charles
Forrester, Jay
Hopper, Grace Murray
Hyatt, Gilbert
Jobs, Steven
Mauchly, John William
Sutherland, Ivan Edward
Wozniak, Stephen

Dentistry
Wells, Horace

Ecology
Carson, Rachel Louise
Leopold, Aldo

Energy and Power Industries
Boulton, Matthew
Drake, Edwin Laurentine
Edison, Thomas
Ferranti, Sebastian Ziani de
Leclanché, Georges
Newcomen, Thomas
Papin, Denis
Swan, Sir Joseph Wilson
Westinghouse, George

Engineering
Armstrong, Edwin
Baird, John Logie
Benz, Karl Friedrich
Bessemer, Sir Henry
Brindley, James
Brunel, Isambard Kingdom
Brunel, Sir Marc Isambard
Campbell-Swinton, Alan Archibald
Daimler, Gottlieb Wilhelm
Darby, Abraham
Diesel, Rudolf
Ferranti, Sebastian Ziani de
Hollerith, Herman
Hyatt, Gilbert
Lebon, Philippe
Leclanché, Georges
Marconi, Guglielmo
Mestral, George de
Newcomen, Thomas
Nipkow, Paul Gottlieb
Otis, Elisha Graves
Otto, Nikolaus August
Parsons, Sir Charles Algernon
Polhem, Christopher
Poncelet, Jean-Victor
Savery, Thomas
Sikorsky, Igor
Sperry, Elmer Ambrose
Stephenson, George
Sutherland, Ivan Edward
Swan, Sir Joseph Wilson
Swinburne, James
Telford, Thomas
Tesla, Nikola
Thomson, Elihu
Trevithick, Richard
Watson-Watt, Robert Alexander
Watt, James
Westinghouse, George
Whitehead, Robert
Wright, Wilbur and Orville
Young, James
Zeppelin, Ferdinand
Zworykin, Vladimir

Exploration
Magellan, Ferdinand
Polo, Marco

Food Technology
Appert, Nicolas
De Laval, Carl Gustav Patrik
Pennington, Mary Engle

Genetics Research
Avery, Oswald T.
Crick, Francis
De Vries, Hugo
King, Mary-Claire
McClintock, Barbara
Mendel, Johann Gregor
Morgan, Thomas Hunt
Venter, J. Craig

Geography
Mercator, Gerardus
Ortelius, Abraham

Geology
Agricola, Georgius
Holmes, Arthur

Home Economics
Richards, Ellen Swallow

Horology
Harrison, John

Immunology
Matzinger, Polly Celine Eveline

Marine Biology
Carson, Rachel Louise

Mathematics
Ampère, André-Marie
Archimedes
Babbage, Charles
Hero of Alexandria
Kelvin, Baron (William Thomson)

Leibniz, Gottfried Wilhelm
Napier, John
Newton, Sir Isaac
Pappus
Pascal, Blaise
Poncelet, Jean-Victor
Von Neumann, John

MEDICINE
Apgar, Virginia
Banting, Sir Frederick G.
Barton, Clara
Best, Charles Herbert
Caldicott, Helen
Elion, Gertrude Belle ("Trudy")
Hopkins, Donald (Public Health)
Jenner, Edward
Kolff, Willem J.
Lister, Joseph
Love, Susan
Medawar, Peter Brian

METALLURGY
Agricola, Georgius
Swedenborg, Emanuel

METEOROLOGY
Abbe, Cleveland
Ackerman, Thomas P.
Daniell, John Frederic

MICROBIOLOGY
Pasteur, Louis
Salk, Jonas Edward

MINERALOGY
Agricola, Georgius

MOLECULAR BIOLOGY
Crick, Francis
Gilbert, Walter
Gurdon, Sir John Bertrand

NATURAL HISTORY
Darwin, Charles Robert
Pliny the Elder

NEUROBIOLOGY
Greenfield, Susan

NEUROPATHOLOGY
Freud, Sigmund

NURSING
Barton, Clara

OCEANOGRAPHY
Cousteau, Jacques-Yves
Franklin, Benjamin

OPTICS
Zeiss, Carl Friedrich

PATHOLOGY
Florey, Howard Walter

PHILOSOPHY
Leibniz, Gottfried Wilhelm

PHOTOGRAPHY
Eastman, George

PHYSICS
Alvarez, Luis Walter
Ampère, André-Marie
Archimedes
Becquerel, Antoine-Henri
Crookes, Sir William
Dalton, John
Einstein, Albert
Fermi, Enrico
Franklin, Benjamin
Frisch, Otto Robert
Gabor, Dennis
Galilei, Galileo
Goddard, Robert Hutchings
Hertz, Heinrich Rudolf
Hounsfield, Godfrey Newbold
Huygens, Christiaan
Joliot-Curie, Irène
Joule, James Prescott
Kapitsa, Pyotr Leonidovich
Kelvin, Baron (William Thomson)
Maiman, Theodore
Marconi, Guglielmo
Maxwell, James Clerk
Meitner, Lise
Newton, Sir Isaac
Oppenheimer, J. Robert
Ørsted, Hans Christian
Pascal, Blaise
Planck, Max
Popov, Alexander Stepanovich
Röntgen, Wilhelm Conrad
Shockley, William
Tesla, Nikola
Volta, Count Alessandro
Wheatstone, Sir Charles
Zworykin, Vladimir

PHYSIOLOGY
De Vries, Hugo
Einthoven, Willem
Pavlov, Ivan Petrovich
Swedenborg, Emanuel

PLASTICS INDUSTRY
Hyatt, John Wesley
Swinburne, James

POTTERY
Spode, Josiah

PSYCHOBIOLOGY
Sperry, Roger

PSYCHOLOGY
Binet, Alfred
Freud, Sigmund
Piaget, Jean
Skinner, B. F.

Rubber Industry
Goodyear, Charles
Hancock, Thomas

Steel and Iron Industry
Bessemer, Sir Henry
Darby, Abraham
Kelly, William
Krupp, Alfred
Wilkinson, John

Textile Industry
Arkwright, Sir Richard
Carothers, Wallace Hume
Hargreaves, James

Jacquard, Joseph-Marie
Kay, John
Mestral, George de
Perkin, William Henry
Singer, Isaac Merrit

Theology
Swedenborg, Emanuel

Toxicology
Silbergeld, Ellen Kovner

Transportation Industry
Benz, Karl Friedrich
Daimler, Gottlieb Wilhelm

Diesel, Rudolf
Dunlop, John Boyd
Lanchester, Frederick William
Lear, William
Michaux, Pierre
Otis, Elisha Graves
Sikorsky, Igor
Starley, James
Telford, Thomas

Xerography
Carlson, Chester

Entries by Country of Birth

Australia
Caldicott, Helen
Florey, Howard Walter

Austria
Freud, Sigmund
Frisch, Otto Robert
Meitner, Lise
Mendel, Johann Gregor

Belgium
Baekeland, Leo Hendrik
Ortelius, Abraham

Brazil
Medawar, Peter Brian

Canada
Avery, Oswald Theodore
Banting, Sir Frederick G.

Croatia
Tesla, Nikola

Denmark
Ørsted, Hans Christian

England
Arkwright, Sir Richard
Babbage, Charles
Bakewell, Robert
Bessemer, Sir Henry
Boulton, Matthew
Brindley, James
Brunel, Isambard Kingdom
Cooke, Sir William Fothergill
Crick, Francis
Crookes, Sir William
Dalton, John
Daniell, John Frederic
Darby, Abraham
Darwin, Charles Robert
Davy, Sir Humphry
Ferranti, Sebastian Ziani de
Franklin, Rosalind Elsie
Greenfield, Susan
Gurdon, Sir John Bertrand
Hancock, Thomas
Hargreaves, James
Harrison, John
Hodgkin, Dorothy Crowfoot
Holmes, Arthur
Hounsfield, Godfrey Newbold
Jenner, Edward
Joule, James Prescott
Kay, John
Kipping, Frederic Stanley
Lanchester, Frederick William
Lawes, Sir John Bennet
Lister, Joseph
Newcomen, Thomas
Newton, Sir Isaac
Parsons, Sir Charles Algernon
Perkin, William Henry
Playfair, Lyon
Sanger, Frederick
Savery, Thomas
Shockley, William
Soddy, Frederick
Spode, Josiah
Starley, James
Stephenson, George
Swan, Sir Joseph Wilson
Thomson, Elihu
Trevithick, Richard
Tull, Jethro
Wheatstone, Sir Charles
Whitehead, Robert
Wilkinson, John

France
Ampère, André-Marie
Appert, Nicolas
Becquerel, Antoine-Henri
Binet, Alfred
Brunel, Sir Marc Isambard
Cousteau, Jacques-Yves
Diesel, Rudolf
Jacquard, Joseph-Marie
Joliot-Curie, Irène

Lavoisier, Antoine-Laurent
Leblanc, Nicolas
Lebon, Philippe
Leclanché, Georges
Lumière, Auguste and Louis
Michaux, Pierre
Niepce, Joseph
Papin, Denis
Pascal, Blaise
Pasteur, Louis
Poncelet, Jean-Victor

Germany
Agricola, Georgius
Benz, Karl Friedrich
Berliner, Emile
Bunsen, Robert Wilhelm Eberhard von
Chain, Sir Ernst Boris
Daimler, Gottlieb Wilhelm
Domagk, Gerhard
Einstein, Albert
Haber, Fritz
Hertz, Heinrich Rudolf
Krupp, Alfred
Leibniz, Gottfried Wilhelm
Nipkow, Paul Gottlieb
Otto, Nikolaus August
Planck, Max
Röntgen, Wilhelm Conrad
Zeiss, Carl Friedrich
Zeppelin, Ferdinand

Greece
Hero of Alexandria
Pappus

Hungary
Gabor, Dennis
Von Neumann, John

Ireland
Kelvin, Baron (William Thomson)

Italy
Fermi, Enrico
Galilei, Galileo
Marconi, Guglielmo
Pliny the Elder
Polo, Marco
Volta, Count Alessandro

Netherlands
De Vries, Hugo
Einthoven, Willem
Huygens, Christiaan
Kolff, Willem J.
Mercator, Gerardus

Portugal
Magellan, Ferdinand

Russia
Kapitsa, Pyotr Leonidovich
Pavlov, Ivan Petrovich
Popov, Alexander Stepanovich
Zworykin, Vladimir

Scotland
Baird, John Logie
Campbell-Swinton, Alan Archibald
Dunlop, John Boyd
Fleming, Sir Alexander
Maxwell, James Clerk
Napier, John
Swinburne, James
Telford, Thomas
Watson-Watt, Robert Alexander
Watt, James
Young, James

Sicily
Archimedes

Sweden
Alfvén, Hannes
De Laval, Carl Gustav Patrik
Nobel, Alfred Bernhard
Polhem, Christopher
Swedenborg, Emanuel

Switzerland
Mestral, George de
Piaget, Jean

Ukraine
Sikorsky, Igor

United States
Abbe, Cleveland
Ackerman, Thomas P.
Aiken, Howard Hathaway
Allen, Paul
Alvarez, Luis Walter
Apgar, Virginia
Armstrong, Edwin
Barton, Clara
Best, Charles Herbert
Carlson, Chester
Carothers, Wallace Hume
Carson, Rachel Louise
Daily, Gretchen
Drake, Edwin Laurentine
Eastman, George
Edison, Thomas
Elion, Gertrude Belle ("Trudy")
Farnsworth, Philo
Forrester, Jay
Franklin, Benjamin
Gilbert, Walter
Goddard, Robert Hutchings
Goodyear, Charles
Hollerith, Herman
Hoobler, Icie Gertrude Macy
Hopkins, Donald
Hopper, Grace Murray
Hyatt, Gilbert
Hyatt, John Wesley
Jobs, Steven

Kelly, William
King, Mary-Claire
Lancefield, Rebecca Craighill
Lear, William
Leopold, Aldo
Love, Susan
Lucid, Shannon W.
Maiman, Theodore
Marcy, Geoffrey
Matzinger, Polly Celine Eveline
Mauchly, John William

McClintock, Barbara
Mead, Margaret
Morgan, Thomas Hunt
Oppenheimer, J. Robert
Otis, Elisha Graves
Pennington, Mary Engle
Quimby, Edith H.
Richards, Ellen Swallow
Rowland, F. Sherwood
Sagan, Carl Edward
Salk, Jonas Edward

Silbergeld, Ellen Kovner
Singer, Isaac Merrit
Skinner, B. F.
Sperry, Elmer Ambrose
Sperry, Roger
Sutherland, Ivan Edward
Venter, J. Craig
Wells, Horace
Westinghouse, George
Wozniak, Stephen
Wright, Wilbur and Orville

Entries by Country of Major Scientific Activity

Australia
Caldicott, Helen

Austria
Freud, Sigmund
Mendel, Johann Gregor

Belgium
Ortelius, Abraham

Canada
Avery, Oswald Theodore
Banting, Sir Frederick G.
Best, Charles Herbert

Denmark
Ørsted, Hans Christian

England
Arkwright, Sir Richard
Babbage, Charles
Bakewell, Robert
Bessemer, Sir Henry
Boulton, Matthew
Brindley, James
Brunel, Isambard Kingdom
Brunel, Sir Marc Isambard
Chain, Sir Ernst Boris
Cooke, Sir William Fothergill
Crick, Francis

Crookes, Sir William
Dalton, John
Daniell, John Frederic
Darby, Abraham
Darwin, Charles Robert
Davy, Sir Humphry
Ferranti, Sebastian Ziani de
Florey, Howard Walter
Franklin, Rosalind Elsie
Frisch, Otto Robert
Gabor, Dennis
Greenfield, Susan
Gurdon, Sir John Bertrand
Hancock, Thomas
Hargreaves, James
Harrison, John
Hodgkin, Dorothy Crowfoot
Holmes, Arthur
Hounsfield, Godfrey Newbold
Jenner, Edward
Joule, James Prescott
Kay, John
Kipping, Frederic Stanley
Lanchester, Frederick William
Lawes, Sir John Bennet
Medawar, Peter Brian
Newcomen, Thomas
Newton, Sir Isaac
Parsons, Sir Charles Algernon
Perkin, William Henry

Playfair, Lyon
Sanger, Frederick
Savery, Thomas
Soddy, Frederick
Spode, Josiah
Starley, James
Stephenson, George
Swan, Sir Joseph Wilson
Swinburne, James
Telford, Thomas (also built roads in Scotland, Wales, etc.)
Trevithick, Richard
Tull, Jethro
Wheatstone, Sir Charles
Whitehead, Robert
Wilkinson, John

France
Ampère, André-Marie
Appert, Nicolas
Becquerel, Antoine-Henri
Binet, Alfred
Cousteau, Jacques-Yves
Jacquard, Joseph-Marie
Joliot-Curie, Irène
Lavoisier, Antoine-Laurent
Leblanc, Nicolas
Lebon, Philippe
Leclanché, Georges

Lumière, Auguste and Louis
Michaux, Pierre
Niepce, Joseph
Papin, Denis
Pascal, Blaise
Pasteur, Louis
Poncelet, Jean-Victor

GERMANY
Agricola, Georgius
Benz, Karl Friedrich
Bunsen, Robert Wilhelm Eberhard von
Daimler, Gottlieb Wilhelm
Diesel, Rudolf
Domagk, Gerhard
Einstein, Albert
Haber, Fritz
Hertz, Heinrich Rudolf
Krupp, Alfred
Leibniz, Gottfried Wilhelm
Nipkow, Paul Gottlieb
Otto, Nikolaus August
Planck, Max
Röntgen, Wilhelm Conrad
Zeiss, Carl Friedrich
Zeppelin, Ferdinand

GREECE
Hero of Alexandria
Pappus

ITALY
Archimedes
Galilei, Galileo
Marconi, Guglielmo
Pliny the Elder
Polo, Marco
Volta, Count Alessandro

NETHERLANDS
De Vries, Hugo
Einthoven, Willem
Huygens, Christiaan
Mercator, Gerardus

RUSSIA
Kapitsa, Pyotr Leonidovich
Pavlov, Ivan Petrovich
Popov, Alexander Stepanovich

SCOTLAND
Baird, John Logie
Campbell-Swinton, Alan Archibald
Dunlop, John Boyd
Fleming, Sir Alexander
Kelvin, Baron (William Thomson)
Lister, Joseph
Maxwell, James Clerk
Napier, John
Watson-Watt, Robert Alexander
Watt, James
Young, James

SPAIN
Magellan, Ferdinand

SWEDEN
Alfvén, Hannes
De Laval, Carl Gustav Patrik
Meitner, Lise
Nobel, Alfred Bernhard
Polhem, Christopher
Swedenborg, Emanuel

SWITZERLAND
Einstein, Albert
Mestral, George de
Piaget, Jean

UNITED STATES
Abbe, Cleveland
Ackerman, Thomas P.
Aiken, Howard Hathaway
Allen, Paul
Alvarez, Luis Walter
Apgar, Virginia
Armstrong, Edwin
Baekeland, Leo Hendrik
Barton, Clara
Berliner, Emile
Carlson, Chester
Carothers, Wallace Hume
Carson, Rachel Louise
Daily, Gretchen
Drake, Edwin Laurentine
Eastman, George
Edison, Thomas
Einstein, Albert
Elion, Gertrude Belle ("Trudy")
Farnsworth, Philo
Fermi, Enrico
Forrester, Jay
Franklin, Benjamin
Gilbert, Walter
Goddard, Robert Hutchings
Goodyear, Charles
Hollerith, Herman
Hoobler, Icie Gertrude Macy
Hopkins, Donald
Hopper, Grace Murray
Hyatt, Gilbert
Hyatt, John Wesley
Jobs, Steven
Kelly, William
King, Mary-Claire
Kolff, Willem J.
Lancefield, Rebecca Craighill
Lear, William
Leopold, Aldo
Love, Susan
Lucid, Shannon W.
Maiman, Theodore
Marcy, Geoffrey
Matzinger, Polly Celine Eveline
Mauchly, John William
McClintock, Barbara
Mead, Margaret
Morgan, Thomas Hunt
Oppenheimer, J. Robert

Otis, Elisha Graves
Pennington, Mary Engle
Quimby, Edith H.
Richards, Ellen Swallow
Rowland, F. Sherwood
Sagan, Carl Edward
Salk, Jonas Edward
Shockley, William
Sikorsky, Igor
Silbergeld, Ellen Kovner
Singer, Isaac Merrit
Skinner, B. F.
Sperry, Elmer Ambrose
Sperry, Roger
Sutherland, Ivan Edward
Tesla, Nikola
Thomson, Elihu
Venter, J. Craig
Von Neumann, John
Wells, Horace
Westinghouse, George
Wozniak, Stephen
Wright, Wilbur and Orville
Zworykin, Vladimir

Entries by Year of Birth

Third Century b.c.–Fourth Century a.d.
Archimedes
Hero of Alexandria
Pappus
Pliny the Elder
Polo, Marco

1400–1599
Agricola, Georgius
Galilei, Galileo
Magellan, Ferdinand
Mercator, Gerardus
Napier, John
Ortelius, Abraham

1600–1699
Darby, Abraham
Harrison, John
Huygens, Christiaan
Leibniz, Gottfried Wilhelm
Newcomen, Thomas
Newton, Sir Isaac
Papin, Denis
Pascal, Blaise
Polhem, Christopher
Savery, Thomas
Swedenborg, Emanuel
Tull, Jethro

1700–1799
Ampère, André-Marie
Appert, Nicolas
Arkwright, Sir Richard
Babbage, Charles
Bakewell, Robert
Boulton, Matthew
Brindley, James
Brunel, Sir Marc Isambard
Dalton, John
Daniell, John Frederic
Davy, Sir Humphry
Franklin, Benjamin
Hancock, Thomas
Hargreaves, James
Jacquard, Joseph-Marie
Jenner, Edward
Kay, John
Lavoisier, Antoine-Laurent
Leblanc, Nicolas
Lebon, Philippe
Niepce, Joseph
Ørsted, Hans Christian
Poncelet, Jean-Victor
Spode, Josiah
Stephenson, George
Telford, Thomas
Trevithick, Richard
Volta, Count Alessandro
Watt, James
Wilkinson, John

1800–1849
Abbe, Cleveland
Barton, Clara
Benz, Karl Friedrich
Bessemer, Sir Henry
Brunel, Isambard Kingdom
Bunsen, Robert Wilhelm Eberhard von
Cooke, Sir William Fothergill
Crookes, Sir William
Daimler, Gottlieb Wilhelm
Darwin, Charles Robert
De Laval, Carl Gustav Patrik
De Vries, Hugo
Drake, Edwin Laurentine
Dunlop, John Boyd
Edison, Thomas
Goodyear, Charles
Hyatt, John Wesley
Joule, James Prescott
Kelly, William
Kelvin, Baron (William Thomson)
Krupp, Alfred
Lawes, Sir John Bennet
Leclanché, Georges
Lister, Joseph
Maxwell, James Clerk
Mendel, Johann Gregor
Michaux, Pierre
Nobel, Alfred Bernhard

Otis, Elisha Graves
Otto, Nikolaus August
Pasteur, Louis
Pavlov, Ivan Petrovich
Perkin, William Henry
Playfair, Lyon
Richards, Ellen Swallow
Röntgen, Wilhelm Conrad
Singer, Isaac Merrit
Starley, James
Swan, Sir Joseph Wilson
Wells, Horace
Westinghouse, George
Wheatstone, Sir Charles
Whitehead, Robert
Young, James
Zeiss, Carl Friedrich
Zeppelin, Ferdinand

1850–1899
Armstrong, Edwin
Avery, Oswald Theodore
Baekeland, Leo Hendrik
Baird, John Logie
Banting, Sir Frederick G.
Becquerel, Antoine-Henri
Berliner, Emile
Best, Charles Herbert
Binet, Alfred
Campbell-Swinton, Alan Archibald
Carothers, Wallace Hume
Diesel, Rudolf
Domagk, Gerhard
Eastman, George
Einstein, Albert
Einthoven, Willem
Ferranti, Sebastian Ziani de
Fleming, Sir Alexander
Florey, Howard Walter
Freud, Sigmund
Goddard, Robert Hutchings
Haber, Fritz
Hertz, Heinrich Rudolf

Hollerith, Herman
Holmes, Arthur
Hoobler, Icie Gertrude Macy
Joliot-Curie, Irène
Kapitsa, Pyotr Leonidovich
Kipping, Frederic Stanley
Lancefield, Rebecca Craighill
Lanchester, Frederick William
Leopold, Aldo
Lumière, Auguste
Lumière, Louis
Marconi, Guglielmo
Meitner, Lise
Morgan, Thomas Hunt
Nipkow, Paul Gottlieb
Parsons, Sir Charles Algernon
Pennington, Mary Engle
Piaget, Jean
Planck, Max
Popov, Alexander Stepanovich
Quimby, Edith H.
Sikorsky, Igor
Soddy, Frederick
Sperry, Elmer Ambrose
Swinburne, James
Tesla, Nikola
Thomson, Elihu
Watson-Watt, Robert Alexander
Wright, Orville
Wright, Wilbur
Zworykin, Vladimir

1900–1909
Aiken, Howard Hathaway
Alfvén, Hannes
Apgar, Virginia
Carlson, Chester
Carson, Rachel Louise
Chain, Sir Ernst Boris
Farnsworth, Philo
Fermi, Enrico
Frisch, Otto Robert

Gabor, Dennis
Hopper, Grace Murray
Lear, William
Mauchly, John William
McClintock, Barbara
Mead, Margaret
Mestral, George de
Oppenheimer, J. Robert
Skinner, B. F.
Von Neumann, John

1910–1919
Alvarez, Luis Walter
Cousteau, Jacques-Yves
Crick, Francis
Elion, Gertrude Belle ("Trudy")
Forrester, Jay
Hodgkin, Dorothy Crowfoot
Hounsfield, Godfrey Newbold
Kolff, Willem J.
Medawar, Peter Brian
Salk, Jonas Edward
Sanger, Frederick
Shockley, William
Sperry, Roger

1920–1929
Franklin, Rosalind Elsie
Maiman, Theodore
Rowland, F. Sherwood

1930–1939
Caldicott, Helen
Gilbert, Walter
Gurdon, Sir John Bertrand
Hyatt, Gilbert
Sagan, Carl Edward
Sutherland, Ivan Edward

1940–1949
Ackerman, Thomas P.
Hopkins, Donald
King, Mary-Claire

Love, Susan
Lucid, Shannon W.
Matzinger, Polly Celine
    Eveline
Silbergeld, Ellen Kovner
Venter, J. Craig

1950–1959
Allen, Paul
Greenfield, Susan
Jobs, Steven
Marcy, Geoffrey
Wozniak, Stephen

1960–1969
Daily, Gretchen

Chronology

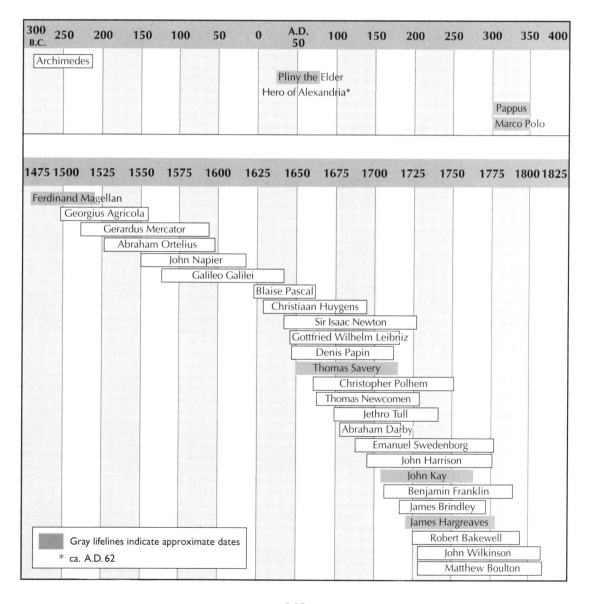

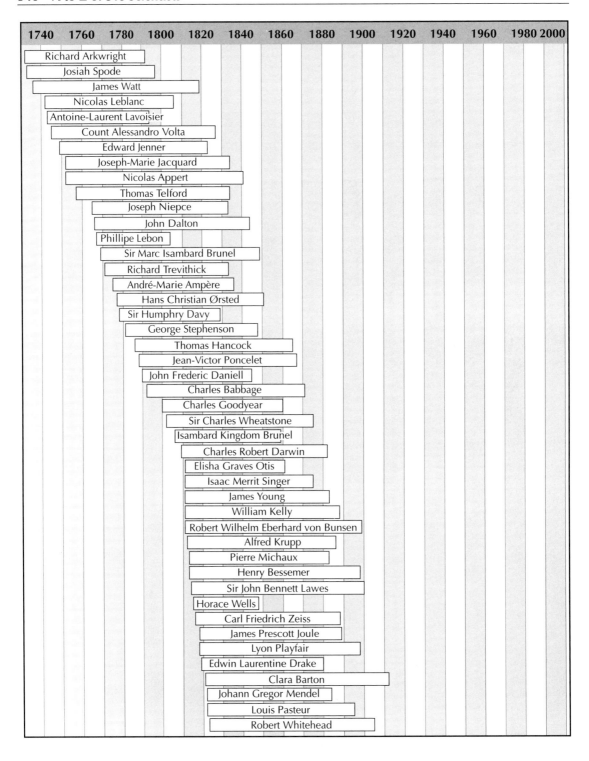

Chronology   347

| 1740 | 1760 | 1780 | 1800 | 1820 | 1840 | 1860 | 1880 | 1900 | 1920 | 1940 | 1960 | 1980 | 2000 |

William Thomson, Baron Kelvin
Joseph Lister
Sir Joseph Wilson Swan
James Starley
James Clerk Maxwell
Nikolaus August Otto
Sir William Crookes
Alfred Bernhard Nobel
Gottlieb Wilhelm Daimler
John Wesley Hyatt
William Henry Perkin
Cleveland Abbe
Ferdinand Zeppelin
George Leclanché
John Boyd Dunlop
Ellen Swallow Richards
Karl Friedrich Benz
Carl Gustav Patrik De Laval
Wilhelm Conrad Röntgen
George Westinghouse
Thomas Edison
Hugo De Vries
Ivan Petrovich Pavlov
Emile Berliner
Antoine Henri Becquerel
Thomson Elihu
Sir Charles Algernon Parsons
George Eastman
Sigmund Freud
Nikola Tesla
Heinrich Rudolf Hertz
Alfred Binet
Rudolf Diesel
Max Planck
James Swinburne
Alexander Stepanovich Popov
Willem Einthoven
Herman Hollerith
Elmer Ambrose Sperry
Paul Gottlieb Nipkow
Auguste Lumière
Alan Archibald Campbell-Swinton
Leo Hendrik Baekeland
Frederic Stanley Kipping
Sebastian Ziani de Ferranti

Timeline	Scientist
1864–	Louis Lumière
1866–	Thomas Hunt Morgan
1867–	Wilbur Wright
1868–	Frederick William Lanchester
1871–	Orville Wright
1872–	Mary Engle Pennington
1874–	Guglielmo Marconi
1877–	Oswald Theodore Avery
1877–	Frederick Soddy
1878–	Lise Meitner
1879–	Albert Einstein
1881–	Sir Alexander Fleming
1882–	Robert Hutchings Goddard
1887–	Aldo Leopold
1888–	John Logie Baird
1889–	Igor Sikorsky
1889–	Vladimir Zworykin
1890–	Edwin Armstrong
1890–	Arthur Holmes
1891–	Sir Frederick G. Banting
1891–	Edith H. Quimby
1892–	Robert Alexander Watson-Watt
1892–	Icie Gertrude Macy Hoobler
1894–	Pyotr Leonidovich Kapitsa
1895–	Gerhard Domagk
1895–	Rebecca Craighill Lancefield
1896–	Wallace Hume Carothers
1896–	Jean Piaget
1897–	Irène Joliot-Curie
1898–	Howard Walter Florey
1899–	Charles Herbert Best
1900–	Howard Hathaway Aiken
1900–	Dennis Gabor
1901–	Enrico Fermi
1901–	Margaret Mead
1902–	William Lear
1902–	Barbara McClintock
1903–	John Von Neumann
1904–	J. Robert Oppenheimer
1904–	Otto Robert Frisch
1904–	B. F. Skinner
1906–	Chester Carlson
1906–	Philo Farnsworth
1906–	Sir Ernst Boris Chain
1906–	Grace Murray Hopper

Chronology

Timeline: 1740–2000

- Rachel Louise Carson
- John William Mauchly
- George de Mestral
- Hannes Alfvén
- Virginia Apgar
- William Shockley
- Dorothy Crowfoot Hodgkin
- Jacques-Yves Cousteau
- Luis Walter Alvarez
- Willem J. Kolff
- Roger Sperry
- Jonas Edward Salk
- Peter Brian Medawar
- Francis Crick
- Gertrude Belle (Trudy) Elion
- Jay Forrester
- Frederick Sanger
- Godfrey Newbold Hounsfield
- Rosalind Elsie Franklin
- Theodore Maiman
- F. Sherwood Rowland
- Walter Gilbert
- Sir John Bertrand Gurdon
- Carl Edward Sagan
- Helen Caldicott
- Gilbert Hyatt
- Ivan Edward Sutherland
- Donald Hopkins
- Ellen Kovner Silbergeld
- Mary-Claire King
- J. Craig Venter
- Polly C. E. Matzinger
- Thomas Ackerman
- Susan Love
- Stephen Wozniak
- Susan Greenfield
- Paul Allen
- Geoffrey Marcy
- Steven Jobs
- Gretchen Daily

Gray lifelines indicate approximate dates

BIBLIOGRAPHY

Abir-Am, P. G., and Dorinda Outram, eds. *Uneasy Careers and Intimate Lives: Women in Science 1789–1979*. New Brunswick, N.J.: Rutgers University Press, 1987.

American Men and Women of Science, 1995–96: A Biographical Directory of Today's Leaders in Physical, Biological, and Related Sciences. New York: R. R. Bowker, 1994.

Asimov, Isaac. *Asimov's Biographical Encyclopedia of Science and Technology: The Lives and Achievements of 1510 Great Scientists from Ancient Times to the Present Chronologically Arranged*. New York: Doubleday, 1982.

Association of Women in Mathematics. "Women in Mathematics." Available online. URL: http://www.awm-math.org.

Bailey, Brooke. *The Remarkable Lives of 100 Women Healers and Scientists*. Holbrook, Mass.: Bob Adams, 1994.

Bailey, M. J. *American Women in Science*. Santa Barbara, Calif.: ABC-CLIO, Inc., 1994.

Barrett, Eric C., and David Fisher, eds. *Scientists Who Believe: Twenty-one Tell Their Own Stories*. Chicago: Moody Press, 1984.

The Biographical Dictionary of Scientists. New York: P. Bedrick Books, 1983–85.

Biography: Scientists, Mathematicians & Inventors. Available on-line. URL: http://www.teaneckschools.org/mscl/pages/biogsci.html.

Bridges, Thomas C., and Hubert H. Tiltman. *Master Minds of Modern Science*. New York: L. MacVeagh, 1931.

Cassutt, Michael. *Who's Who in Space: The First 25 Years*. Boston: G. K. Hall, 1987.

Concise Dictionary of Scientific Biography. New York: Scribner's, 1981.

Current Biography. New York: H. W. Wilson, 1940– .

Daintith, John, Sarah Mitchell, Elizabeth Tootill, and Derek Gjertsen, eds. *Biographical Encyclopedia of Scientists*. Philadelphia: Institute of Physics Publishing, 1994.

Darrow, Floyd L. *Masters of Science and Invention*. New York: Harcourt, 1923.

Dash, Joan. *The Triumph of Discovery: Women Scientists Who Won the Nobel Prize*. Englewood Cliffs, N.J.: Julian Messner, 1991.

Debus, A. G., ed. *World Who's Who in Science*. Chicago: Marquis Who's Who, 1968.

Defries, Amelia D. *Pioneers of Science*. London: Routledge and Sons, 1928.

Dictionary of American Biography. New York: Scribner's, 1928– .

Dictionary of Scientific Biography. New York: Scribner's, 1970–1980.

Elliott, Clark A. *Biographical Dictionary of American Science: The Seventeenth through the Nineteenth Centuries*. Westport, Conn.: Greenwood Press, 1979.

Eric's Treasure Trove of Scientific Biographies. Available on-line. URL: http://www.treasure-troves.com

Feldman, Anthony. *Scientists and Inventors.* New York: Facts On File, 1979.

Gaillard, Jacques. *Scientists in the Third World.* Lexington: University Press of Kentucky, 1991.

Grinstein, Louise S., Rose K. Rose, and Miriam H. Rafailovich, eds. *Women in Chemistry and Physics: A Biobibliographic Sourcebook.* Westport, Conn.: Greenwood Press, 1993.

Herzenberg, Caroline L. *Women Scientists from Antiquity to the Present: An Index.* West Cornwall, Conn.: Locust Hill, 1986.

Howard, Arthur Vyvyan. *Chamber's Dictionary of Scientists.* New York: Dutton, 1961.

Hsiao, T. C., ed. *Who's Who in Computer Education and Research: U.S. Edition.* Latham, N.Y.: Science and Technology Press, 1975.

Index of Inventors. Available on-line. URL: http://www.invent.org/book/book-text/indexbyname.html.

Ireland, Norma O. *Index to Scientists of the World from Ancient to Modern Times: Biographies and Portraits.* Boston: Faxon, 1962.

James, Laylin K., ed. *Nobel Laureates in Chemistry, 1901–1992.* Washington, D.C.: American Chemical Society/Chemical Heritage Foundation, 1993.

Jones, Bessie Zaban, ed. *The Golden Age of Science: Thirty Portraits of the Giants of 19th-Century Science by Their Scientific Contemporaries.* New York: Simon and Schuster, 1966.

Kass-Simon, Gabriele, and Patricia Farnes. *Women of Science: Righting the Record.* Bloomington: Indiana University Press, 1990.

Kessler, James H., Jerry S. Kidd, Renee A. Kidd, Katherine Morin, and Tracy More, eds. *Distinguished African American Scientists of the 20th Century.* Phoenix, Ariz.: Oryx Press, 1996.

Krapp, Kristine M. *Notable Black American Scientists.* Detroit, Mich.: Gale Research, 1998.

Makers of Modern Science: A Twentieth Century Library Trilogy. New York: Scribner's, 1953.

McGraw-Hill Modern Men of Science: 426 Leading Contemporary Scientists. New York: McGraw-Hill, 1966–1968.

McGraw-Hill Modern Scientists and Engineers. New York: McGraw-Hill, 1980.

McGrayne, Sharon Bertsch. *Nobel Prize Women in Science: Their Lives, Struggles, and Momentous Discoveries.* Secaucus, N.J.: Carol Publishing Group, 1993.

McMurray, Emily J., and Donna Olendorf, eds. *Notable Twentieth Century Scientists.* Detroit, Mich.: Gale Research, 1995.

Miles, Wyndham D. ed. *American Chemists and Chemical Engineers.* Washington, D.C.: American Chemical Society, 1976–1994.

Millar, David, Ian Millar, John Millar, and Margaret Millar, eds. *The Cambridge Dictionary of Scientists.* New York: Cambridge University Press. 1996.

Murray, Robert H. *Science and Scientists in the Nineteenth Century.* New York: Macmillan, 1925.

National Inventors Hall of Fame. Available on-line. URL: http://www.invent.org/inventure.html.

The Nobel Foundation. "Nobel Laureates." Available on-line. URL: http://www.nobel.com.

North, J. *Mid-Nineteenth-Century Scientists.* Oxford, N.Y.: Pergamon Press, 1969.

Olby, Robert C., ed. *Late Eighteenth Century European Scientists.* Oxford, N.Y.: Pergamon Press, 1966.

A Passion to Know: 20 Profiles in Science. New York: Scribner's, 1984.

Pelletier, Paul A. *Prominent Scientists: An Index to Collective Biographies.* New York: Neal-Schuman, 1994.

Porter, Roy, ed. *The Biographical Dictionary of Scientists*. New York: Oxford University Press, 1994.

Schlessinger, Bernard S., and June H. Schlessinger. *Who's Who of Nobel Prize Winners*. Phoenix, Ariz.: Oryx Press, 1991.

Science, Technology & Society. Available on-line. URL: http://www.personal.u-net.com/~nchadd/home.htm.

Scientific Biographies. Available on-line. URL: http://home.achilles.net/~jtalbot/bio/biographies.html.

Scott, Michael Maxwell. *Stories of Famous Scientists*. London: Barker, 1967.

Shearer, B. F., and B. S. Shearer, eds. *Notable Women in the Physical Sciences*. Westport, Conn.: Greenwood Press, 1997.

Siedel, Frank, and James M. Siedel. *Pioneers in Science*. Boston: Houghton, 1968.

St. Andrews, History of Science. Available on-line. URL: http://www-history.mcs.st-andrews.ac.uk/history/.

STS Links, North Carolina State University. Available on-line. URL: http://www.ncsu.edu/chass/mds/stslinks.html.

Uglow, Jennifer S. *International Dictionary of Women's Biography*. New York: Continuum, 1885.

University of California at Los Angeles. "Women in Science and Engineering." Available on-line. URL: http://www.physics.ucla.edu/~cwp/Phase2/.

Unterburger, Amy L., ed. *Who's Who in Technology*. Detroit: Mich.: Gale Research Inc., 1989.

Van Sertima, Ivan, ed. *Blacks in Science: Ancient and Modern*. New Brunswick, N.J.: Transaction Books, 1983.

Van Wagenen, Theodore F. *Beacon Lights of Science: A Survey of Human Achievement from the Earliest Recorded Times*. New York: Thomas Y. Crowell, 1924.

Weisberger, Robert A. *The Challenged Scientists: Disabilities and the Triumph of Excellence*. New York: Praeger, 1991.

Who's Who in Science and Engineering, 1994–1995. New Providence, N.J.: Marquis Who's Who, 1994.

Who's Who in Science in Europe. Essex, England: Longman, 1994.

Who's Who of British Scientists, 1980–81. New York: St. Martin's, 1981.

Williams, Trevor I., ed. *A Biographical Dictionary of Scientists*. Fourth edition. Glasgow: HarperCollins, 1994.

Youmans, William J., ed. *Pioneers of Science in America: Sketches of Their Lives and Scientific Work*. New York: Arno Press, 1978.

Yount, Lisa. *A to Z of Women in Science and Math*. New York: Facts On File, 1998.

Zuckerman, Harriet. *Scientific Elite: Nobel Laureates in the United States*. New York: Free Press, 1979.

INDEX

Note: Page numbers in **boldface** indicate main topics. Page numbers in *italic* refer to illustrations.

A

Abbe, Cleveland **1–3**, *2*
Abbe, Ernst 323, 324
Abel Wolman Award 272
Aberdeen, University of 277
Académie des Sciences (France) 47, 141, 142, 176, 180, 238, 251, 318
Academy of Moral and Political Sciences (France) 40
Academy of Science (U.K.) 80
Academy of Sciences (USSR) 8
accelerators 11
Acheson-Lilienthal Report 225
Ackerman, Thomas P. **3–4**
activism 49–50, 155–157, 239, 277
Adams, Mark 300
Adelaide, University of 49, 99
Advice to a Young Scientist (Medawar) 203
Aerial Flight (Lanchester) 170, 171
aeronautical engineers 317–319, 324–326
aeronautics
    aerial combat 171
    aircraft 175, 269–271, 317–319, 324–326
    astronauts 184–185
    automatic pilot system 174–175
    flight simulation 102, 285–286
    gyroscopes 278
    rocketry 116–117
The Age of the Earth (Holmes) 135
Agricola, Georgius *4*, **4–5**

Agricultural Academy of Hohenheim 260
Agricultural College (Nebraska) 101
agriculture 27–28, 123–124, 173–174, 296–297
AIDS treatment 92
Aiken, Howard Hathaway **5–7**
aircraft 175, 269–271, 317–319, 324–326
aircraft stability and control analyzer (ASCA) 102
Albert Lasker Awards 116, 199, 280
Albert Medal 47, 61, 174, 242
Aldo Leopold Leadership Program 68
Alfvén, Hannes *7*, **7–8**
Allen, Paul **9–10**
allopurinol 91
Altdorf, University of 179
alternating current (AC) 96, 97, 292–293
Alvarez, Luis Walter *10*, **10–11**
Ambrose, James 140
American Academy of Achievement 165
American Academy of Arts and Sciences 138
American Association for Cancer Research 163
American Association for the Advancement of Science 194, 211
American Association of Humanistic Psychology 50

American Association of University Women 57, 258
American Balsa Company 240
American Chemical Society 25, 137, 164, 240
American Civil War 31–32
American Diabetes Foundation 39
American Electrochemical Society 25
American Forestry Association 181
American Heart Association Achievement Award 170
American Home Economics Association 258
American Humanist Association 275
American Institute of Electrical Engineers 293, 328
American Institute of Refrigeration 240
American Meteorological Society 3
American Philosophical Society 104
American Physical Society 3–4, 190, 269
American Radium Society 256
American Red Cross 30–32
American Society of Anesthesiologists 14
American Society of Biological Chemists 116
American Society of Chemical Engineering 25
American Society of Mechanical Engineers 269, 278

American Society of Refrigerating
   Engineers 241
ammonia synthesis 123–124
Ampère, André Marie **11–13,** *12,*
   227
Amsterdam, University of 77
A. M. Turing Award 286
Analytical Society 23
Anderson, Carl 224
Andersonian Institution 245
And Keep Your Powder Dry (Mead)
   200, 201
anesthesia 13–14, 308–310
Angers, University of 231
animal breeding 27–28
Anschutz-Kaempfe, Hermann 278
Anthropological Association (U.S.)
   201
anthropology 200–201
antibacterial agents 79–80, 97–101
Apgar, Virginia *13,* **13–14**
"An Apparatus for Detecting and
   Recording Electrical Oscillations"
   (Popov) 252
Appert, Nicolas **14–15**
Appia, Louis 31, 32
Apple computers 9, 149–151,
   315–316
Applications d'analyse et de géométrie
   (Poncelet) 251
AquaLung 61
aqueducts 291, 292
Archimedes **15–16,** 233
Argentina, identification of children
   162
Ariel (bicycle) 281, 282
Arkwright, Sir Richard **16–17,** 42
Armstrong, Edwin **17–19**
ARPA 285
arsenic 47, 98
*Ars Nova ad Aquam Ignis Adminiculo
   Efficacissime Elevandam* (Papin)
   232
artificial gas 177–178
artificial organs 164–165
Artificial Organs (Kolff) 165
artillery shells 37
*The Art of Preserving All Kinds of
   Animal and Vegetable Substances
   for Several Years* (Appert) 15
"The Art of the Soluble" (Medawar)
   203
Ashland, James N. 87

Association for Computing
   Machinery 286
astronauts 184–185
astronomers 113–114, 141–142,
   191–192, 233, 263–264
astrophysics 7–8
Atanasoff, John 194–195
atmospheric chemistry 261–262
Atmospheric Radiation (Abbe) 2
atomic bomb *See* Manhattan Project
Atomic Energy Commission (U.S.)
   205, 225, 256, 303
Atomic Energy Research
   Establishment (U.K.) 109
atomic theory of matter 70–71
atomic weights 47, 64, 65
"Attraction and Repulsion resulting
   from Radiation" (Crookes) 65
Audubon Society 181
Australian National University 101
Austrian Order of Francis Joseph 314
Automatic Sequence Controlled
   Calculator (ASCC) 6
automobiles 33–34, 68–70, 170, 171
Avery, Oswald Theodore **19–21,** 63
aviation *See* aeronautics

B
Babbage, Charles 6, **23–24,** 148
bacteriology
      Oswald Avery **19–21,** 63
      Gerhard Domagk **79–80**
      Sir Alexander Fleming 58, 59,
         97–99, 100
      Rebecca Craighill Lancefield
         169–170, *170*
Baekeland, Leo Hendrik *24,*
   24–25, 289, 290
Baird, John Logie **25–27,** 93, 94,
   219
Bakelite 24–25
Bakewell, Robert **27–28**
Balinese Character (Mead) 200
Ballantine Medal 190
Banting, Frederick G. **28–30,** 37,
   38, 39
Bardeen, John 268, 269
Barnard College 200
Barnard Medal 260
baronetage 290
baronies 74, 101, 183
Barsky Award 272
Barton, Clara **30–32,** *31*

Basel Institute for Immunology 193
BASIC 9, 316
Battelle Memorial Institute 53
batteries 71, 72, 178–179, 286, 287,
   301
Bauer, John 91
Baumgartner Prize 131, 260
BBC (British Broadcasting
   Corporation) 26, 27
Beckman Instruments 269
Becquerel, Antoine-Henri *32,*
   32–33, 260
behaviorism 273–275
The Behavior of Organisms (Skinner)
   275
Beit Memorial Fellowship 266
Bell, Alexander Graham 35, 36
Bell Helicopter Company 285
Benz, Karl Friedrich **33–34,** 68, 69,
   70, 230
Berg, Paul 114, 115
Berlin, University of 108, 123, 130,
   204, 244, 302
Berlin Academy 180
Berlin Academy of Sciences 301
Berliner, Emile **34–36**
Bern, University of 88
Berzelius Medal 59
Bessemer, Henry **36–37,** 158, 159,
   166
Best, Charles Herbert 28, 29–30,
   37–39
beta decay theory 95, 96
Beth Israel Hospital 183
Beyond Freedom and Dignity
   (Skinner) 275
Bibliothèque Royale 141
bicycles 82, 83, 209–210, 281–283,
   317
Billingham, Rupert Everett
   202–203
Billings, John Shaw 133
BINAC 194, 195
Binet, Alfred **39–40,** 242
Binney, Edward William 322
biochemists 58–59, 136–137,
   266–267, 299–300
Biogen N.V. 114
biologists 67–68, 202–203
biophysics 255–256
Biotechnology Industry
   Organization 300
Birks, John W. 3, 4

Birmingham, University of 108, 109, 202
birth defects 14
Bissell, George 81
blackbody radiation 244
Blackett, Patrick 108
Blenkinsop, John 283
Block, Felix 11
Bohr, Niels 108, 109
bone china 280, 281
boneshakers 209–210, 282
Bonn, University of 131
Boone, Gary 143, 144
Bosch, Carl 124
Boston Society of Natural History 210
botany 59–61
Boulton, Matthew **41–42,** 177–178, 308, 315
Boyle, Robert 231
Boyneburg, Johann Christian von 179, 180
brain and central nervous system 119–120, 239, 278, 279–280
Brainpower: Working out the Human Mind (Greenfield) 120
Brain Story (TV series) 120
braking system 310
Branly, Edouard 252
Brattain, Walter 268, 269
breast cancer 162, 163, 183–184
Breda, College of 141
Brent, Leslie 202–203
Bressa Prize 131
Bridgeport, University of 271
bridges 43, 44, 73, 291–292, 314, 315
Bridges, Calvin B. 211
A Brief Account of the Most Famous Inventions (Polhem) 248
Briggs, Henry 214
Brigham Young University 93
Brindley, James **42–43**
Bristol Iron Company 73
British Thompson-Houston Company (BTH) 111–112
Broca's Brain (Sagan) 264
Brookhaven National Laboratory 261
Brudersaus Machinenfabrik 69
Brunel, Isambard Kingdom **43–44**
Brunel, Marc Isambard 43, **44–46,** 60

Bryn Mawr College 210
Buckley Solid State Physics Prize 190, 269
Budapest, University of 302
Budapest Technical University 111
A Bugs Life (film) 150
Bunsen, Robert Wilhelm Eberhard von 46, **46–47,** 179, 245
Bürge, Joost 214
Burnet, Frank Macfarlane 203
Burroughs Wellcome 91–92
Busch, Adophus 78–79
Business Week 190
Butler, R. Paul 191–192

C

calculating machines 6, 23–24, 179, 236
calculus 179, 180, 216
Caldicott, Helen 49, **49–50**
California, University of, at Berkeley 10, 11, 67, 77, 136, 161, 192, 224, 263, 316
California, University of, at Davis 193
California, University of, at Irvine 193, 261, 262
California, University of, at Los Angeles 184, 191
California, University of, at San Diego 8, 193, 299
California, University of, at Santa Cruz 191
California Institute of Technology 52, 120–121, 211, 224, 268, 269, 279, 285, 286
Cambridge University 23, 58, 63, 74, 100, 105, 109, 114, 121, 124, 155, 160, 193, 196, 197, 216, 223, 234, 266, 267
cameras *See* photography
Campbell, Robert V. D. 6
Campbell-Swinton, Alan Archibald **51–52,** 219, 326
canals 42–43, 280, 281, 291–292
cancer 91–92, 162, 163, 183–184, 255–256
cannons 37, 166–167, 314
C. A. Parsons Company 234
Capital City Commercial College 55
Carleton College 161
Carlson, Chester **52–54**

Carnegie, Andrew 37
Carnegie Institution 191, 198
Carnegie-Mellon University 285
Carothers, Wallace Hume 54, **54–56**
Carson, Rachel Louise **56–58,** 57
Carson National Forest 181
Carter, Jimmy 138
Carter Center (Emory University) 138
cartography 206–207, 227–228
Catalogus Auctorum (Ortelius) 228
Catch Me Who Can (locomotive) 296
cathode ray tubes 51, 94
Catoptrica (Hero) 129
CAT scanning 140–141
CDC *See* Centers for Disease Control and Prevention
Celera Genomics 300
celluloid 144–145
census taking 133–134
Centers for Disease Control and Prevention 137–138
Central College for Women 136
central nervous system *See* brain and central nervous system
cephalosporin C 101
CFCs *See* chlorofluorocarbons
Chain, Ernst Boris **58–59,** 98, 99, 100, 132
Charcot, Jean-Martin 39, 107
Chargaff, Erwin 63
Charité Hospital (Berlin) 58
Charles I, king of Spain (later Holy Roman Emperor Charles V) 187, 188
Charlton works 97
Chatham College (Pennsylvania College for Women) 56
chemical engineering 321–322
Chemical Heritage Foundation 300
Chemical News 65
Chemical Society (France) 242
Chemical Society (Germany) 242
Chemical Society (U.K.) 47, 164, 267
The Chemistry of Cooking and Cleaning (Richards) 258
chemists
    Leo Baekeland 24, **24–25,** 289, 290

Robert Bunsen 46, **46–47,** 179, 245
Wallace Carothers 54, **54–56**
William Crookes 64, **64–65**
John Dalton 70, **70–71**
John Frederic Daniell **71–72**
Humphrey Davy **74–75,** 283, 301, 309
Gertrude Elion **90–92,** *91*
Rosalind Franklin 63, **104–106**
Walter Gilbert **114–116,** *115*
Fritz Haber **123–124**
Dorothy Hodgkin *131,* **131–132**
Frederic Stanley Kipping **163–164**
Antoine-Laurent Lavoisier **171–173,** *172*
Nicolas Leblanc **175–176**
Philippe Lebon **177–178**
Joseph Niepce **217–218**
Alfred Nobel 79, **219–221**
Louis Pasteur 15, 183, **237–238**
Mary Engle Pennington **239–241**
William Henry Perkin 163, **241–242**
Lyon Playfair **245–246,** 321
Ellen Swallow Richards **257–258,** *258*
F. Sherwood Rowland **261–262**
Frederick Soddy **275–277,** *276*
Joseph Wilson Swan **286–288,** 289
Chester College of Science 65
Chibnall, Charles 266
Chicago, University of 6, 11, 96, 136, 137, 261, 263, 279
Chicago World Columbian Exposition (1893) 293
children 13–14, 200–201, 242–243
Children's Hospital Medical Center (Boston) 50
China 249–250
china 280–281
chlorofluorocarbons (CFCs) 261, 262
choline 37, 38
Chomsky, Noam 275
chromosome theory of heredity 210–211
Church, Arthur H. 241
Cincinnati Weather Bulletin 1
cinematography 145, 185–186
ciphers 245–246, 311, 312
circumnavigation of the world 187–188
City and Guilds College (U.K.) *See* Imperial College of Science and Technology
City College of New York 265
civil engineering 291–292
Clark, Edward 273
Clarke, Chapman and Company 234
Clark University 117
Cleveland Abbe Award 3
Cleveland Clinic Foundation 165
Clifton College 289
Clifton Suspension Bridge 43
Clinton Liberal Institute 31
clocks (horology) 127–128, 141, 142
cloning 120–121
Clowes Award 163
Coal Utilization Research Association 105
COBOL 6, 138, 139
coherer 252
coin stamping machines 41–42
coke-firing process 72–73
Cole, Rufus 20
Colgate University 20
Collection (Pappus) 232–233
Collège de France 12, 39–40, 251, 326
College of the City of New York 1
Collège Saint-Louis 237
Collier Trophy 175, 271
Collins, Francis 300
Collip, James Bertram 28, 30, 38
Colorado, University of 136, 137, 189
Colton, Garner 309
Columbia School of Mines 132–133
Columbia University 13, 14, 18, 19, 20, 96, 116, 169, 170, 200, 211, 255, 256
combustion 172–173
Coming of Age in Samoa (Mead) 200, 201
Comité National de l'Union des Femmes Françaises 152
communications
  gramophone 35–36
  radio broadcasting 17–19, 174–175
  telegraph industry 59–61, 71, 72, 160–161, 179, 190–191, 251–253, 312
  telephone 35
  television 25–27, 51–52, 93–95, 218–219, 326–328
Compagnie du Chemin de l'Est 178
Companion of the British Empire 267
compiler 138, 139
Compton, Arthur 225
computers
  Apple 9, 149–151, 315–316
  EMI-DEC 1100 140
  first compiler 138, 139
  first digital electronic 194–196
  integrated circuit microprocessors 142–144
  interactive graphics 284–285
  languages 138, 139
  Mark 5–7, 138–139
  mechanical 23–24, 179, 236
  memory 101, 102, 302, 303
  Microsoft 9–10
  system dynamics 101, 103
  transistors 268–269
  virtual reality 285–286
  Whirlwind 101, 102
Computer Society 7
concertina 312
condensers 306–308
conditioned reflex theory 239
Congressional Medals *See* U.S. Congress
Conklin Medal 121
Conservation Foundation 181
Constructio (Napier) 214
continental drift 134, 135
Continental Edison Company 293
A Continuation of New Experiments (Boyle, Papin) 231
Cooke, William Fothergill **59–61,** 312
Cooper Union (Cooper Institute) 35
Copenhagen, University of 226
Copernicus 113, 114

Copley Medal 21, 47, 65, 72, 75, 104, 128, 161, 211, 301
copying process 52–54
Corday-Morgan Medal and Prize 267
Corey Medal 239
Cormack, Allan M. 141
Cornell University 198, 263, 277
Cosmical Electrodynamics (Alfvén) 8
The Cosmic Connection (Sagan) 264
Cousteau, Jacques-Yves **61–62**
Cousteau Amazon (TV show) 62
Cousteau Odyssey (TV show) 62
Craig, John 307
cream separator 75.76
Creighton, Harriet 198
Crick, Francis 19, **62–64**, 105–106, 114, 115, 199
Crimean War 37
Cristallotechnie (Leblanc) 176
A Critique of the Theory of Evolution (Morgan) 211
Crompton Parkinson, Ltd. 52
Crookes, William 64, **64–65**
Crookes tubes 51, 64, 65
Crutzen, Paul 3, 4, 261, 262
cryogenics 155–157
crystallization 176
crystallography *See* X-ray crystallography
Crystal Palace Exposition (1854) 229
Cumberland's Quaker School 70
Curie, Marie 33
Curie, Pierre 33
Cyrus B. Comstock Award 269
cystic fibrosis 49–50

D

Daedalus Hyperboreus 288
Daguerre, Louis 217, 218
Daily, Gretchen **67–68**
Daimler, Gottlieb Wilhelm 33, 34, **68–70**, 230
Daimler-Motoren Gesellschaft (DMG) 69
Dalton, John 70, **70–71**
Dana Farber Cancer Institute 184
Daniel Guggenheim Medal 271, 319
Daniell, John Frederic **71–72**
Danish Society 157
Danish Society for the Promotion of Natural Science 227

Darby, Abraham **72–73**
DARPA 285
Darwin, Charles Robert **73–74**
Darwin Medal 211
data processing *See* information processing
Davy, Humphry **74–75,** 283, 301, 309
Davy Medal 47, 65, 164, 242
DDT 57–58
De Anza College 316
De Casibus Perplexis (Leibniz) 179
De Cupro (Swedenborg) 288
deed stamps 36
De Ferro (Swedenborg) 288
DeForest, Lee 18, 19
De Laval, Carl Gustav Patrik **75–76**
Deming, W. Edwards 171
De novis quibusdam machinis (Papin) 232
dentistry 308–310
De peste libri III (Agricola) 5
De prima ac simplici institutione grammatica (Agricola) 5
De Principio Individui (Leibniz) 179
De re metallica libris XII (Agricola) 4, 5
"Der leer der specifieke energieen" (Einthoven) 89–90
Desargues, Girard 236
The Descent of Man, and Selection in Relation to Sex (Darwin) 74
Deutz, Gasmotoren-Fabrik 69
De Vries, Hugo **59–61,** 211
dew-point hygrometer 71
DGG (Deutsche Grammophon Gesellschaft) 35
diabetes 28–30, 37–39
Dialogue on the Two Chief World Systems, Ptolemaic and Copernican (Galileo) 114
Dialogues Concerning Two New Sciences (Galileo) 114
Die Konstruktion des Sitenglavanometers (Einthoven) 90
Die Physik im Kampf um die Weltanschauung (Planck) 244
Diesel, Rudolf **77–79**, 230
Difference Engines 23–24
digestive system 238, 239
dinosaur extinction 11

Dioptra (Hero) 129
Dirac, Paul Adrien Maurice 95
direct current (DC) 96, 97, 292–293
dirigibles 324–326
displacement law 276
Divisament dou Monde (Polo) 249–250
DNA (deoxyribonucleic acid) 19, 20–21, 62–64, 104–106, 114–116, 266–267
Dochez, Alphonse Raymond 20
docks, harbor 44
"Does the Inertia of a Body Depend on Its Energy Content?" (Einstein) 88
Domagk, Gerhard **79–80**
Doty, Robert W. 279
"The Double Bond" (Carothers) 55
Dow Corning Corporation 164
Doyle, Bernard W. 145
The Dragons of Eden (Sagan) 264
Drake, Edwin Laurentine **80–82**
Draper, John 316
Dr. Susan Love's Breast Book (Love) 184
Du Cros, W. H. 82, 83
Duisburg, University of 207
Duncan, Thomas 294
Dunlop, John Boyd **82–83**
DuPont Company 54, 55, 56
Durham, University of 60
dyes 241–242
Dynamical Theory of the Electrical Field (Maxwell) 197
dynamite 219–221
DYNAMO programs 103
dynamos 97, 234

E

earthenware 280–281
Eastbourne College 276
East India Company 41
Eastman, George 25, **85–86**
Eckener, Hugo 325
Eckert, J. Presper 194, 195
École Centrale des Arts et Manufactures 178
École d'Application de l'Artillerie et du Génie 250, 251
École de Chirurgie 176
École de la rue de la Grange-aux-Belles 242

École des Ponts et Chaussées 177
École Navale 61
École Normale Supérieure 237, 238
École Polytechnique 32, 33, 250–251
École Polytechnique Fédérale de Lausanne 208
Ecological Society of America 181
ecology 56–58, 180–181, 258
Eddyville Iron Works 158
The Edge of the Sea (Carson) 57
Edinburgh, University of 60, 74, 82, 135, 136, 165, 196, 245, 246
Edinburgh Royal Infirmary 182
Edison, Thomas Alva 35, 36, **86–87,** 87, 186, 286, 287, 292, 293, 294, 310, 311
Edison Medal 328
EDVAC 303
Egerton, Francis 42
Eggs (Pennington) 240
The Ego and the Id (Freud) 107
Ego: The Neuroscience of Emotion (Greenfield) 120
Ego: The Neuroscience of the Self (Greenfield) 120
Ehrlich, Paul and Anne 67–68
Eight Minutes to Midnight (film) 50
"Ein Beitrag zur Chemotherapie der bakteriellen Infektionen" (Domagk) 80
Einstein, Albert **87–89**
Einthoven, Willem 89, **89–90**
electrical engineers
 Sebastian de Ferranti **96–97**
 Gilbert Hyatt **142–144**
 Paul Gottlieb Nipkow 26, **218–219**
 Joseph Wilson Swan **286–288**
 James Swinburne **289–290**
 Nikola Tesla **292–293,** 310–311
 Elihu Thomson **293–295**
 Vladimir Zworykin 94, **326–328,** 327
Electric and Musical Industries, Ltd. (EMI) 27, 35, 36, 52, 140, 141
electricity
 batteries 71, 72, 178–179, 286, 287, 301
 generation and distribution 96–97, 292–295
 hedgehog transformer 289

light bulbs 86–87, 286–288
one-fluid theory 103–104
resistance 311–312
electric locomotive 277
electric motors 292–293
electrocardiogram (ECG) 89, 90
Electrochemical Institute (St. Petersburg) 252
electrochemistry 75
electrodynamics 11–13, 130–131
electromagnetism 196, 197, 225–227
Electronic Frontier Foundation 316
electronics engineers 284–286
electron microscopes 111, 112, 328
electrophorus 301
electrostatic copying 52–54
Elementary Treatise on Chemistry (Lavoisier) 173
elements, discovery of 47, 64, 65, 74, 75, 173, 204
Elements of Agricultural Chemistry (Davy) 75
Elements of Chemical Philosophy (Davy) 75
elevators 228–229
Eli Lilly 30
Elion, Gertrude Belle ("Trudy") **90–92,** 91
Elliott Cresson Medal 260, 306, 318
embossing machine 36
EMI *See* Electric and Musical Industries, Ltd.
EMI-DEC 1100 computer 140
Emory University 138
employee benefits 166–167, 324
energy and power *See also* electricity
 gas lighting and heating 177–178
 petroleum 80–82
 steam engines 41–42, 214–215, 231–232, 233–235, 314–315
 steam pump 267–268
 waterpower 247, 248
Engineering and Science Hall of Fame 92, 139
engineers *See also* aeronautical engineers; electrical engineers; mechanical engineers
 Edwin Armstrong **17–19**
 John Logie Baird **25–27,** 93, 94, 219
 Karl Benz **33–34,** 68, 69, 70, 230

Sir Henry Bessemer **36–37,** 158, 159, 166
James Brindley **42–43**
Isambard Kingdom Brunel **43–44**
Sir Marc Isambard Brunel 43, **44–46,** 60
Alan Archibald Campbell-Swinton **51–52,** 219, 326
Gottlieb Daimler 33, 34, **68–70,** 230
Abraham Darby **72–73**
Herman Hollerith **132–134**
Philippe Lebon **177–178,** 217–218
Georges Leclanché **178–179**
Guglielmo Marconi 18, 51, **190–191,** 251, 252
George de Mestral **207–209**
Elisha Otis **228–229**
Nikolaus August Otto 69, **229–230**
Jean-Victor Poncelet **250–251**
Thomas Savery 214, 215, 232, **267–268**
Elmer Ambrose Sperry **277–278**
George Stephenson **283–284,** 296
Ivan Sutherland **284–286**
Thomas Telford **291–292**
Richard Trevithick **295–296**
Robert Alexander Watson-Watt **305–306**
James Watt **206–308**
George Westinghouse 293, **310–311**
Robert Whitehead **312–314**
James Young **321–322**
ENIAC 194, 195, 303
Enrico Fermi Award 205, 225, 303
Enrico Fermi Prize 96
Entropy (Swinburne) 290
Environmental Defense Fund 272
environmental movement 56–58, 61–62, 258
environmental toxicology 271–272
An Essay on Teeth (Wells) 308
eugenics 269
euthenics 257–258
Euthenics: Science of Controllable Environment (Richards) 258
Everson, George 93–94

"Evidence for Multiple Companions to Upsilon Andromedae" (Marcy, et al.) 192
evolution 73–74, 211
Evolution and Adaptation (Morgan) 211
"Existence of Electromagnetic-Hydrodynamic Waves" (Alfvén) 8
exobiology 263–264
Experimental Project Aiming to Describe What Happens When the Cone Comes in Contact with a Plane (Pascal) 236
Experiments and Observations on Electricity, Made at Philadelphia in America (Franklin) 104
explorers 187–188, 249–250
Extraordinary Facts Relating to the Vision of Colours (Dalton) 70
extraterrestrial life 263–264

F

Faculty of Sciences (University of Paris) 33, 151–152, 251
Failla, Gioacchino 255, 256
Faleiro, Rui 187, 188
Fanny and John Hertz Foundation Award for Applied Physical Science 190
Faraday House Electrical Engineering College 140
Faraday Medal 97, 120, 328
Faraday Society 287, 290
Farman, Joe 262
farming *See* agriculture
Farnsworth, Philo 27, **93–95**, 219
Faulkner Breast Center 184
Federal Aviation Administration (U.S.) 175
Federal Institute of Technology (Switzerland) 88, 123, 259, 302
Fermat, Pierre de 235, 237
Fermi, Enrico 8, 95, **95–96**, 225
Ferranti, Sebastian Ziani de **96–97**
fertilizers 123–124, 173–174
Fettes College 51
Fish and Wildlife Coordination Act (1934) 181
Fleming, Alexander 58, 59, **97–99**, 100
Fleming, John Ambrose 289
flight simulation 102, 285–286

Florence, University of 96
Florey, Howard Walter 59, 98, 99, **99–101**
Flow-Matic computer language 138
fluorescence 33
flu vaccine 264, 265
flying shuttle (weaving) 157–158
FM radio 19
Food Materials and Their Adulterations (Richards) 258
food technology and science
  cream separator and milking machine 75–76
  dye toxicity 271, 272
  pasteurization 237, 238
  preservation 14–15
  refrigeration and inspection 239–241
Fordham University 183
Forrester, Jay **101–103**
Forward Gas Engine Company 171
"The Foundation of the General Theory of Relativity" (Einstein) 88
"Fourier's expansions of functions in trigonometrical series" (Thomson) 160
Frances Lou Kallman Award 67
Francie, Thomas, Jr. 264, 265
Frankfurt, University of 80
Franklin, Benjamin 103, **103–104**
Franklin, Rosalind Elsie 63, **104–106**
Franklin Institute 8, 19, 25, 103, 190, 260, 306, 318
Franklin Medal 19, 25
Frank M. Hawks Memorial Award 175
Fraser, Claire 300
Frederic Stanley Kipping Award 164
Free Academy *See* College of the City of New York
Freeman, Derek 201
Freud, Sigmund **106–107**
Friedrich-Wilhelm University 58
Frisch, Otto Robert **107–109**, *108*, 203, *204*, 204–205
frozen foods 240
Fuchs, Ephraim 193

G

Gabor, Dennis **111–113**, *112*
Gagnan, Emile 61

Galápagos Islands 74
Galilei, Galileo **113–114**
Galvani, Luigi 301
galvanometers 89, 90
Galvin, Paul 175
Game Management (Leopold) 181
Garvan Medal 137, 240
gas engine 229–230
gases 70–71, 196, 197
gas lighting and heating 177–178
Gasmotoren-Fabrik Deutz 69
Gassner, Carl 179
Gates, Bill 9–10
General Electric Company 293, 294
genetics 59–61, 161–163, 197–200, 205–206, 210–211
  DNA 19, 20–21, 62–64, 104–106, 114–116, 266–267
  human genome mapping 116, 266, 267, 299–300
Geneva, Treaty of 31, 32
Geneva University 243
geography 206–207, 227–228
Geological Society (U.K.) 245
Geological Society of America 135
Geological Society of London 135
geology 134–136
geometry 232–233, 236, 250–251
George III, king of Great Britain 128
George Washington University 103
Gergonne, Joseph 251
German Television Society 219
German University (Prague) 88
germ theory of disease 237, 238
Ghent, University of 24–25
Giessen, University of 260
Giffard, Henri 325
Gilbert, Joseph Henry 173–174
Gilbert, Walter **114–116**, *115*
Glaser, Donald 11
Glasgow, University of 26, 160, 183, 276, 307
Glasgow Royal Infirmary 202
Global 2000 Guinea Worm Eradication Program 138
Goddard, Robert Hutchings 116, **116–117**
Gonet, Alfred 208
Goodwin, Hannibal 86
Goodyear, Charles 83, **117–119**, 125

Göttingen, University of 46, 47, 95, 223, 302
Gower, Earl 42, 43
Graf Zeppelin (dirigible) 325
gramophones 35–36
Grand Croix dans l'Ordre National du Mérite (France) 62
Grand Cross of the Civil Order of Health (Spain) 80
Grand Trunk Canal 43
Graphical User Interface (GUI) 285
gravitation, theory of 216–217
Gray's Inn 296
Great Britain, SS 43, 44
Great Eastern, PSS 43, 44
Great Exhibition (1851) 245
Great Western, PS 43, 44
Great Western Railway 43, 44, 60
Greenfield, Susan **119–120**
Greifswald, University of 79
Groningen, University of 164
Grosvernor Gallery Company 97
Growing Up in New Guinea (Mead) 200
Guettard, Jean-Étienne 172
guidance systems 278
guinea worm eradication 138
Gulf Stream, study of 104
Gum Elastic and Its Varieties (Goodyear) 119
Gurdon, John Bertrand **120–121**
gyroscopes 277–278

H

Haber, Fritz **123–124**
Hahn, Otto 108, 204, 205
Haldane, J. B. S. 121
Halle, University of 77
Hall of Fame of Great Americans 87, 311
Haloid Company 53–54
Hamburg, University of 108, 302
Hancock, Thomas 119, **124–126**
Hargreaves, James **126–127**
Harrison, John **127–128**
Harrison, William 128
Harry M. Goode Award 7, 196
Harvard College 223
Harvard Mark computers 5–7, 138–139
Harvard Medical School 49, 50
Harvard School of Public Health 137

Harvard University 6–7, 55, 114, 115, 138–139, 263, 275, 279, 285
Hayward, Nathaniel 118
heart, artificial 165
heart disease 89–90
heart-lung machine 164, 165
Heatley, Norman 59, 100
hedgehog transformer 289
Heidelberg, University of 47, 77, 123
helicopters 269–271
heliographs 217–218
helium 155–156
Helmholtz, Hermann von 130
heparin 37, 38
Herbert A. Sober Memorial Award 116
heredity *See* genetics
Hero of Alexandria **129–130,** 234
herpesviruses 91–92
Hertz, Heinrich Rudolf 130, **130–131**
Hess, Harry 135
Hewlett-Packard 149, 316
Hieron II, king of Sicily 15, 16
Higgins, Elmer 56
Hillier, James 328
Hillman, William 282
Hindenburg (dirigible) 326
histaminase 37, 38
Historia Naturalis (Pliny the Elder) 246, 247
Historical Museum of Gotland 248
A History of the Discovery and Application of Nitrous Oxide Gas, Ether and Other Vapours to Surgical Operations (Wells) 309
Hitchings, George 91–92
Hoagland Laboratory 20
Hodgkin, Dorothy Crowfoot 131, **131–132**
Hofmann Medal 242
Hollerith, Hermann **132–134**
Holley Medal 269
Holmes, Arthur **134–136**
holography 111–113
Home Sanitation (Richards) 258
home economics 257–258
Honorary Chief Doy-gei-tau 99
Hoobler, Icie Gertrude Macy **136–137**
Hooke, Robert 231
Hopkins, Donald

Hopper, Grace Murray 6, **138–139,** *139*
Horologium Oscillatorium (Huygens) 142
horology 127–128, 141, 142
Hospital for Sick Children (Toronto) 29
Hounsfield, Godfrey Newbold **140–141**
House of Appert 15
House of Nobles (Sweden) 288
Houston, Edwin J. 294
Howard N. Potts Award 103, 195
Howe, Elias 273
Huckins, Olga Owens 57
Hughes Aircraft Company 189
Hughes Medal 287, 306
human genome mapping 116, 266, 267, 299–300
Humanist of the Year Award 50
The Human Mind Explained (Greenfield) 120
Hunter College 91
Huygens, Christiaan **141–142,** *142,* 231, 232
Hyatt, Gilbert **142–144**
Hyatt, John Wesley 86, **144–145**
hydrostatics 15–16
hygrometer, dew-point 71

I

IBM (International Business Machines) 6, 9, 10, 132, 134, 150, 285
iconoscopes 94, 326
identification through genetics 161–163
If You Love This Planet: A Plan to Heal the Earth (Caldicott) 50
I. G. Farbenindustrie 79–80
Igor I. Sikorsky International Trophy 271
Illinois, University of 55
image dissectors 94
"The Immunological Relationships of *Streptococcus viridans* and Certain of Its Chemical Fractions" (Lancefield) 170
immunology 192–194
Imperial College (London) 59, 134
Imperial College of Science and Technology (City and Guilds College) 140, 163

Imperial Medical Academy (St. Petersburg) 239
incandescent light bulbs 86, 87, 286–288
Indiana University at Bloomington 275
industrial chemistry 219–221
Industrial Revolution
    canals 42–43
    flying shuttle (weaving) 157–158
    Jacquard loom 147–148
    spinning machines 16–17, 126–127
    steam engines 41–42, 214–215, 267, 308
    steelmaking 158–159
information processing 132–134
Ingenhousz, Jan 148
An Inquiry into the Causes and Effects of the Variolae Vaccinae . . . (Jenner) 149
Institute of Chemistry (U.K.) 163, 174
The Institute of Genomic Research (TIGR) 300
Institute of Neuroscience (California) 119
Institute of Radio Engineers 19, 269, 328
Institute of Radium 151–152
Institute of Theoretical Physics (Copenhagen) 108
Institution of Electrical Engineers (U.K.) 52, 97, 287, 290
Institut J. J. Rousseau 243
insulin 28–30, 37–39, 131, 132, 266
integrated circuit microprocessors 142–144
Intel 143
intelligence tests 39–40
interactive graphics 284–285
internal combustion engine 33–34, 229–230
"The Internal Secretion of the Pancreas" (Banting, Best) 38
International Aerospace Hall of Fame 175
International Bureau of Education (Geneva) 243
International Business Machines *See* IBM
International Diabetes Foundation 39

International Psychoanalytic Association 107
International Society for Differentiation 121
International Telephone and Telegraph Company (ITT) 94
International Union of Physiological Sciences 39
International Year of Peace Award (Australia) 50
The Interpretation of Dreams (Freud) 107
Intracellular Pangenesis (De Vries) 77
Introduction to Nuclear Physics (Fermi) 96
Introduction to the Study of Chemical Philosophy (Daniell) 72
inventors
    Emile Berliner 34–36
    Matthew Boulton 41–42, 177–178, 308, 315
    Chester Carlson 52–54
    Sir William Fothergill Cooke 59–61, 312
    Gustav De Laval 75–76
    George Eastman 25, 85–86
    Thomas Edison 35, 36, 86–87, 87, 186, 286, 287, 292, 293, 294, 310, 311
    Philo Farnsworth 27, 93–95, 219
    Charles Goodyear 83, 117–119, 125
    Thomas Hancock 119, 124–126
    James Hargreaves 126–127
    John Wesley Hyatt 86, 144–145
    Joseph-Marie Jacquard 147–148
    John Kay 16, 157–158
    William Kelly 158–159
    Alfred Krupp 166–167
    Frederick Lanchester 170–171
    William Lear 174–175
    Pierre Michaux 209–210
    Thomas Newcomen 41, 73, 214–215, 231, 232, 267, 268, 307
    Denis Papin 231–232
    Isaac Merrit Singer 272–273
    James Starley 281–283
    Elihu Thomson 293–295

Inventors Hall of Fame (U.S.) *See* National Inventors Hall of Fame
ionosphere 305
iron *See* steel and iron
Iron Bridge 315
isotopes 275, 276–277
Istituto Superiore di Sanità 59
Italian Scientific Study 131

J
Jacquard, Joseph Marie **147–148**
Janeway Medal 256
Jarvik, Robert 165
Jean Brachet Memorial Prize 121
Jellinek, Emil 69
Jena, University of 179
Jenner, Edward **148–149**
Joanneum Technical University 292
Jobs, Steven **149–151,** 315–316
John Burroughs Medal 57
John D. and Catherine T. MacArthur Foundation 138, 199, 272
John Potts Award 195
Johns Hopkins University 14, 56, 194, 210, 272
Joliot-Curie, Irène 151, **151–152**
Joliot-Curie, Jean Frédéric 151, 151–152
Joule, James Prescott 152, **152–153,** 245
Journal (Darwin) 74
Journey to the Centres of the Brain (Greenfield) 120
Journey to the Centres of the Mind: Towards a Science of Consciousness (Greenfield) 120
Judgment and Reasoning in the Child (Piaget) 243
Judson, Horace 63

K
Kaiser Wilhelm Institute 58, 88, 123, 124, 204
kaleidophone 312
Kansas, University of 261
Kapitsa, Pyotr Leonidovich **155–157,** 156
Kapor, Mitch 316
Karlsruhe Lyzeum 33
Karlsruhe Technical College 130
Kassel, University of 47

Kay, John 16, **157–158**
K-electron capture 11
Kelly, William **158–159**
Kelvin, Baron (William Kelvin Thomson) 97, 152, 153, **159–161,** 160
Kennedy, John F. 246
Kentucky, University of 210
Khan, Kublai 249
kidney, artificial 164–165
Kiel, University of 79, 244
Kiev Polytechnic Institute 269
Kimber Genetics Award 199
kinetoscope 186, 326
King, Mary-Claire **161–163**
King's College (London) 71–72, 183, 196, 312
Kipping, Frederic Stanley **163–164**
Kirchoff, Gustav 47
knighthoods 17, 30, 37, 46, 59, 61, 65, 74, 80, 99, 121, 161, 183, 203, 217, 235, 242, 246, 248, 306, 312
Kodak cameras 85–86
Kolff, Willem J. **164–165**
Korad Corporation 189
Kornei, Otto 53, 54
Kreel, Louis 140
Krupp, Alfred **166–167**
Kundt, August 259

L

Laboratorium Mecanicum 248
La Caze Prize 131
lac repressors 114, 115
Ladies' Home Journal 14
Lady Godiva (sewing machine) 282
Lallement, Pierre 209
Lancefield, Rebecca Craighill **169–170,** 170
Lanchester, Frederick William **170–171**
Langen, Eugen 230
Langley Medal 117, 318
The Language and Thought of the Child (Piaget) 243
L'Année Psychologique 40
The Laser Odyssey (Maiman) 190
lasers 112, 188–190
Lasker Awards 116, 199, 280
"La Sortie des Ouvriers de l'Usine Lumière" (film) 186
Laue, Max von 105
Lavoisier, Antoine-Laurent **171–173,** 172

Lavoisier Medal 242
Lawes, John Bennet **173–174**
Lawrence, Ernest Orlando 225
Lay Down Your Arms (Suttner) 221
Lear, William **174–175**
Learjet Inc. 175
learning 242–243
Leblanc, Nicolas **175–176**
Lebon, Phillipe **177–178**
Leclanché, Georges **178–179**
Lefèvre, Alexander 209
Legion of Honor (France) 18, 59, 119, 239, 287, 311, 314, 319
Legion of Merit 139
Leibniz, Gottfried Wilhelm 141, **179–180,** 236
Leiden, University of 77, 89, 96, 141, 164, 165
Leipzig University 5, 179
Lemelson/MIT Lifetime Achievement Award 92
Lenoir, Etienne 230
Leopold, Aldo **180–181**
Leo Szilard Lectureship Award 3–4
Les enfants anormaux (Binet) 40
Les idées sur les enfants (Binet) 40
L'étude expérimentale de l'intelligence (Binet) 40
Liebig, Justus von 183, 245
Life 165
light 88, 130–131, 141, 142, 216, 217
light bulbs 86–87, 286–288
Lilienthal, Otto 317, 318
Lille University 238
Linde, Carl von 78
Lister, Joseph **182–183**
lithography 218
Locomotion (locomotive) 284
locomotives 277, 283–284, 295–296
Lodge, Oliver 252
logarithm, invention of 213–214
London, University of 38, 98, 105–106, 108, 112, 163, 276, 305
London Electric Supply Corporation Ltd. 97
London Hospital 100
longitude problem 127–128
Longstaff Medal 164, 242
looms 147–148, 158
Louisa Horwitz Gross Prize 116
Louvain, University of 206–207
Love, Susan 183, **183–184**

Lovelace, Ada 24
Lovelock, James 261–262
Lucid, Shannon **184–185**
Luftschiff Zeppelin No. 1 (LZ1) (dirigible) 325
Lumière, Auguste and Louis **185–186**
luminescence 33
Lunar Society 41

M

MacArthur Foundation 138, 199, 272
Macintosh, Charles 124–125
Macintosh computer 150
mackintoshes 125
MacLeod, Colin M. 21
Macleod, John James Rickard 28, 29–30, 37–38
Macmillan, Kirkpatrick 209
MacRoberts Award 141
Magellan, Ferdinand **187–188**
magneto-hydrodynamics 7–8
Maiman, Theodore **188–190**
Maize & Burns 229
Manchester Literary and Philosophical Society 70–71
Manhattan Project 10, 11, 95, 96, 109, 223, 225, 256, 303
maps 206–207, 227–228
Marburg, University of 232
Marcellus Hartley Medal 2
Marconi, Guglielmo 18, 51, **190–191,** 251, 252
Marconi Company 18, 27, 52
Marcy, Geoffrey **191–192**
Marine Biological Laboratory (Woods Hole) 56
marine biology 56–58
Marine Steam Turbine Company 52, 234
Marischal College 196
Mark computers 5–7, 138–139
Marsburg, University of 47
Maryland, University of 56, 272
Maschinenbau Gesellschaft 69
Massachusetts General Hospital 309
Massachusetts Institute of Technology (MIT) 11, 101–103, 257–258, 268, 285, 294
mass and energy 88
masticator 125

The Mathematical Foundations of Quantum Mechanics (Von Neumann) 302
Mathematical Principles of Natural Philosophy (Principia) (Newton) 180, 216, 217
mathematicians
 André Ampère **11–13,** *12,* 227
 Archimedes **15–16,** 233
 Charles Babbage 6, **23–24,** 148
 Hero of Alexandria **129–130,** 234
 Baron Kelvin 97, 152, 153, **159–161,** *160*
 Gottfried Wilhelm Leibniz 141, **179–180,** 236
 John Napier **213–214**
 Isaac Newton 142, 179, 180, *216,* **216–217**
 Papus **232–233**
 Blaise Pascal *235,* **235–236**
 Jean-Victor Poncelet **250–251**
 Alessandro Volta 178, **301**
 John Von Neumann **301–303,** *302*
Matteucci Medal 131
Matzinger, Polly Celine Eveline **192–194,** *193*
Mauchly, John William **194–196**
Maxam, Allan 114
Max Planck Medal 205
Maxwell, James Clerk *196,* **196–197**
May, George 219
Maybach, Wilhelm 68, 69, 230
Mayor, Michel 191
McCarty, Maclyn 21
McClintock, Barbara **197–200,** *198*
McGill University (Montreal, Canada) 276
Mead, Margaret **200–201,** *201*
Measurement of the Circle (Archimedes) 15
mechanical engineers 77–79, 233–235, 247–248, 306–308
mechanical equivalent of heat 152–153
mechanics 113–114
Mechanics (Hero) 129
Mechanics Institute (Manchester) 313

The Mechanism of Mendelian Heredity (Morgan, Bridges, Muller, Sturtevant) 211
Medal of Merit (U.S.) 96, 268, 306
Medal of Science (U.S.) 328
Medawar, Peter Brian **202–203**
Medical Research Council (U.K.) 202, 266
Meitner, Lise 107, 108–109, **203–205,** *204*
Memoir of a Thinking Radish (Medawar) 203
Memoir on the Mathematical Theory of Electrodynamic phenomena, uniquely Deduced from Experience (Ampère) 12–13
Menai Suspension Bridge 291, 292
Mendel, Gregor 76–77, **205–206,** 211
Mercator, Gerardus 206–207, 227, 228
Mercedes (automobile) 69–70
Merrill-Palmer School for Motherhood and Child Development 136–137
Mestral, George de **207–209**
metallurgy 4–5, 288–289
Meteorological Essays (Daniell) 71
meteorologists 1–3, 3–4, 71–72
Meterological Observations and Essays (Dalton) 70
"A Method of Reaching Extreme Altitudes" (Goddard) 117
Metrica (Hero) 129
Michaux, Pierre **209–210**
Michelin, André and Edouard 83
Michigan, University of 265
microbiology 237–238, 264–265
Micro Computer, Inc. 143
microscopes 323–324
Microsoft 9–10
military strategy 170, 171
milking machine 75.76
Mindwaves: Thoughts on Intelligence, Identity and Consciousness (Greenfield) 119
miner's safety lamps 74, 75, 283, 286, 287
mining and smelting 4–5, 248, 267–268, 283, 288
Minnesota, University of 275
Mirifici logarithmorum canonis descriptio (Napier) 214

Missile Envy: The Arms Race and Nuclear War (Caldicott) 50
Missouri, University of, at Columbia 198
Model Instrumentation and Telemetry Systems (MITS) 9
Modern Medicine Award 137
molecular biology 62–64, 114–116, 120–121
Molina, Mario 261, 262
Montagu, Mary Wortley 148
Monthly Weather Review 2
Morehouse College 137
Morgan, Thomas Hunt **210–211**
Morgenstern, Oskar 303
Morris Leibmann Memorial Prize 269, 328
Morse, Samuel 59, 60
Morton, William T. G. 309
Moses and Monotheism (Freud) 107
motion, laws of 216–217
motion picture industry 185–186
motorcycles 69, 210
Mount Holyoke College 13
Muller, Hermann Joseph 211
Munich, University of 130, 163, 243–244, 260
Münster, University of 79
Murdoch, William 177–178
Museum of Natural History (France) 32, 33
Museum of Natural History (U.S.) 200, 201
The Mutation Theory (De Vries) 77
Myers, Gene 300

N

Napier, John **213–214**
Napoleon 301
NASA 3, 4, 185, 264
National Academy (U.S.) 138
National Academy of Engineering (U.S.) 286
National Academy of Sciences (U.S.) 2, 87, 170, 199, 201, 211, 262, 269
National Aeronautics and Space Administration (NASA) 3, 4, 185, 264
National Book Award 57
National Breast Cancer Coalition (U.S.) 183
National Conservatory of Arts and Crafts (France) 32, 33

National Foundation–March of Dimes (U.S.) 14
National Foundation for Scientific Research (Switzerland) 243
National Geographic Society 62
National Human Genome Research Institute (U.S.) 300
National Institute for Medical Research (U.K.) 203
National Institutes of Health (U.S.) 193–194, 272, 279, 299
National Inventors Hall of Fame (U.S.) 92, 95, 103, 119, 165, 175, 190, 209, 230
National Medal of Science (U.S.) 92, 199, 271, 280
National Medal of Technology (U.S.) 139, 316
National Physical Laboratory (U.K.) 305
National Research Council (U.S.) 138, 265
National Women's Hall of Fame (U.S.) 92
natural history 246–247
Nature's Services: Societal Dependence on Natural Ecosystems (Daily) 68
Naval Ordnance Development Award 139
Nebraska, University of 101
neoprene 54, 55
nervous system *See* brain and central nervous system
Neuchâtel, University of 242–243
neurobiology 119–120
neuropathology 106–107
neurospecificity 278, 279
neutrons 11
newborn health 13–14
New College (Manchester) 70
Newcomen, Thomas 41, 73, **214–215,** 231, 232, 267, 268, 307
New Experiments Concerning Vacuums (Pascal) 236
The New Horse Hoeing Husbandry (Tull) 297
New Mexico Game Protection Association 181
A New System of Chemical Philosophy (Dalton) 71
Newton, Isaac 142, 179, 180, *216,* **216–217**
New York Academy of Sciences 170
New York City Memorial Hospital for Cancer and Allied Diseases 255
New York Times, The 5
New York University 91, 119, 265
NeXT Company 150
Nichols Medal 25
Niels Bohr International Gold Medal 157
Niepce, Joseph **217–218**
Nipkow, Paul Gottlieb 26, **218–219**
nitrogen, discovery of 173
nitroglycerine 220–221
nitroprussides 245
nitrous oxide 74, 75, 308–310
Nobel, Alfred Bernhard 79, **219–221**
Nobel Foundation 219, 221
Nobel Institute of Theoretical Physics 204
Nobel Prize in chemistry 114–116, 123–124, 131–132, 151–152, 261–262, 266–267, 275–277
Nobel Prize in physics 7, 8, 10–11, 32–33, 87–89, 111–113, 157, 190–191, 243–245, 268–269
Nobel Prize in physiology or medicine 28–30, 37–39, 58–59, 62–64, 79–80, 89–92, 97–99, 99–101, 106, 140–141, 197–200, 202–203, 210–211, 238–239, 259–261, 280
The Nomenclature of Petrology (Holmes) 135
Norland Achievement Award 137
Northrop Aircraft Company 195
Notre Dame of Maryland 183
Nottingham University (University College in Nottingham) 163
Nova (TV show) 263, 264
"Nova Methodus pro Maximus et Minimis" (Leibniz) 180
nuclear disintegration 275, 276
nuclear energy 277
nuclear fission 95, 96, 107, 108–109, 203, 204–205
Nuclear Madness: What You Can Do! (Caldicott, Herrington, Stiskin) 50
nuclear medicine 255–256
nuclear physics 151
nuclear weapons 3–4, 49–50, 223, 225, 264 *See also* Manhatten Project
"Nuclear Winter: Global Consequences of Multiple Nuclear Explosions" (Turco, Toon, Ackerman, Pollack, Sagan) 3, 264
nuclear winter theory 3–4, 264
nursing 30–32
nutrition 132, 136–137
nylon 54, 55

O

Oakes, James 321, 322
Oberlin College 279
object-oriented programming (OOP) 150
"Observations on the Contractile Tissue of the Iris" (Lister) 182
"Observations on the Muscular Tissue of the Skin" (Lister) 182
obstetrics (childbirth, infant health) 13–14
oceanography 61–62, 103–104
O'Connor, Daniel Basil 265
Ohio Wesleyan University 261
oil industry *See* petroleum industry
Oklahoma, University of 184
Oklahoma Medical Research Foundation 184
Olivier, René and Aimé 210
Olmütz University 205
Olsen, Don 165
"On a Heuristic Point of View about the Creation and Conversion of Light" (Einstein) 88
"On Depletion of Antarctic Ozone" (Solomon et al.) 262
100 Most Important Americans of the Twentieth Century 165
"On Evaporation" (Dalton) 71
On Floating Bodies (Archimedes) 16
"On Protein Synthesis" (Crick) 62, 63, 64
"On Some Chemical Agencies of Electricity" (Davy) 75
"On the Absorption of Gases by Water" (Dalton) 71
"On the Antibacterial Action of Cultures of a Penicillium . . ." (Fleming) 99, 100

"On the Constitution of Mixed Gases" (Dalton) 71
"On the Dynamical Theory of Heat" 160
"On the Electrodynamics of Moving Bodies" (Einstein) 88
"On the Expansion of Gases by Heat" (Dalton) 71
"On the Force of Steam" (Dalton) 71
On the Mechanical Equivalent of Heat (Joule) 153
On the migration of birds (Jenner) 149
"On the Motion of Small Particles Suspended in a Stationary Liquid According to the Molecular Kinetic Theory of Heat" (Einstein) 88
On the Origin of Species by Means of Natural Selection (Darwin) 74
On the Production of Heat by Voltaic Electricity (Joule) 152
"On the Relation of Metallic Powders to Electrical Oscillations" (Popov) 252
On the Sphere and the Cylinder (Archimedes) 16
On the Subject of Metals (Agricola) 4, 5
Oppenheimer, J. Robert **223–225,** *224*
Opticks (Newton) 217
optics 323–324
"Optimum Human Population Size" (Ehrlich, Ehrlich, Daily) 68
Order of Leopold 311
Order of Merit (U.K.) 65, 132, 235, 267
Order of Merit of the Rising Sun (Japan) 80
Order of Orange-Nassau (Netherlands) 165
Order of the Crown (Italy) 311
organic chemistry 241–242
Organic Chemistry (Kipping, Perkin) 163
"Organic Derivatives of Silicon" (Kipping) 164
organ transplants 193, 202, 203
The Origins of Intelligence in Children (Piaget) 243
ornithology 148, 149

Ørsted, Hans Christian 12, **225–227,** *226*
Ørsted Medal 227
Ortelius, Abraham **227–228**
O. Tingley & Company 229
Otis, Elisha Graves **228–229**
Otto, Nikolaus August 69, **229–230**
Outstanding African American 138
overpopulation 67–68
Owens College 163
Oxford University 59, 100, 101, 119, 120, 121, 132, 202, 203, 276, 277, 296
oxygen, discovery of 172, 173
ozone layer 261–262

P
Padua, University of 301
Papin, Denis, **231–232**
Pappus **232–233**
paraffin 321–322
Paraffin Light and Mineral Oil Company 322
Pardee, Arthur M. 55
Paris, University of (Sorbonne, Faculty of Sciences) 12, 33, 40, 60, 151–152, 243, 251
Paris Academy of Sciences (Académie Parisienne) 131, 236, 237, 238, 301
Parsons, Charles Algernon **233–235**
particle physics 11
Pascal, Blaise 235, **235–237**
Pasteur, Louis 15, 183, **237–238**
pasteurization 237, 238
pathology 99–101
Patterson, Tim 10
Paul Ehrlich Centenary Prize 59
Paul Ehrlich Gold Medal and Prize 21, 80
Pavia, University of 301
Pavlov, Ivan Petrovich **238–239**
Pease, Edward 284
pendulums 113, 141, 142
penicillin 58–59, 97–101
Penn, John 282
Pennington, Mary Engle **239–241**
Pennsylvania, University of 100, 104, 194, 240, 303
Pennsylvania Gazette 104
Pennsylvania State University 4

Penrose Medal 135
Pensées (Pascal) 235, 237
Perkin, William Henry 163, **241–242**
Perkin Medal 25, 145, 242
Personal Narrative of the Origin and Progress of the Caoutchouc or India Rubber Manufacture in England (Hancock) 126
pesticides 57–58
Petrograd Naval College 269
Petrograd Polytechnic Institute 155
Petrographic Methods and Calculations (Holmes) 135
petroleum industry 80–82, 321–322
Pew fellowships 68
Phelps, Orson C. 273
Philadelphia Health Department 240
philosophy 179–180
Philosophy of Physics (Planck) 244
phonographs 35, 86, 87
photography 24–25, 85–86, 145, 185, 217–218, 286, 287
photon theory of light 88
Physical Foundations of Radiology (Quimby) 256
Physicians for Social Responsibility 49, 50
physicists
    Luis Alvarez 10, **10–11**
    André Ampère **11–13,** *12,* 227
    Archimedes **15–16,** 233
    Henri Becquerel 32, **32–33,** 260
    William Crookes 64, **64–65**
    John Dalton 70, **70–71**
    Albert Einstein **87–89**
    Enrico Fermi 8, 95, **95–96,** 225
    Benjamin Franklin 103, **103–104**
    Otto Frisch **107–109,** *108,* 203, *204,* 204–205
    Dennis Gabor **111–113,** *112*
    Robert Hutchings Goddard 116, **116–117**
    Heinrich Rudolf Hertz 130, **130–131**
    Godfrey Hounsfield **140–141**

Christiaan Huygens **141–142,** *142,* 231, 232
James Prescott Joule *152,* **152–153,** 245
Pyotr Kapitsa **155–157,** *156*
Baron Kelvin 97, 152, 153, **159–161,** *160*
Theodore Maiman **188–190**
Guglielmo Marconi 18, 51, **190–191,** 251, 252
James Clerk Maxwell *196,* **196–197**
Lise Meitner 107, 108–109, **203–205,** *204*
Isaac Newton 142, 179, 180, 216, **216–217**
J. Robert Oppenheimer **223–225,** *224*
Hans Christian Ørsted 12, **225–227,** *226*
Blaise Pascal *235,* **235–236**
Max Planck **243–245,** *244*
Alexander Popov **251–253**
Wilhelm Conrad Röntgen *259,* **259–261**
William Shockley **268–269**
Nicola Tesla **292–293,** 310–311
Charles Wheatstone 59, 60–61, 245–246, *311,* **311–312**
Vladimir Zworykin 94, **326–328,** *327*
Physikalisch Technische Reischsanstalt 108
physiology 59–61, 89–90, 238–239
Piaget, Jean **242–243**
The Pine Cone 181
Pisa, University of 95, 113
Pittsburgh, University of 265, 326
Pixar 149, 150
Placzek, George 108
A Plaine Discovery of the Whole Revelation of St. John (Napier) 213
Planck, Max **243–245,** *244*
plant hybridization experiments 205–206
plasma theory 7–8
plastics industry 24–25, 144–145, 289–290
Playfair, Lyon **245–246,** 321
A Plea for Terrestrial Physics (Abbe) 2

Pliny the Elder **246–247**
Pliny the Younger 246, 247
plows 297
Pneumatic Institute 75
Pneumatics (Hero) 129–130
pneumatic tires 82–83
Polhem, Christopher **247–248,** 288
polio vaccine 264–265
Pollack, James B. 3–4
Polo, Marco **249–250**
polymers 24–25, 55, 118, 144–145, 290
Polytechnic Institute of Copenhagen 227
Polytechnikum (Germany) 33
Poncelet, Jean-Victor **250–251**
Pont Cysylltau Aqueduct 291, 292
Poor Richard's Almanac (Franklin) 104
Pope, Franklin L. 87
Popov, Alexander Stepanovich **251–253**
Popper, Karl 275
population biology 67–68
Portsmouth Dock 45
pottery 280–281
Poultry Historical Society 241
Prague, University of 292
Prandl, Antonin 76
pregnancy and childbirth 136–137
"Preparatory Studies for Deductive Methods in Storm and Weather Prediction" (Abbe) 2
Presbyterian Hospital 13
Presidential Certificate of Merit (U.S.) 271
Presidential Medal of Freedom (U.S.) 62, 265, 303
Price Waterhouse Information Technology Leadership Award for Lifetime Achievement 286
Priestley, Joseph 172
Princes and Peasants: Smallpox in History (Hopkins) 137
Princeton University 87, 117, 225, 261, 268, 302
Principia (Newton) 180, 216, 217
Principia Rerum Naturalium (Swedenborg) 288
Principles of Physical Geology (Holmes) 135, 136
"Prismatic analysis of electric light" (Wheatstone) 312

The Private Life of the Brain (Greenfield) 120
P. R. Mallory Company 53
"*Production artificielle d'éléments radioactifs*" (Joliot-Curie) 151
Project for a Scientific Psychology (Freud) 107
Project SAGE (Semi-Automated Ground Environment) 101, 102
Protonsil Rubrum 79–80
Prussian Academy of Sciences 88
Prussian Order of the Crown 260
psychoanalysis 106–107
psychobiology 278–280
psychologists 39–40, 106–107, 239–238, 242–243, 273–275
public health 137–138
Puffing Devil (locomotive) 295
Pulitzer Prize 264
pulley blocks 45
punch card tabulator 132–134, *133*
punch-card weaving system 147–148
Pure Food and Drug Act (1906) 240

Q

quantum mechanics 95, 302
quantum physics 223
quantum theory 243, 244
Queen Anne Act 127, 128
Queen Elizabeth Hospital (Boston) 49
"*Quelques remarques sur la mécanisme de l'articulation du coude*" (Einthoven) 89
Quimby, Edith H. **255–256,** *256*
Quincy Radio Laboratory 175

R

Rabdolgia (Napier) 214
rabies vaccination 238
radar 11, 305–306
Radcliffe Observatory 64
radiation 33, 49–50, 244
radioactive decay 275, 276
radio broadcasting 17–19, 174–175
Radiological Society of North America 256
radiology 255–256, 259, 260, 294
radiometer 64, 65
Radio Shack 9
Radio Society of Great Britain 52
radio waves 130–131, 305–306 *See also* communications

railroads 43, 44, 240, 277, 283–284, 295–296, 310
random access memory (RAM) 101, 102
rayon 286, 287
RCA (Radio Corporation of America) 18, 19, 27, 35, 36, 94, 175, 327, 328
Reale Scuola Normale Superiore 95
Red Cross 30–32
Reed College 149
Reflections on the Romance of Science (Sagan) 264
refrigeration 239–240
Regent Street Polytechnic Institute 98
Regnum Animale (Swedenborg) 289
Reith, J. C. W. 26
relativity, theories of 88, 244
resistance, electrical 311–312
retortion 321–322
Reuleaux, Franz 230
rheostats 312
Rice Institute 77
Richards, Alfred 100
Richards, Ellen Swallow **257–258,** 258
Riggs, John 309
Rittenhouse Medal 264
river blindness 138
RNA (ribonucleic acid) 106, 115
roads 291–292
Rochas, Alphonse Beau de 230
Rockefeller Foundation 59, 100, 243, 302
Rockefeller University (Rockefeller Institute for Medical Research) 20, 169, 170, 203, 328
Rocket (locomotive) 283, 284
rocketry 116–117
Roebuck, John 41, 308
Rogers-Low, Barbara 132
Rome, University of 96
Röntgen, William Conrad 259, **259–261**
Roscoe, Henry 47
Rosing, Boris 326
Rothamsted Experimental Station (RES) 173–174
Rowland, F. Sherwood **261–262**
Roxbury India Rubber Company 118
Royal Academy of Sciences 172
Royal Albert bridges 44
Royal Astronomical Society (U.K.) 8, 23
Royal College of Chemistry (London) 64
Royal Collège of Franche-Comté 237
Royal College of Mines (Sweden) 288
Royal College of Science (U.K.) 241
Royal College of Surgeons (U.K.) 98, 182
Royal Institution (U.K.) 52, 75, 119, 120, 121, 245
Royal Italian Academy 191
Royal Medal 65, 121, 161, 203, 242
Royal Mint (U.K.) 41–42, 217
Royal School of Como 301
Royal Society of Arts (U.K.) 52, 174
Royal Society of Chemical Industry 242
Royal Society of Edinburgh 161
Royal Society of London 52, 148, 149
  Albert Medal 47, 61
  Copley Medal 21, 47, 65, 72, 75, 104, 128, 161, 211, 301
  Corey Medal 239
  Davy Medal 47, 65, 164, 242
  Faraday Medal 120
  fellowship 24, 37, 41, 44, 46, 47, 65, 71, 75, 80, 97, 101, 104, 121, 135, 141, 142, 153, 155, 163, 180, 231, 267, 287, 290, 306, 308
  Gold Medal 71
  Hughes Medal 287, 306
  presidents 65, 75, 101, 161, 183, 217
  Royal Medal 65, 121, 161, 203, 242
  Rumford Medal 72, 131, 260
Royal Technical College (Glasgow) 26
rubber industry 82–83, 117–119, 124–126
Rumford Medal 72, 131, 260
Russian-Baltic Railroad Car Works 270
Russian Physical and Chemical Society 252, 253
Rutherford, Ernest 276

S
Sabin, Albert Bruce 265
Sachs, Julius von 77
Sagan, Carl Edward 3–4, **263–264,** 264
SAGE (Semi-Automated Ground Environment) air defense system 101, 102
St. Andrews, University of 213
St. Augustine's College 96
St. Mary's Hospital Medical School 98
St. Petersburg Institute of Technology 326
St. Petersburg University 252
Salk, Jonas **264–265**
Salk Institute for Biological Sciences 64, 265
Salpêtrière Hospital (Paris) 39
A Sand County Almanac, and Sketches Here and There (Leopold) 180, 181
San Francisco State University 191
Sanger, Frederick 114, 115, **266–267**
Sarnoff, David 94, 327
Saturn, rings of 197
Savery, Thomas 214, 215, 232, **267–268**
Scandinavian Society of Immunology 193
Schaffert, Roland M. 53
Schawlow, Arthur L. 189
Schedules of Reinforcement (Skinner) 275
Schockley, William
School of Mines (U.K.) 245
Schott, Otto 324
Schwab, Robert "Swampy" 193
Science and Human Behavior (Skinner) 275
Scientist of the Year (California) 192
Sculley, John 150
The Sea Around Us (Carson) 57
seed drill 296–297
Select Methods in Chemical Analysis (Crookes) 65
selenium 218, 219
Seminario Benzi 301
Senefelder, Alois 218
sewing machines 272–273, 282
Sex and Temperament in Three Primitive Societies (Mead) 200

sexuality, theories of 106–107
Sheffield, University of 100
ships 43, 44, 234–235, 278
Shockley, William **268–269**
Sikorsky, Igor **269–271**
Silbergeld, Ellen Kovner 271, 271–272
Silent Spring (Carson) 57–58
The Silent World (Cousteau) 62
silicon chemistry 163–164
Simon, Théodore 39, 40
SIMPLE (Simulation of Industrial Management Problems with Lots of Equations) 103
Singer, Isaac Merrit **272–273**
single-chip integrated circuit computer architecture 143–144
6-mercaptopurine (6-MP) 91
Sixty Feet Down (film) 61
Sketchpad 285
skin grafts 202, 203
Skinner, B. F. **273–275,** 274
Sloan, Alfred 103
smallpox 137–138, 148–149
Smith, Hamilton O. 300
Smithsonian Institution 117, 318
Sobrero, Ascanio 220
Society for Developmental Biology 121
Society for the Encouragement of National Industry 15
Society of American Bacteriologists 170
Society of Chemical Industry (U.S.) 145, 242
Society of the Arts (U.K.) 87
soda (sodium carbonate) production 175–176
Soddy, Frederick **275–277,** 276
Solar Spots and Terrestrial Temperature (Abbe) 2
Solomon, Susan 262
Sorbonne *See* Paris, University of
The Soul in Nature (Ørsted) 227
space flight 116–117, 184–185
Space Medal of Honor (U.S.) 184
space shuttle 185
spectroscopy 46, 47, 312
Sperry, Elmore Ambrose **277–278**
Sperry, Roger **278–280**
Sperry Rand 195
spinning machines 16–17
Spode, Josiah **280–281**

Sprengle, Hermann 287
Sproull, Robert 285, 286
Stanford-Binet Intelligence Test 39, 40
Stanford University 67, 68, 189, 269
Starley, James **281–283**
Starley, John Kemp 281, 282–283
Starry Messenger (Galileo) 114
Star Wars (film) 150
State University of New York 183, 299
Statistical Society of London 23
steam engines 41–42, 214–215, 233–235, 295–296, 306–308, 314–315
steam pumps 214–215, 267–268
Stearn, Charles 287
steel and iron industries 36–37, 72–73, 158–159, 166–167, 314–315
Stephenson, George **283–284,** 296
stereochemistry 237
stereoscope 312
"Stereoscopie door kleurverschil" (Einthoven) 89
Stern, Otto 108, 155
stock price ticker 87
The Stork and the Plow: The Equity Solution to the Human Dilemma (Ehrlich, Ehrlich, Daily) 68
Story of My Childhood (Barton) 32
A Story of the Red Cross (Barton) 32
Strasbourg, University of 259, 260
Strassmann, Fritz 108
Strathclyde, University of (Anderson's University/Institute) 321, 322
streptococcus research 169–170
Strutt, Robert 134–135
Studies in Hysteria (Freud) 107
"Studies on the Chemical Nature of the Substance Inducing Transformation of Pneumococcal Types . . ." (Avery, MacLeod, McCarty) 21
Sturgeon, William 152
Sturtevant, A. H. 211
Stuttgart Polytechnic 69
sulfanilamides 80
Sun Laboratories (Sun Microsystems) 286
superfluidity 155, 156

superphosphates fertilizer 173–174
surgery 182–183
Susan G. Komen Foundation Award 163
Sutherland, Ivan Edward **284–286**
Suttner, Bertha von 221
Sutton, Granger 300
Swade, Doran 23
Swan, Joseph Wilson **286–288,** 289
Swedenborg, Emanuel **288–289**
Swedenborgian Church 288–289
Swinburne, James **289–290**
Syme, James 182, 183
Systema Saturnium (Huygens) 141
system dynamics 101, 103
Systems, Man, and Cybernetics Society 103

T

Tabulating Machine Company 134
Talbot, Marion 258
Tarkio College 55
Technical Institute (Karlsruhe) 123
Technische Hochschule (Berlin) 111
Technische Hochschule (Munich) 78
telegraph industry 59–61, 71, 72, 160–161, 179, 190–191, 251–253, 312
telephones 35
telescopes 113, *113,* 114
television 25–27, 51–52, 93–95, 218–219, 326–328
Telford, Thomas **291–292**
temperature scale 160
Tesla, Nicola **292–293,** 310–311
Texas Instruments 143
textile industry
  dyes 241–242
  fibers 54, 56, 286, 287
  sewing machine 272–273
  spinning 16–17, 126–127
  Velcro 207–209
  weaving 147–148, 157–158
Thames Tunnel 43, 44, 45–46
Theatrum Orbis Terrarum (Ortelius) 227, 228
theology 288–289
"Theorie und Konstruktion eines rationellen Wäremotors" (Diesel) 78

The Theory of Games and Economic Behavior (Von Neumann) 303
thermochemistry 172, 173
thermodynamics 123, 152–153
Thermodynamics of Technical Gas Reactions (Haber) 123
Thiebaut, J. A. 179
Thimonnier, Barthélemy 273
Thompson, "Sneaker" 99
Thomson, Elihu **293–295**
Thomson, Robert William 82, 83
Thomson, William Kelvin *See* Kelvin, Baron
Thomson-Ferranti alternator 97
Thorn EMI *See* Electric and Musical Industries, Ltd. (EMI)
Three Essays on the Theory of Sexuality (Freud) 107
ticker tape 87
Time 95, 190
Tin Toy (film) 150
tires 82–83
tobacco mosaic virus 106
Tolman, Justin 93, 94
tomography 140–141
Toon, Owen B. 3–4
Toronto, University of 29, 30, 37, 38, 39
torpedos 312–314
Torpedo School (Russian Navy) 252
Townes, Charles H. 189
Townshend, James 81
toxicology 271–272
Toy Story (film) 150
Traité de la Lumière (Huygens) 142
Traité des propriétés projectives des figures (Poncelet) 250, 251
Transactions of the Edinburgh Geological Society 135
transformers 289
transistors 268–269
transportation industry
    automobiles 33–34, 68–70, 170, 171
    bicycles 209–210, 281–283, 317
    bridges 43, 44, 73, 291–292, 314, 315
    canals 42–43, 280, 281, 291–292
    diesel engine 77–79
    elevators 228–229
    internal combustion engine 33–34, 229–230
    motorcycles 69, 210
    pneumatic tires 82–83
    railroads 43, 44, 240, 277, 283–284, 295–296, 310
    roads 291–292
    steam engines 41–42, 214–215, 233–235, 295–296, 306–308, 314–315
    tunnels 43–44, 45–46
Treatise on Meteorological Apparatus (Abbe) 2
"Treatise on Meteorological Apparatus and Methods" (Abbe) 2
Treatise on the Equilibrium of Liquids (Pascal) 237
Trevithick, Richard **295–296**
Trinity College 234
Trowbridge, W. P. 133
Trueblood, Kenneth 132
TRW Electronics 189
TTAPS team 3–4
tuberculosis 80
Tull, Jethro **296–297**
tunnels 43–44, 45–46
turbines 233–235
Turbinia (ship) 234
Turco, Richard P. 3–4
Turin Royal Academy 131
Turner, Josiah 282

U

"Ultra-Short Electromagnetic Waves" (Alfvén) 8
"Undersea" (Carson) 56
The Undersea World of Jacques Cousteau 62
Under the Sea Wind (Carson) 57
Union Carbide and Carbon Company 25
UNIVAC 194, 195
University College (Dundee) 305
University College (London) 62, 97, 182, 203, 276, 321
University College Hospital Medical School 58
University College of Wales 276
Uppsala, University of 8, 248, 288
Urban Dynamics (Forrester) 103
Ursinus College 194
U.S. Army 18, 302

U.S. Census Bureau 133–134, 195
U.S. Congress
    Gold Medal 87, 117
    Medal of Merit 96
    Space Medal of Honor 184
U.S. Department of Agriculture 240
U.S. Department of Defense 139, 285
U.S. Department of Energy 4, 96
U.S. Fish and Wildlife Service (Bureau of Fisheries) 56–57
U.S. Forest Products Laboratory 181
U.S. Forest Service 181
U.S. Navy 6–7, 101, 102, 138–139, 278
U.S. Office of Scientific Research and Development 275
U.S. Patent Office 31, 143–144
U.S. Veterans Administration 256
U.S. Weather Bureau 1–2
Utah, University of 93, 165
Utrecht, University of 89

V

vaccines 148–149, 238, 264–265
vacuum, existence of absolute 236
vacuum tubes 194
Valentine Medal 165
Variation in Animals and Plants Under Domestication (Darwin) 74
Vassar College 138, 257
Velcro 207–209
Venter, J. Craig **299–300**
Vetlesen Prize 136
Victor Talking Machine Company 36
Vienna, University of 106–107, 108, 137, 204, 205
Vienna Academy of Sciences 131, 260
Vienna General Hospital 107
View of Chemical Laws (Ørsted) 226
virtual reality 285–286
viruses 106
Viscoloid Company 145
vitamins 132, 136, 137
Volta, Alessandro 178, **301**
voltaic cells 71, 72
voltaic pile 301
Von Neumann, John **301–303**, 302
vulcanization 117–119, 125

W

Walden Two (Skinner) 274, 275
Wallace, Alfred Russel 74
Warner-Lambert Award for Distinguished Women in Science 272
Washington, University of 3, 162
Washington State University 9
waterpower 247, 248
Watson, James 19, 62, 63, 64, 105–106, 114, 115
Watson-Watt, Robert Alexander **305–306**
Watt, James 41–42, 177–178, 214, 215, 295, **306–308**, *307*, 315
wattmeters 294
weak force 96
weather reporting 1–2
weaving 147–148, 157–158
Wedgwood, Josiah 42, 43
Weiss, Paul A. 279
Wellcome Cancer Research Campaign Institute 121
Wells, Horace **308–310**
Western Ontario, University of 29
Westinghouse, George 293, **310–311**
Westinghouse Corporation 311, 326
Wheatstone, Sir Charles 59, 60–61, 245–246, *311*, **311–312**
Whirlwind computer 101, 102
Whitehead, Robert **312–314**
Whitehouse, E. O. W. 161
Whitman College 255
Wilderness Society 181
Wilkins, Maurice 62, 63, 64, 105–106
Wilkinson, John **314–315**
William of Cleves, Duke 206–207
Wilson, Allan C. 161
Wimperis, H. E. 305–306
Winsor, Frederic 178
Wisconsin, University of 5, 181
Wolf Prize (Israel) 199
Wollaston Medal 135
Woman of the Year in science 14
women, role of 200–201
Women's Medical College of Pennsylvania 240
Wood, Frances 249
Woods Hole Marine Biological Laboratory 56
Worcester Polytechnic Institute 117
workers' benefits 166–167, 324
World Dynamics (Forrester) 103
World Peace Council 152
World War I 79, 117, 124, 246, 270, 278, 305, 325
World War II 18, 61, 62, 101, 117, 138–139, 194–195, 246, 302–303, 306, 328 *See also* Manhattan Project
Worsley Canal (Duke's Canal) 42
Wozniak, Stephen 149–150, **315–316**
Wright, Wilbur and Orville *317*, **317–319**, *318*, *319*
Wright-Fleming Institute of Microbiology 99
Wright Trophy 271
W. T. Henley Telegraph Works Company 52
Würzburg, University of 77, 259–260

X

xerography 52–54
Xerox Company 52, 54
X-ray crystallography 105–106, 131–132
X rays 33, 51, 255–256, 259, 260, 294

Y

Yale University 138, 181, 240
Yerkes Laboratories 279
Yomah Oil Company 135
Young, James **321–322**

Z

Zeiss, Carl Friedrich **323–324**
Zeiss, Roderick 324
Zeppelin, Ferdinand **324–326**
Zurich, University of 88, 242
Zworykin, Vladimir 94, **326–328**, *327*
Zworykin Award 286

Library Media Center
Watertown High School